thelightingpress.com

ARMY SMARTBOOK

7th Edition (AODS7)

(AODS7) The Army Operations & Doctrine SMARTbook, 7th Ed.

FM 3-0 Multidomain OPERATIONS

The Lightning Press
Norman M Wade

The Lightning Press

2227 Arrowhead Blvd.
Lakeland, FL 33813
24-hour Voicemail/Fax/Order: 1-800-997-8827
E-mail: SMARTbooks@TheLightningPress.com
www.TheLightningPress.com

(AODS7) The Army Operations & Doctrine SMARTbook, 7th Ed.

Multidomain Operations

Copyright © 2023 Norman M. Wade

ISBN: 978-1-935886-91-4

All Rights Reserved

No part of this book may be reproduced or utilized in any form or other means, electronic or mechanical, including photocopying, recording or by any information storage and retrieval systems, without permission in writing by the publisher. Inquiries should be addressed to The Lightning Press.

Notice of Liability

The information in this SMARTbook and quick reference guide is distributed on an "As Is" basis, without warranty. While every precaution has been taken to ensure the reliability and accuracy of all data and contents, neither the author nor The Lightning Press shall have any liability to any person or entity with respect to liability, loss, or damage caused directly or indirectly by the contents of this book. If there is a discrepancy, refer to the source document. This SMARTbook does not contain classified or sensitive information restricted from public release.

"The views presented in this publication are those of the author and do not necessarily represent the views of the Department of Defense or its components."

SMARTbook is a trademark of The Lightning Press.

About our cover photo: Trained and Ready! 173rd Airborne Brigade paratroopers conduct reconnaissance and provide security while on a patrol during Exercise Bayonet Ready 22 at the Joint Multinational Readiness Center in Hohenfels Training Area, Germany. The brigade routinely trains alongside NATO allies and partners to build our partnerships and strengthen the alliance. (U.S. Army photo by Cpl. Uriel Ramirez.)

Photos Credits. All other photos courtesy Dept. of the Army and/or Dept. of Defense and credited individually where applicable.

Printed in India, 2023 by Citicap Channels Limited - connect@printprinters.com

View, download FREE samples and purchase online:
www.TheLightningPress.com

(AODS7) Notes to Reader

The Army's **primary mission** is to organize, train, and equip its forces to conduct prompt and sustained land combat to defeat enemy ground forces and seize, occupy, and defend land areas. Army forces shape operational environments, counter aggression on land during crisis, prevail during large-scale ground combat, and consolidate gains.

Army forces achieve objectives through the **conduct of operations**. Operations vary in many ways. They occur in all kinds of physical environments and vary in scale of forces involved and duration.

Multidomain operations are the combined arms employment of joint and Army capabilities to create and exploit relative advantages that achieve objectives, defeat enemy forces, and consolidate gains on behalf of joint force commanders. Multidomain operations are the Army's contribution to joint campaigns, spanning the **competition continuum**.

The Army provides forces capable of **transitioning to combat operations**, fighting for information, producing intelligence, adapting to unforeseen circumstances, and defeating enemy forces. Army forces employ capabilities from multiple domains in a **combined arms approach** that creates complementary and reinforcing effects through multiple domains while preserving combat power to maintain options for the joint force commander.

Combat power is the total means of destructive and disruptive force that a military unit/formation can apply against an enemy at a given time. It is the **ability to fight**. The complementary and reinforcing effects that result from synchronized operations yield a powerful blow that overwhelms enemy forces and creates friendly momentum. Army forces deliver that blow through a combination of five dynamics: leadership, firepower, information, mobility, and survivability.

The **warfighting functions** contribute to **generating and applying** combat power. Well sustained units able to move and maneuver bring combat power to bear against the opponent. Joint and Army indirect fires complement and reinforce organic firepower in maneuver units. Survivability is a function of protection tasks that focus friendly strengths against enemy weaknesses. Information contributes to the disruption and destruction of enemy forces. Intelligence determines how and where to best apply combat power against enemy weaknesses. C2 enables leadership, the most important qualitative aspect of combat power.

SMARTbooks - DIME is our DOMAIN!

Recognized as a "whole of government" doctrinal reference standard by military, national security and government professionals around the world, SMARTbooks comprise a comprehensive professional library designed for all levels of Service.

SMARTbooks can be used as quick reference guides during actual operations, as study guides at education and professional development courses, and as lesson plans and checklists in support of training. Visit **www.TheLightningPress.com**!

Introduction-1

AODS7: The Army Operations & Doctrine SMARTbook (Multidomain Operations)

FM 3-0 (Oct '22)
See following pages for an overview.

(Chapters 1 & 2)

"BSS6: The Battle Staff SMARTbook"

Chap 1: Operations (FM 3-0, 2022)

The Army's primary mission is to organize, train, and equip its forces to conduct prompt and sustained land combat to defeat enemy ground forces and seize, occupy, and defend land areas. It supports four strategic roles for the joint force. Army forces shape operational environments, counter aggression on land during crisis, prevail during large-scale ground combat, and consolidate gains.

Multidomain operations are the combined arms employment of joint and Army capabilities to create and exploit relative advantages that achieve objectives, defeat enemy forces, and consolidate gains on behalf of joint force commanders (JFC). Employing Army and joint capabilities makes use of all available combat power from each domain to accomplish missions at least cost. Multidomain operations are the Army's contribution to joint campaigns, spanning the competition continuum.

The Army provides forces capable of transitioning to combat operations, fighting for information, producing intelligence, adapting to unforeseen circumstances, and defeating enemy forces. Army forces employ capabilities from multiple domains in a combined arms approach that creates complementary and reinforcing effects through multiple domains, while preserving combat power to maintain options for the joint force commander. Creating and exploiting relative advantages require Army forces to operate with endurance and in depth.

Chap 2: Combat Power (Generating & Applying)

Combat power is the total means of destructive and disruptive force that a military unit/formation can apply against an enemy at a given time. It is the ability to fight. The complementary and reinforcing effects that result from synchronized operations yield a powerful blow that overwhelms enemy forces and creates friendly momentum. Army forces deliver that blow through a combination of five dynamics: are leadership, firepower, information, mobility, and survivability.

The warfighting functions contribute to generating and applying combat power. Well sustained units able to move and maneuver bring combat power to bear against the opponent. Joint and Army indirect fires complement and reinforce organic firepower in maneuver units. Survivability is a function of protection tasks, the protection inherent to Army platforms, and schemes of maneuver that focus friendly strengths against enemy weaknesses. Intelligence determines how and where to best apply combat power against enemy weaknesses. C2 enables leadership, the most important qualitative aspect of combat power.

"TLS6: The Leader's SMARTbook"

Information

The Six Warfighting Functions

| ADP 6-0 | ADP 3-90 | ADP 2-0 | ADP 3-19 | ADP 4-0 | ADP 3-37 |
| (Chapter 3) | (Chapter 4) | (Chapter 5) | (Chapter 6) | (Chapter 7) | (Chapter 8) |

Chap 3: Command & Control (ADP 6-0)
The command and control warfighting function is the related tasks and a system that enable commanders to synchronize and converge all elements of combat power. The primary purpose of the command and control warfighting function is to assist commanders in integrating the other elements of combat power to achieve objectives and accomplish missions.

Chap 4: Movement and Maneuver (ADP 3-90 & others)
The movement and maneuver warfighting function is the related tasks and systems that move and employ forces to achieve a position of relative advantage over the enemy and other threats. Direct fire and close combat are inherent in maneuver. The movement and maneuver warfighting function includes tasks associated with force projection related to gaining a position of advantage over the enemy. Movement is necessary to disperse and displace the force as a whole or in part when maneuvering. Maneuver is the employment of forces in the operational area.

Chap 5: Intelligence (ADP 2-0)
The intelligence warfighting function is the related tasks and systems that facilitate understanding the enemy, terrain, and civil considerations. This warfighting function includes understanding threats, adversaries, and weather. It synchronizes information collection with the primary tactical tasks of reconnaissance, surveillance, security, and intelligence operations. Intelligence is driven by commanders and is more than just collection. Developing intelligence is a continuous process that involves analyzing information from all sources and conducting operations to develop the situation.

Chap 6: Fires (ADP 3-19)
The fires warfighting function is the related tasks and systems that create and converge effects in all domains against the threat to enable actions across the range of military operations. These tasks and systems create lethal and nonlethal effects delivered from both Army and Joint forces, as well as other unified action partners.

Chap 7: Sustainment (ADP 4-0)
The sustainment warfighting function is the related tasks and systems that provide support and services to ensure freedom of action, extend operational reach, and prolong endurance. The endurance of Army forces is primarily a function of their sustainment. Sustainment determines the depth and duration of Army operations. It is essential to retaining and exploiting the initiative. Sustainment provides the support necessary to maintain operations until mission accomplishment.

Chap 8: Protection (ADP 3-37)
The protection warfighting function is the related tasks and systems that preserve the force so the commander can apply maximum combat power to accomplish the mission. Preserving the force includes protecting personnel (combatants and noncombatants) and physical assets of the United States and multinational military and civilian partners, to include the host nation. The protection warfighting function enables the commander to maintain the force's integrity and combat power. Protection determines the degree to which potential threats can disrupt operations and then counters or mitigates those threats.

FM 3-0 Operations (Oct' 22): Overview

Continued from previous page

The 2022 version of FM 3-0 establishes multidomain operations as the Army's operational concept. Conceptually, multidomain operations reflect an evolutionary inflection point, building on the incremental changes in doctrine as the operational environment has changed over the last forty years. In practice, however, these conceptual changes will have revolutionary impacts on how the Army conducts operations in the coming decades. The 2017 version of FM 3-0 introduced many multidomain considerations and ideas. This version of FM 3-0 codifies the multidomain approach to operations in terms of the combined arms employment of capabilities from multiple domains. The multidomain operations concept draws from previous Army operational concepts, including AirLand Battle, Full Spectrum Operations, and Unified Land Operations.

Multidomain operations are the Army's contribution to unified action, conducted by Army echelons in an operational environment consisting of five domains and three dimensions, and the strategic contexts of competition, crisis, and armed conflict. It concludes with a description of multidomain operations through guiding principles of war, tenets, and imperatives that enable Army forces to accomplish missions, defeat enemy forces, and meet objectives.

Full Spectrum Operations accounted for the operations that Army forces conducted outside the bounds of armed conflict. This version of FM 3-0 updates this material for the present by describing how Army forces operate during competition below armed conflict and during crisis. It goes a step further by describing how these operations set conditions for success during armed conflict.

Unified Land Operations emphasized the integration and synchronization of Army, joint, and other unified action partners during operations. This version of FM 3-0 retains the focus on large-scale combat operations. It also builds on the importance of integrating joint and multinational capabilities and expands the combined arms approach with a focus on creating complementary and reinforcing effects with capabilities from multiple domains.

The nature of war remains unchanged. The model for understanding an operational environment, specifically the physical, information, and human dimensions, reinforces the Clausewitzian idea that war is an act of force to compel the enemy's will. In other words, physical action can influence human perceptions, behavior, and decision making. Although there are new capabilities in space and cyberspace, Army forces use them just as they employ any other capability—to accomplish missions on land.

Summary of major changes:

- Establishes multidomain operations as the Army's operational concept.
- Organizes chapters around the range of military operations that occur along the competition continuum in the context of competition, crisis, and conflict. Continues focus on large-scale combat operations.
- Emphasizes understanding an operational environment through three dimensions (physical, information, human) and five physical domains (land, air, maritime, space, cyberspace).
- Develops different Tenets (Agility, Convergence, Endurance, and Depth) and new Imperatives applicable to operations.
- Adopts the joint definition of Combat Power and clarifies relationship with Warfighting Functions.
- Adjusts the Operational Framework to assigned areas, main effort/supporting effort, and deep/close/rear operations; eliminates "decisive, shaping, and sustaining operations" component; emphasizes "continuous consolidation of gains" during all operations.
- Establishes the Theater Strategic level of warfare as a fourth distinct level separate from national strategic.

Continued from previous page

- Adds a chapter on the unique requirements of maritime environments and a contested deployment appendix.
- Establishes a 9th form of contact – Influence: interactions intended to shape perceptions, behaviors, and decisions.
- METT-TC mnemonic updated to METT-TC(I), adding informational considerations to the mission variables.

FM 3-0 Operations (Oct '22) Logic Map

Peer threats contest the joint force in all domains through several methods:
Information warfare   Preclusion   Isolation   Sanctuary   Systems warfare

The U.S. joint force addresses these threats through

Unified action:
The synchronization, coordination, and/or integration of the activities of governmental and nongovernmental entities with military operations to achieve unity of effort (JP 1, Volume 1).

The Army's contribution to unified action is

Multidomain operations:
The combined arms employment of joint and Army capabilities to create and exploit relative advantages that achieve objectives, defeat enemy forces, and consolidate gains on behalf of joint force commanders.

Army forces conduct multidomain operations throughout an operational environment that consists of 5 domains and 3 dimensions

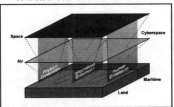

and executed by Army echelons during competition, crisis, and conflict across the range of military operations

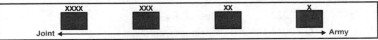

Joint ← → Army

respecting the

Principles of war
Maneuver   Objective   Offensive   Surprise   Economy of force
Mass   Unity of command   Security   Simplicity

reflecting the

Tenets of operations
Agility   Convergence   Endurance   Depth

adhering to

Imperatives of operations:
- See yourself, see the enemy, and understand the operational environment
- Account for being under constant observation and all forms of enemy contact
- Create and exploit relative physical, information, and human advantages in pursuit of decision dominance
- Make initial contact with smallest element possible
- Impose multiple dilemmas on the enemy
- Anticipate, plan, and execute transitions
- Designate, weight, and sustain the main effort
- Consolidate gains continuously
- Understand and manage the effects of operations on units and Soldiers

executed through

Offensive operations   Defensive operations   Stability operations   Defense support of civil authorities

to accomplish missions, defeat enemy forces, and consolidate gains that meet joint and national objectives.

Ref: FM 3-0 (Oct '22), Introductory figure. FM 3-0 logic chart.

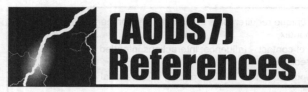

(AODS7) References

The following references were used to compile The Army Operations & Doctrine SMARTbook. All references are available to the general public and designated as "approved for public release; distribution is unlimited." The Army Operations & Doctrine SMARTbook does not contain classified or sensitive material restricted from public release.

Army Doctrinal Publications (ADPs)

ADP 1-02	Aug 2018	Terms and Military Symbols
ADP 2-0	Jul 2019	Intelligence
ADP 3-0	Jul 2019	Operations
ADP 3-07	Jul 2019	Stability
ADP 3-19	Jul 2019	Fires
ADP 3-28	Jul 2019	Defense Support of Civil Authorities
ADP 3-37	Jul 2019	Protection
ADP 3-90	Jul 2019	Offense and Defense
ADP 4-0	Jul 2019	Sustainment
ADP 5-0	Jul 2019	The Operations Process
ADP 6-0	Jul 2019	Mission Command: Command and Control of Army Forces

Field Manuals (FMs) &

FM 3-0*	Oct 2022	Operations
FM 3-34	Apr 2014	Engineer Operations
FM 3-55	May 2013	Information Collection
FM 3-52	Oct 2016	Airspace Control
FM 3-90-1	Mar 2013	Offense and Defense, Volume 1
FM 3-90-2	Mar 2013	Reconnaissance, Security, and Tactical Enabling Tasks, Volume 2
FM 6-0*	May 2022	Commander and Staff Organization and Operations
FM 4-0*	Jul 2019	Sustainment

Army Techniques Publications (ATPs)

ATP 3-35	Mar 2015	Army Deployment and Redeployment
ATP 2-01.3*	Mar 2019	Intelligence Preparation of the Battlefield

Joint Publications (JPs)

JP 3-0*	Jun 2022	Joint Campaigns and Operations
JP 5-0*	Dec 2020	Joint Planning

* Denotes new/updated reference since previous edition.

(AODS7) Table of Contents

Chap 1: Operations (FM 3-0, 2022)

I. Foundations of Operations ... 1-1
- I. Army Operations ... 1-1
  - Operations (Definition) ... 1-1
  - Operational Categories ... 1-1
- II. Multidomain Operations ... 1-2
- III. Challenges for Army Forces ... 1-4
- IV. Lethality: Overcoming Challenges ... 1-6
- V. War and Warfare ... 1-6
  - A. The Nature of War ... 1-7
  - B. Principles of War ... 1-8
  - C. Characteristics of Warfare ... 1-9
  - D. Methods of Warfare (Conventional / Irregular) ... 1-9
  - E. Offense, Defense, and Stability ... 1-11
  - F. Large-Scale Combat Operations ... 1-11
  - G. Combined Arms ... 1-11
  - H. Levels of Warfare ... 1-12
  - I. Army Strategic Contexts ... 1-14
  - J. Consolidating Gains (Continuously) ... 1-16
- VI. Understanding an Operational Environment ... 1-17
  - A. Domains ... 1-18
  - B. Dimensions ... 1-20
  - C. Operational and Mission Variables ... 1-22
    - Informational Considerations ... 1-22

II. Strategic Environment ... 1-23
- I. Strategic Environment ... 1-23
  - A. China and Russia ... 1-23
  - B. North Korea and Iran ... 1-23
  - C. Non-state Actors ... 1-23
  - D. Threats / Threat Methods ... 1-24
    - Information Warfare ... 1-25
    - Systems Warfare ... 1-25
    - Preclusion ... 1-26
    - Isolation ... 1-27
    - Sanctuary ... 1-27
- II. Unified Action and Army Forces ... 1-28
  - A. Joint Operations and Activities ... 1-28
  - B. Multinational Operations ... 1-28
  - C. Interagency Coordination and Interorganizational Cooperation ... 1-28
  - D. Conventional & Special Operations Forces Integration ... 1-28

E. Joint Interdependence 1-36
F. Army Force Posture.. 1-36
G. Army Echelons.. 1-32
H. Domain Interdependence 1-34
III. Why the Army Needs Mulitdomain Operations........................... 1-29

III. Fundamentals of Operations 1-37
I. Multidomain Operations: the Army's Operational Concept........... 1-37
II. Tenets 1-38
   A. Agility.. 1-38
   B. Convergence.. 1-39
      - Integration.. 1-40
      - Synchronization 1-41
      - Achieving Convergence 1-42
   C. Endurance 1-42
   D. Depth 1-43
III. Imperatives 1-44
IV. Operational Approach.. 1-46
   A. Defeating Enemy Forces in Detail 1-46
   B. Defeat and Stability Mechanisms....................................... 1-47
      - Defeat Mechanisms (Destroy, Dislocate, Isolate, Disintegrate)........... 1-48
      - Stability Mechanisms (Compel, Control, Influence, Support() 1-47
   C. Risk.. 1-49
V. The Elements of Operational Art... 1-50
VI. Strategic Framework 1-52
   - Strategic Support Area 1-52
   - Joint Security Area 1-52
   - Extended Deep Area 1-52
VII. Operational Framework.. 1-54
   A. Assigned Areas (Area of Operations, Zone, Sector)........... 1-54
   B. Deep, Close, and Rear Operations..................................... 1-58
      Support Area Operations 1-58
   C. Main Effort, Supporting Effort, & Reserve.......................... 1-62

IV(A). COMPETITION (Below Armed Conflict) 1-63
I. Overview of Operations During Competition 1-63
II. Adversary Methods During Competition 1-64
III. Preparation for Large-Scale Combat Operations 1-66
   A. Set the Theater 1-66
   B. Build Allied and Partner Capabilities and Capacity............. 1-67
   C. Interoperability... 1-67
   D. Protect Forward-Stationed Forces...................................... 1-68
   E. Prepare to Transition and Execute Operation Plans............ 1-68
   F. Train and Develop Leaders 1-68
IV. Interagency Coordination 1-69
V. Competition Activities 1-70
VI. Relative Advantages During Competition................................ 1-72
VII. Consolidating Gains During Competition 1-73
VIII. Transition to Crisis and Armed Conflict 1-74

IV(B). Operations During CRISIS...1-77
I. Overview of Operations During Crisis...1-77
II. Adversary Methods During Crisis 1-78
III. Operations Security 1-79
IV. Relative Advantages During Crisis 1-79
V. Army Support to the Joint Force during Crisis 1-80
VI. Force Projection 1-82
VII. Consolidating Gains 1-86
VIII. Transition to Competition or Armed Conflict..............................1-86

IV(C). Operations during ARMED CONFLICT...........................1-86
I. Armed Conflict and Large-Scale Combat Operations....................1-86
  A. Enemy Approaches to Armed Conflict 1-88
    - China 1-90
    - Russia...1-91
  B. Relative Advantages During Armed Conflict 1-89
  C. Operating as Part of the Joint Force...1-94
    - Doctrinal Template of Depths & Frontages 1-95
    - Establishing Command and Control 1-96
    - Notional Echelon Roles & Responsibilities 1-97
  D. Applying Defeat Mechanisms 1-106
  E. Enabling Operations 1-107
II. Defensive Operations 1-108
  A. Purpose and Conditions for the Defense..................................1-108
  B. Types of Defensive Operations...1-110
    - Area Defense..1-110
    - Mobile Defense...1-110
    - Retrograde..1-110
  C. Characteristics of the Defense..1-112
  D. Enemy Offense 1-112
  E. Defensive Operational Framework Considerations 1-114
III. Offensive Operations 1-118
  A. Purpose and Conditions for the Offense 1-118
  B. Characteristics of the Offense...1-118
  C. Enemy Defense 1-118
  D. Types of Offensive Operations 1-119
    - Movement to Contact..1-119
    - Attack..1-119
    - Exploitation 1-119
    - Pursuit...1-119
  E. Offensive Operational Framework Considerations 1-123
  F. Transition to Defense and Stability..1-126
  G. Transition to Post-Conflict Competition 1-126

V. Army Operations in Maritime Environments 1-127
I. Overview of the Maritime Environment...1-127
II. Operational Framework 1-120
III. Operational Approach 1-132

VI. Contested Deployment...1-139
I. Force Projection and Threat Capabilities.......................................1-139
II. Movement Phase...1-140
III. RSOI During Contested Deployments..1-143
  - Homeland Defense 1-144
  - Defense Support Of Civil Authorities (DSCA) 1-144

Table of Contents - 3

Chap 2

Combat Power (Generating & Applying)

I. Combat Power 2-1
I. Warfighting Functions 2-1
- Command and Control Warfighting Function 2-2
- Movement and Maneuver Warfighting Function 2-2
- Intelligence Warfighting Function 2-2
- Fires Warfighting Function 2-3
- Sustainment Warfighting Function 2-3
- Protection Warfighting Function 2-3
II. Dynamics of Combat Power 2-4
A. Leadership 2-4
B. Firepower 2-5
C. Information 2-5
D. Mobility...2-5
E. Survivability..2-6

II. Leadership (as a Dynamic of Combat Power) 2-7
I. The Art of Command and the Commander............................2-7
- Commander Presence on the Battlefield 2-9
- Imperative: Understand and Manage the Effects of Operations on Units
  and Soldiers 2-9
II. Applying the Art of Command 2-8
A. Commander's Intent 2-9
B. Initiative..2-9
C. Drive the Operations Process...2-10
D. Discipline 2-10
E. Accepting Risk to Create and Exploit Opportunities 2-11
F. Command and Control During Degraded or Denied Communications....2-12

III. Information (as a Dynamic of Combat Power).......................2-13
I. Information 2-13
II. Information Operations (IO)..2-13
III. Information (as one of the Joint Functions) 2-14
A. Joint Force Capabilities, Operations, and Activities for
   Leveraging Information...2-14
B. Information-Related Capabilities (IRCs) 2-15
C. Cyberspace Operations (CO) and Electromagentic Warfare (EW) 2-16
D. Information Use Across the Competition Continuum (Examples)...........2-18

4 - Table of Contents

Command & Control Warfighting Function

Chap 3

Command and Control Warfighting Function....................................3-1
I. The Command and Control Warfighting Function 3-1
II. Mission Command..3-4
III. Principles of Mission Command 3-6
IV. Command and Control System 3-8

I. The Operations Process 3-9
I. The Operations Process 3-9
II. Principles of the Operations Process..3-10
    A. Drive the Operations Process 3-10
        - Understand 3-11
        - Visualize 3-11
        - Describe..3-11
        - Direct 3-11
        - Lead 3-11
        - Assess 3-11
    B. Build and Maintain Situational Understanding 3-14
    C. Apply Critical and Creative Thinking...3-14
III. Activities of the Operations Process 3-12
        - Plan...3-13
        - Prepare 3-13
        - Execute 3-12
        - Assess..3-13
IV. Army Planning Methodologies 3-15
    A. The Army Design Methodology (ADM) 3-15
    B. The Military Decision-Making Process (MDMP) 3-15
    C. Troop Leading Procedures (TLP) 3-15
V. Imperatives: Command & Control (C2) 3-16
VI. Integrating Processes...3-20
        - Intelligence Preparation of the Battlefield (IPB) 3-20
        - Information Collection 3-20
        - Targeting 3-21
        - Risk Management 3-21
        - Knowledge Management 3-21
VII. Battle Rhythm...3-22

II. Command and Support Relationships 3-23
I. Chain of Command...3-23
II. Joint Command Relationships...3-24
III. Army Command & Support Relationships 3-25
III. Other Relationships 3-28

III. Command Posts 3-29
I. Command Posts 3-29
II. Types of Command Posts...3-30
III. CP by Echelon and Type of Unit..3-32

IV. Army Airspace Command and Control (A2C2).......................3-33
I. Airspace Coordinating Measures (ACMs)..3-34
II. Key Positions and Responsibilities 3-36

Table of Contents - 5

Chap 4

Movement & Maneuver Warfighting Function

Movement & Maneuver Warfighting Function 4-1
- Imperatives: Movement & Maneuver 4-2
- I. Offense and Defense...4-6
  - The Tactical Level of War 4-6
  - The Offense..4-6
  - The Defense..4-6
  - Enabling Operations...4-7
- II. Stability Operations..4-8
  - Primary Army Stability Tasks...4-8
- III. Force Projection/Deployment Operations.....................................4-10
  - Force Projection 4-10
  - Deployment Operations 4-10
  - Reception, Staging, Onward Movement, and Integration (RSOI) 4-10

I. Tactics and Tactical Mission Tasks (ADP 3-90) 4-11
- I. The Tactical Level of War..4-11
- II. The Art and Science of Tactics 4-12
- III. Tactical Mission Tasks 4-12
  - Tactical Doctrinal Taxonomy 4-13
  - A. Actions by Friendly Forces...4-14
  - B. Effects on Enemy Forces 4-15
  - C. Mission Symbols..4-16

II. Reconnaissance 4-17
  - Reconnaissance Objective..4-17
- I. Reconnaissance 4-18
  - A. Route Reconnaissance...4-19
  - B. Zone Reconnaissance 4-19
  - C. Area Reconnaissance...4-19
  - D. Reconnaissance in Force 4-19
  - E. Special Reconnaissance..4-19
- II. Reconnaissance Fundamentals 4-20

III. Security Operations 4-21
- I. Security Operations Tasks 4-21
  - A. Screen..4-21
  - B. Guard 4-22
  - C. Cover 4-24
  - D. Area Security 4-24
- II. Fundamentals of Security Ops 4-23

IV. Mobility and Countermobility 4-25
- I. Mobility...4-25
  - Assured Mobility 4-26
  - Aspects of Mobility 4-28
- II. Countermobility..4-30
  - Obstacle Planning 4-32
  - Obstacle Control Measures...4-33
- III. Engineer Support...4-34

6 - Table of Contents

Chap 5

Intelligence Warfighting Function

Intelligence Warfighting Function..5-1
- Fighting for Intelligence 5-1
- I. Intelligence Overview...5-2
  - Logic Map...5-3
- II. Intelligence Warfighting Function Tasks..5-4
- III. Intelligence Support to Commanders and Decisionmakers..........5-5
- IV. Intelligence Integrating Functions..5-6
  - Intelligence Preparation of the Battlefield (IPB) 5-6
  - Information Collection 5-7
- V. Types of Intelligence Products...5-8
  - A. Intelligence Estimate..5-8
  - B. Intelligence Summary 5-8
  - C. Intelligence Running Estimate 5-8
  - D. Common Operational Picture (COP) 5-8

I. The Intelligence Process 5-9
- The Joint Intelligence Process...5-9
- The Army Intelligence Process 5-9
- Commander's Guidance...5-10
- I. Intelligence Process...5-10
  - A. Plan and Direct 5-12
  - B. Collect..5-13
  - C. Produce 5-13
  - D. Disseminate...5-13
- II. Requirements Management 5-11
- III. Intelligence Process Continuing Activities 5-14
  - A. Analyze 5-14
  - B. Assess..5-14

II. Army Intelligence Capabilities 5-15
- I. All-Source Intelligence 5-15
- II. Single-Source Intelligence...5-18
  - A. The Intelligence Disciplines 5-16
    - Counterintelligence (CI)..5-16
    - Geospatial Intelligence 5-16
    - Human Intelligence (HUMINT)...5-16
    - Measurement and Signature Intelligence (MASINT) 5-17
    - Open-Source Intelligence (OSINT).....................................5-17
    - Signals Intelligence (SIGINT) 5-17
    - Technical Intelligence (TECHINT).......................................5-17
  - B. Complementary Intelligence Capabilities.............................5-18
  - C. Processing, Exploitation, and Dissemination (PED).............5-18

Table of Contents - 7

Chap 6

Fires
Warfighting Function

Fires Warfighting Function 6-1
   I. Fires Warfighting Function 6-1
   II. Fires Overview 6-2
      - Fires Logic Diagram 6-3
   III. Fires Across the Domains 6-4

I. I. Execute Fires Across the Domains 6-5
   I. Surface-to-Surface Fires 6-5
      - Army Surface-to-Surface Capabilities 6-7
         - Rockets 6-7
         - Missiles 6-7
         - Cannon Artillery 6-7
         - Mortars 6-7
   II. Air-to-Surface Fires 6-6
      - Fixed-Wing Aircraft 6-6
      - Rotary-Wing Aircraft 6-6
      - UAS 6-6
      - Air Force Assets 6-8
         - Air Interdiction 6-9
         - Close Air Support (CAS) 6-9
   III. Surface-to-Air Fires 6-10
      - High-to-Medium Air Defense (HIMAD) 6-10
      - Short-Range Air Defense (SHORAD) 6-11
   IV. Space Operations 6-12
   V. Special Operations 6-12
   VI. Cyberspace Operations and Electronic Warfare 6-13
      - Cyberspace Electromagnetic Activities (CEMA) 6-23
   VII. Information Operations (IO) 6-14

III. Integrate Army, Multinational and Joint Fires 6-15
   I. Fires in the Operations Process 6-15
      A. Integrating Fires into Planning 6-15
      B. Fires Preparation 6-16
      C. Fires Assessment 6-16
      D. Airspace Planning and Integration 6-17
   II. Integrating Multinational Fires 6-18
   III. Army Targeting Process (D3A) 6-20
      A. Decide 6-20
      B. Detect 6-20
      C. Deliver 6-20
      D. Assess 6-20
      - Operations Process & Targeting Relationship 6-21
   IV. Joint Targeting 6-22
   V. Air and Missile Defense Planning/Integration 6-24

8 - Table of Contents

Chap 7

Sustainment Warfighting Function

Sustainment Warfighting Function	7-1
I. Sustainment Warfighting Function	7-1
A. Logistics	7-1
B. Financial Management	7-4
C. Personnel Services	7-4
D. Health Service Support	7-4
II. Sustainment Overview (Underlying Logic)	7-2

I. Sustainment of Unified Land Operations ... 7-5

Operational Context	7-5
I. Operational Reach	7-6
A. Army Pre-Positioned Stocks (APS)	7-6
B. Theater Opening	7-6
- Basing	7-7
C. Theater Closing	7-6
II. Freedom of Action	7-12
A. Sustainment Preparation	7-12
- Negotiations and Agreements	7-13
- Sustainment Preparation of the Operational Environment	7-13
B. Sustainment Execution	7-12
III. Endurance	7-12
- Distribution	7-14

II. Sustain Large-Scale Combat Operations ... 7-15

I. Overview	7-15
II. Sustainment Synchronization	7-16
- Sustainment Rehearsals	7-22
III. Threats to Sustainment Units	7-17
IV. Large-Scale Defensive Operations	7-18
- Sustaining Defensive Operations	7-18
- Sustainment Fundamentals	7-18
- Planning Considerations	7-19
V. Large-Scale Offensive Operations	7-20
- Sustaining Offensive Operations	7-20
- Sustainment Fundamentals	7-21
- Risks during Large-Scale Combat	7-21
VI. Support Area	7-22
VII. Reconstitution Operations	7-24
- Reorganization	7-24
- Regeneration	7-24
- Regeneration Task Force (RTF)	7-25

Chap 8

Protection Warfighting Function

Protection Warfighting Function	**8-1**
I. The Protection Warfighting Function	8-1
II. Protection Overview (and Logic Map)	8-2
III. Primary Protection Tasks	8-4
IV. Survivability (One of the Five Dynamics of Combat Power)	8-6
V. Protection Integration in the Operations Process	8-4
I. Protection Planning	**8-9**
I. Initial Assessments	8-9
A. Threat and Hazard Assessment	8-10
- Threats and Hazards	8-11
B. Criticality Assessment	8-10
D. Vulnerability Assessment	8-14
II. Scheme of Protection Development	8-15
III. Protection Prioritization List	8-16
- Protection Priorities	8-16
- Critical Asset List (CAL) and Defended Asset List (DAL)	8-16
- Criticality	8-16
- Threat Vulnerability	8-17
- Threat Probability	8-17
IV. Protection Cell and Working Group	8-17
III. Protection in Preparation	**8-19**
I. Protection Working Group	8-19
A. Antiterrorism Working Group	8-22
B. Counter Improvised Explosive Device Working Group	8-22
C. Chemical, Biological, Radiological, and Nuclear Working Group	8-22
II. Protection Considerations (Preparation)	8-20
IV. Protection in Execution	**8-23**
I. Execution	8-23
II. Protection in Support of Decisive Action	8-24
V. Protection Assessment	**8-25**
I. Continuous Assessment	8-25
II. Lessons-Learned Integration	8-25
III. Protection Considerations (Assessment)	8-26

10 - Table of Contents

Chap 1
I. Foundations of Operations

Ref: FM 3-0, Operations (Oct. '22), chap. 1.

I. Army Operations

The Army's primary mission is to organize, train, and equip its forces to conduct prompt and sustained land combat to defeat enemy ground forces and seize, occupy, and defend land areas. It supports four strategic roles for the joint force. Army forces shape operational environments, counter aggression on land during crisis, prevail during large-scale ground combat, and consolidate gains. The Army fulfills its strategic roles by providing forces for joint campaigns that enable integrated deterrence of adversaries outside of conflict and the defeat of enemies during conflict or war. The strategic roles clarify the overall purposes for which Army forces conduct multidomain operations on behalf of joint force commanders (JFCs) in the pursuit of a stable environment and other policy objectives. Fulfilling policy objectives requires national-level leaders to orchestrate all instruments of national power throughout the entire government and coalition, in a manner commensurate with national will.

Military operations on land are foundational to operations in other domains because almost all capabilities, no matter where employed, are ultimately based on or controlled from land. While any particular domain may dominate military considerations in a specific context, conflicts are usually resolved on land because that is where people live and make political decisions and where the basis of national power exists.

Operations
Army forces achieve objectives through the conduct of operations. An operation is a sequence of tactical actions with a common purpose or unifying theme (JP 1, Vol 1). Operations vary in many ways. They occur in all kinds of physical environments, including urban, subterranean, desert, jungle, mountain, maritime, and arctic. Operations vary in scale of forces involved and duration. Operations change factors in the physical, information, and human dimensions of an operational environment.

Operational Categories
Army forces meet a diverse array of challenges and contribute to national objectives across a wide range of operational categories, including large-scale combat operations, limited contingency operations, crisis response, and support to security cooperation.

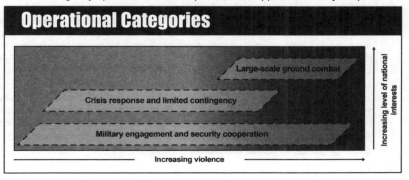

Ref: FM 3-0 (Oct. '22), fig. 1-1. Operational categories and the spectrum of violence.

(Operations) I. Foundations 1-1

Most operations occur on the lower end of the spectrum of violence, and their objectives do not reach the level of vital national interests or national survival. These operations typically shape operational environments in ways that stabilize global security and facilitate conditions that are generally favorable to the United States. They provide valuable options to JFCs because they achieve objectives best supported by persistent presence, often at relatively low cost.

While the overwhelming majority of operations conducted by Army forces occur either below the threshold of armed conflict or during limited contingencies, the focus of Army readiness is on large-scale combat operations. The United States always retains the option to employ greater levels of force when less coercive methods are ineffective, and when a vital interest or national survival is at stake. This requires Army forces to be prepared for the most demanding and dangerous types of operations. Army forces contribute to conventional deterrence through their demonstrated capability, capacity, and will to wage war on land in any environment against any opponent. Credible combat forces make the other instruments of national power more potent, and they help deter the enemy's escalation of violence during other types of operations.

The complex environment in which operations occur demands leaders who understand both the science and art of operations. Understanding the science of operations—such as combat power ratios, weapons ranges, and movement tables—helps leaders improve synchronization and reduce risk. However, there is no way to eliminate uncertainty, and leaders must exercise operational art to make decisions and assume risk. Intangible factors, such as the impact of leadership on morale, using shock effect to defeat enemy forces, and supportive populations are fundamentally human factors that can overcome physical disadvantages and often decide the outcomes of an operation.

Credible combat forces are those able to overcome the advantages peer threats generate within a specific regional context. Enemies typically initiate their aggression under conditions optimal for their success, requiring U.S. forces to respond at a disadvantage. U.S. combat operations typically involve force projection over long distances, providing advantages for enemy forces operating closer to their bases of support. Enemies typically have a degree of popular support cultivated through decades of propaganda and isolation from the free flow of information. This increases the enemy's will to fight and can make local populations hostile to U.S. forces and objectives. Although a combatant command and theater army may accrue a variety of advantages as they set the theater and prepare for armed conflict during periods of competition, Army forces are typically faced with challenges they have to overcome at the onset of hostilities and throughout the conduct or armed conflict.

II. Multidomain Operations

Multidomain operations are the combined arms employment of joint and Army capabilities to create and exploit relative advantages that achieve objectives, defeat enemy forces, and consolidate gains on behalf of joint force commanders. Employing Army and joint capabilities makes use of all available combat power from each domain to accomplish missions at least cost. Multidomain operations are the Army's contribution to joint campaigns, spanning the competition continuum.

Below the threshold of armed conflict, multidomain operations are how Army forces accrue advantages and demonstrate readiness for conflict, deterring adversaries while assuring allies and partners. During conflict, they are how Army forces close with and destroy the enemy, defeat enemy formations, seize critical terrain, and control populations and resources to deliver sustainable political outcomes.

See facing page for further discussion.

Multidomain Operations

Ref: FM 3-0, Operations (Oct. '22), pp. 1-2 to 1-3. See also p. 1-37.

Multidomain operations are the **combined arms** employment of **joint and Army** capabilities to create and exploit **relative advantages** that achieve objectives, defeat enemy forces, and consolidate gains on behalf of joint force commanders.

Employing Army and joint capabilities makes use of all available combat power from each domain to accomplish missions at least cost. Multidomain operations are the Army's contribution to joint campaigns, spanning the **competition continuum**.

Army forces conduct operations in support of joint campaigns which for the most part occur as part of a larger coalition operation. Leaders must understand the interdependencies between their own assigned forces and the forces or capabilities provided by others to generate the complementary and reinforcing effects of combined arms approaches. Army forces employ joint and other unified action partner capabilities to the degree they are available. However, because peer threats can contest the force in all domains, Army forces must be prepared to conduct operations when some or all joint capabilities are unavailable to support mission accomplishment.

All operations are multidomain operations. Army forces employ organic capabilities in multiple domains, and they continuously benefit from air and maritime strategic transportation and space and cyberspace capabilities that they do not control, including global positioning, satellite communications, and intelligence, surveillance, and reconnaissance (ISR). Lower echelons may not always notice the opportunities created by higher echelons or other forces that operate primarily in other domains; however, leaders must understand how the absence of those opportunities affects their concepts of operations, decision making, and risk assessment.

During operations, small advantages can have significant impacts on the outcome of the mission, particularly when they accrue over time. Creating and exploiting relative advantages are therefore necessary for all operations, and they become even more critical when opposing sides are evenly matched. A relative advantage is a location or condition, in any domain, relative to an adversary or enemy that provides an opportunity to progress towards or achieve an objective. Commanders seek and create relative advantages to exploit through action, and they continually assess the situation to identify ways to expand opportunities.

Army leaders are accustomed to creating and exploiting relative advantages through the combined-arms approach that traditionally focuses on capabilities from the land, air, and maritime domains. The proliferation of space and cyberspace capabilities further requires leaders who understand the advantages those capabilities create in their operational environment. The ability to integrate and synchronize space and cyberspace capabilities at the most effective tactical echelon expands options for creating advantages to exploit.

Multidomain operations fracture the coherence of threat operational approaches by destroying, dislocating, isolating, and disintegrating their interdependent systems and formations, and exploiting the opportunities these disruptions provide to defeat enemy forces in detail. Army forces therefore require timely, accurate, relevant, and predictive intelligence to understand threat characteristics, capabilities, objectives, and courses of action. Intelligence initially drives what combinations of defeat mechanisms commanders pursue as they employ the capabilities of their forces in space and time against enemy forces. Army forces combine maneuver and targeting methods to defeat enemy formations and systems. Army forces employ maneuver to close with and destroy enemy formations in close operations. Targeting generally sets priorities for information collection, fires, and other key capabilities to disintegrate enemy networks and systems. Leaders execute the targeting process to create advantages that enable freedom of maneuver and exploit the positional advantages created by maneuver.

(Operations) I. Foundations 1-3

Multidomain Operations

III. Challenges for Army Forces

Ref: FM 3-0, Operations (Oct. '22), pp. 1-3 to 1-5.

The joint force deters most adversaries from seeking to achieve strategic objectives through direct military confrontation with the United States. As a result, adversaries pursue their objectives indirectly through malign activities and armed conflict targeting others in ways calculated to avoid war with the United States. These activities include subversive political and legal strategies, establishing physical presence on the ground to buttress resource claims, coercive economic practices, supporting proxy forces, and spreading disinformation. However, several adversaries have both the ability and the will to conduct armed conflict with the United States under certain conditions, which requires Army forces to be prepared at all times for limited contingencies and large-scale combat operations.

Global and regional adversaries apply all instruments of national power to challenge U.S. interests and the joint force. Militarily, they have extended the battlefield by employing network-enabled sensors and long-range fires to deny access during conflict and challenge friendly forces' freedom of action during competition. These standoff approaches seek to—

- Counter U.S. space, air, and naval advantages to make the introduction of land forces difficult and exploit the overall joint force's mutual dependencies.
- Increase the cost to the joint force and its partners in the event of armed conflict.
- Hold the joint force at risk both in the U.S. and at its overseas bases and contest Army forces' deployment from home station to forward tactical assembly areas overseas.

Adversaries increase risk to the U.S. joint force in order to raise the threshold at which the United States might respond to a provocation with military force. By diluting the joint force's conventional deterrence, adversaries believe they have greater freedom of action to conduct malign activities both within and outside the U.S. homeland. Adversaries exploit this freedom of action through offensive cyberspace operations, disinformation, influence operations, and the aggressive positioning of ground, air, and naval forces to support territorial claims. Adversaries employ different types of forces and capabilities to attack private and government organizations, threaten critical economic infrastructure, and disrupt political processes, often with a degree of plausible deniability that reduces the likelihood of a friendly military response. Conducting these activities in support of policy goals threatens allied cohesion, weakens responses, and creates additional opportunities.

Threat standoff approaches intensify other friendly challenges. These challenges include—

- Gaining and maintaining support of allies and partners.
- Maintaining the continuous information collection needed to determine composition, disposition, strength, and activities of enemy forces.
- Integrating and synchronizing intelligence at all echelons, distributed across large operational areas with diverse requirements.
- Preparing forward-stationed forces to fight and win while outnumbered and isolated.
- Protecting forward-positioned forces and those moving into a theater.
- Minimizing vulnerability to weapons of mass destruction.
- Maintain C2 and sustainment of units distributed across vast distances in noncontiguous areas and outside supporting ranges and distances.
- Maintaining a desirable tempo while defeating fixed and bypassed enemy forces.
- Defeating threat information and irregular warfare attacks against the United States and strategic lines of communications.

1-4 (Operations) I. Foundations

Army forces prepare to conduct operations in contested theaters prior to and during armed conflict, including in the United States. Army forces must account for being under constant observation and the threat's ability to gain and maintain contact in all domains, wherever they are located. Army forces must be ready to deploy on short notice to austere locations and be capable of immediately conducting combat operations. During the initial phases of an operation, Army units may find themselves facing superior threats in terms of both numbers and capabilities. The first deploying units require the capability to defend themselves and continuously collect information on threat activities, as they provide reaction time and freedom of maneuver for follow-on forces. Army units with limited joint support may have to defend while at risk from enemy long-range fires. Forward-stationed forces may defend critical terrain with other coalition forces to delay enemy offensive operations. Some forward-stationed forces may defend joint bases to mitigate the impact of enemy attacks against strategic and operational lines of communications. In both cases, forward-stationed Army forces must be prepared to fight while relatively isolated in the early stages of an enemy attack.

The likelihood of the enemy force's use of massed long-range fires and weapons of mass destruction increases during large-scale combat operations—particularly against command and control (C2) and sustainment nodes, assembly areas, and critical infrastructure. To survive and operate against massed long-range fires and in contaminated environments, commanders ensure as much dispersion as tactically prudent. Army forces seek every possible advantage using dispersion, deception, counterreconnaissance, terrain, cover, concealment, masking, and other procedures to avoid detection and mitigate the impact of enemy fires. In the offense, Army forces maneuver quickly along multiple axes, concentrating only to the degree required to mass effects, and then dispersing to avoid becoming lucrative targets for weapons of mass destruction and enemy conventional fires. Although dispersion disrupts enemy targeting efforts, it increases the difficulty of both C2 and sustainment for friendly forces. Success demands agile units that are able to adjust dispositions rapidly, assume risk, and exploit opportunities when they are available.

The high tempo of large-scale combat operations creates gaps and seams, generating both opportunities and risks as enemy formations disintegrate, disperse, or displace. After generating sufficient combat power for offensive operations, friendly forces may intermingle with or fix and bypass enemy formations. This requires follow-on and supporting units to protect themselves and to defeat enemy remnants in detail within the rear area as part of consolidating gains.

Army forces deploying from the United States and elsewhere face a wide range of threats that are difficult to counter without joint support. The disruptive effects of enemy action may occur at unit home stations, ports of embarkation, while in transit to the theater, and upon arrival at ports of debarkation. Army forces may not have the capability, or the authority, to preempt these attacks, although counterintelligence may aid in early identification of threats. The threat's ability to contest the deployment of forces may degrade combat power available to forward forces and cause unit personnel and equipment to arrive in piecemeal fashion at ports of debarkation.

See pp. 1-39 to 1-44 for more information on deployments contested by threat forces from FM 3-0 (2022).

(Operations) I. Foundations 1-5

Multidomain Operations

IV. Lethality: Overcoming Challenges

Army forces overcome challenges posed by threats and the environment with credible formations able to employ lethal capabilities. Lethality is the capability and capacity to destroy. Employing and threatening the employment of lethal force lies at the core of how Army forces achieve objectives and enable the rest of the instruments of national power to achieve objectives.

Lethality is enabled by formations maneuvering into positions of relative advantage where they can employ weapon systems and mass effects to destroy enemy forces or place them at risk of destruction. The speed, range, and accuracy of weapon systems employed by a formation enhance its lethality. The demands of large-scale combat rapidly deplete available stockpiles and require forces to retain large reserves of ammunition, weapons, and other warfighting capabilities. Leaders multiply the effects of lethal force by employing combinations of capabilities through multiple domains to create, accrue, and exploit relative advantages—imposing multiple dilemmas on enemy forces and overwhelming their ability to respond effectively. Overcoming challenges in the operational environment further requires lethal Army forces that employ all available capabilities to—

- Continuously cultivate landpower networks with allies and partners to facilitate interoperability.
- Be demonstrably prepared for large-scale combat operations to deter conflict on land.
- Employ capabilities in a combined arms manner to create exploitable opportunities.
- Maneuver, mass effects, and preserve combat power to defeat threats to other Service components of the joint force.
- Defend forward-positioned critical joint infrastructure and key terrain.
- Conduct offensive operations to create and exploit opportunities and achieve objectives.
- Consolidate gains during competition, crisis, and armed conflict to enable sustainable political outcomes.

The effective employment of Army forces depends on leaders who understand war, warfare, and the environment within which military forces fight. Gaps in understanding are often causes of failures to achieve sustainable political outcomes with military means.

V. War and Warfare

War is a state of armed conflict between different nations, state-like entities, or armed groups to achieve policy objectives. Wars are fought between nations locally, regionally, or on a global scale. Wars are fought within a nation by a central government against insurgent, separatist, or resistance groups. Armed groups in semiautonomous regions also fight wars to achieve their objectives. Wars range from intense clashes between large military forces—sometimes backed by an official declaration of war—to more subtle hostilities that intermittently breach the threshold of violence.

The object of war is to impose a nation's or group's will on its enemy in pursuit of policy objectives. Regardless of the specific objectives, the decision to wage war represents a major policy decision and changes how Army forces use military capabilities. The nature of war, its principles, and its elements remain consistent over time. However, warfare, the conduct and characteristics of war, reflects changing means and contexts.

1-6 (Operations) I. Foundations

Multidomain Operations

A. The Nature of War

Ref: FM 3-0, Operations (Oct. '22), pp. 1-6 to 1-7.

While the term war has multiple uses depending on the context (for example, the war on drugs or the war on poverty), it is the threat or use of violence to achieve political purposes that distinguishes war in the military context from other human activities. This distinction accounts for three elements of the Army's view of war. War is—

- Fought to achieve a political purpose.
- A human endeavor.
- Inherently chaotic and uncertain.

Note. War, by definition, includes at least two opposing sides. However, not all violence for political gain causes a war. For example, in the current security environment China imposes low levels of violence and new types of violence (including space and cyberspace attacks against government, economic institutions, private industry, and infrastructure) that do not trigger significant military responses. In these cases, China sees itself in a state of war with its adversaries, but its adversaries do not. Such a disparity in perspective is dangerous for those nations opposing China that may endure low levels of violence for long periods, while slowly ceding interests until it is too late to respond effectively. Responding to such situations requires a comprehensive government approach supported by joint and Army forces.

Political Purpose

All U.S. military operations share a common purpose—to achieve or contribute to national policy objectives. As a principle military of war, objective reinforces the proper relationship between military operations and policy. War must always be subordinate to policy and serve a political end. In conjunction with political leaders, military leaders develop strategies to achieve the desired policy outcomes. Policy outcomes often relate to the nation's ability to influence, control, or secure populations, civil infrastructure, natural resources, and access to global commons in all domains.

Human Endeavor

War is shaped by human nature and the complex interrelationships of cognition, emotion, and uncertainty. National sentiments are often targets to be affected or manipulated by one or both sides. Values and ethics are some of the cognitive factors that motivate both the cause for going to war and restrictions in the conduct of war. Fear, passion, camaraderie, grief, and many more emotions affect the resolve of a war's participants. They affect the behavior of combatants, including how and when leaders decide to persevere and when to give up. Individuals react differently to the stress of war; an act that may break the will of one enemy may only serve to stiffen the resolve of another. Human will, instilled through commitment to a cause and leadership, is the driving force of all action in war. The human dimension infuses war with its intangible moral factors.

Inherently Chaotic and Uncertain

War is inherently chaotic and uncertain due to the clash of wills and intense interaction of innumerable factors. Orders are misunderstood, enemy forces do the unexpected, units make wrong turns, unforeseen obstacles appear, the weather changes, and units consume supplies at unexpected rates. This friction affects all military operations, and it must be anticipated by leaders. The chaotic nature of war makes discerning the precise cause and effect of actions difficult, impossible, or delayed. The unintended effects of operations are difficult to anticipate and identify. Such chaos imposes a great deal of uncertainty on all operations and drives the importance of leaders who are skilled at assuming risk.

(Operations) I. Foundations 1-7

B. Principles of War

Ref: FM 3-0, Operations (Oct. '22), pp. 1-7 to 1-8 (table 1-1).

From a U.S. military perspective, war involves nine principles, collectively and classically known as the principles of war. The nine principles of war represent the most important factors that affect the conduct of operations, and they are derived from the study of history and experience in battle.

The principles of war capture broad and enduring fundamentals for the employment of forces in combat. They are not a checklist that guarantees success. Rather, they summarize considerations commanders and their staffs account for during successful operations, applied with judgment in specific contexts. While applicable to all operations, they do not apply equally or in the same way to every situation.

For more information on the principles of war, refer to FM 3-0, app. A.

Maneuver. Place the enemy in a position of disadvantage through the flexible application of combat power.

Objective. Direct every military operation toward a clearly defined, decisive, and attainable goal.

Offensive. Seize, retain, and exploit the initiative.

Surprise. Strike at a time and place or in a manner for which the enemy is unprepared.

Economy of Force. Expend minimum-essential combat power on secondary efforts to allocate the maximum possible combat power on the main effort.

Mass. Concentrate the effects of combat power at the most advantageous place and time to produce decisive results.

Unity of Command. Ensure unity of effort under one responsible commander for every objective.

Security. Prevent the enemy from achieving surprise or acquiring unexpected advantage.

Simplicity. Increase the probability that plans can be executed as intended by preparing clear, uncomplicated plans and orders.

Additional Principles of Joint Operations

- **Restraint**. The purpose of restraint is to prevent the unnecessary use of force. A single act could cause significant military and political consequences; therefore, judicious use of force is necessary.

- **Perseverance**. The purpose of perseverance is to ensure the commitment necessary to attain the national strategic end state. Perseverance involves preparation for measured, protracted military operations in pursuit of the national strategic end state.

- **Legitimacy**. The purpose of legitimacy is to maintain legal and moral authority in the conduct of operations. Legitimacy, which can be a decisive factor in operations, is based on the actual and perceived legality, morality, and rightness of the actions from the various perspectives of interested audiences.

The Army's multidomain operations concept accounts for the constant nature of war and the changing character of warfare. Its balanced approach guides how Army forces operate across the competition continuum given the prevailing characteristics of anticipated operational environments now and in the near future. Doctrine for the conduct of operations begins with a view of war and warfare that includes the—

C. Characteristics of Warfare

Warfare, the conduct and characteristics of war, is affected by changes in technology, national policy, operational concepts, public opinion, and many other factors. Warfare may retain similarities over time, but it inevitably also has great variations. Rapid advances in, and the proliferation of, air, space, and cyberspace capabilities with military applications are changing warfare. Space technology enables persistent overhead surveillance and global communications, navigation, timing, missile warning, and environmental monitoring. Cyberspace technology is integrated into most military capabilities, and it enables near-instantaneous communications and information sharing, creating both opportunities and vulnerabilities that can be exploited by both sides during competition, crisis, and conflict.

D. Methods of Warfare (Conventional/Irregular)

Although the nature and principles of war reflect the continuity of war, the conduct of warfare, like dynamic operational environments, reflects wide variation. Therefore, depending on the situation, strategic actors pursue their objectives in war through different methods of warfare. There are many different methods, but they generally fall into two broad categories: conventional and irregular. Each method of warfare serves the same strategic purpose—to defeat an enemy—but they take fundamentally different approaches to achieving their purpose. Both methods share one characteristic, which is that they involve the use of lethal force to achieve a political end. Warfare rarely fits neatly into any of these subjective categories, and it almost always entails a blend of both methods over the course of a conflict.

Note. These broad categories describe the overall approaches to warfare. Other categories attempt to describe the dominant means used in a particular application, for example "information warfare," "cyber warfare," or "anti-submarine warfare," In these cases, the terms "warfare," "operations," and "activities" are often used interchangeably.

Conventional Warfare

Conventional warfare is a violent struggle for domination between nation-states or coalitions of nation-states. Conventional warfare is generally carried out by two or more military forces through armed conflict. It is commonly known as conventional warfare because it means to fight enemy forces directly, with comparable military systems and organizations. A nation-state's strategic purpose for conducting conventional warfare is to impose its will on an enemy government and avoid imposition of the enemy government's will on it and its citizens. Joint doctrine refers to conventional warfare as "traditional" because it has been understood that way in the West since the Peace of Westphalia (1648), which reserved, for the nation-state alone, a monopoly on the legitimate use of force. However, irregular warfare has a longer history, and it has been just as common as the "traditional" method of warfare in some societies.

Conventional warfare normally focuses on defeating enemy armed forces, enemy warfighting capabilities, and controlling key terrain and populations to decisively influence an enemy government's behavior in favorable ways. During conventional warfare, enemies engage in combat openly against each other and generally employ similar capabilities. Conventional war may escalate to include nation-state use of weapons of mass destruction. Like the other branches of the armed forces, the Army is organized, trained, and equipped primarily to conduct or deter conventional warfare, especially its most lethal manifestation— large-scale combat operations.

See following page for discussion of irregular warfare.

Multidomain Operations

Irregular Warfare
Ref: FM 3-0, Operations (Oct. '22), p. 1-9.

Irregular warfare is the overt, clandestine, and covert employment of military and non-military capabilities across multiple domains by state and non-state actors through methods other than military domination of an adversary, either as the primary approach or in concert with conventional warfare.

Irregular warfare may include the use of indirect military activities to enable partners, proxies, or surrogates to achieve shared or complementary objectives. The main objective of irregular warfare varies with the political context, and it can be successful without being combined with conventional warfare (for example, the Cuban Revolution). While it often focuses on establishing influence over a population, irregular warfare has also historically been an economy of force effort to fix enemy forces in secondary theaters of conflict or to cause enemy leaders to commit significant forces to less critical lines of effort. Two characteristics distinguish irregular warfare from conventional warfare:

- The intent is to erode a political authority's legitimacy and influence or to exhaust its resources and will—not to defeat its armed forces—while supporting the legitimacy, influence, and will of friendly entities engaged in the struggle.
- The nonmilitary instruments of power are more prominent because the military instrument of power alone is insufficient to achieve desired objectives.

JFCs can employ most Army forces and capabilities during irregular warfare. Certain forces and capabilities are irregular warfare focused (for example Army special operations forces), in that they are specifically designed and organized for irregular warfare, but they can also be employed effectively in conventional warfare (for example as combat advisors to host-nation forces). Other forces are irregular warfare capable, in that they are primarily designed and organized for conventional warfare, but they can also be employed effectively in irregular warfare. Historically, the overwhelming majority of Army forces employed to conduct irregular warfare have been conventional forces.

Irregular forces are armed individuals or groups who are not members of the regular armed forces, police, or other internal security forces. The irregular OPFOR can be part of the hybrid threat (HT). The irregular OPFOR component of the HT can be insurgents, guerrillas, or criminals or any combination thereof. The irregular OPFOR can also include other armed individuals or groups who are not members of a governing authority's domestic law enforcement organizations or other internal security forces.

A hybrid threat is the diverse and dynamic combination of regular forces, irregular forces, and/or criminal elements all unified to achieve mutually benefitting effects. Irregular forces are armed individuals or groups who are not members of the regular armed forces, police, or other internal security forces. Irregular forces are unregulated and as a result act with no restrictions on violence or targets for violence. OPFOR5 topics and chapters include irregular and hybrid threat (components, organizations, strategy, operations, tactics), insurgents and guerillas forces, terrorists (motivations, behaviors, organizations, operations and tactics), criminals (characteristics, organizations, activities), noncombatants (armed & unarmed), foreign security forces (FSF) threats, and functional tactics.

1-10 (Operations) I. Foundations

E. Offense, Defense, and Stability

Offense, defense, and stability are inherent elements of conventional and irregular warfare. Divisions and higher echelons typically perform some combination of all three elements in their operations simultaneously. However, the lower the echelon, the more likely it is for that formation to be focused on one element at a time.

An offensive operation is an operation to defeat or destroy enemy forces and gain control of terrain, resources, and population centers (see pp. 1-118 to 1-125). A defensive operation is an operation to defeat an enemy attack, retain key terrain, gain time, and develop conditions favorable for offensive or stability operations (see pp. 1-110 to 1-117). A stability operation is an operation conducted outside the United States in coordination with other instruments of national power to establish or maintain a secure environment and provide essential governmental services, emergency infrastructure reconstruction, and humanitarian relief (see pp. 4-8 to 4-9).

F. Large-Scale Combat Operations *(See p.1-87.)*

The focus of Army readiness is on large-scale combat operations. Large-scale combat operations are extensive joint combat operations in terms of scope and size of forces committed, conducted as a campaign aimed at achieving operational and strategic objectives (ADP 3-0). During ground combat, they typically involve operations by multiple corps and divisions, and they typically include substantial forces from the joint and multinational team. Large-scale combat operations often include both conventional and irregular forces on both sides.

Conflicts encompassing large-scale combat operations are more intense and destructive than limited contingencies, often rapidly amassing heavy casualties. Peer threats employ networks of sensors and long-range massed fires that exploit electromagnetic signatures and other detection methods to create high risk for ground forces, particularly when they are static. Army forces must account for constant enemy observation, including the threat from unmanned systems that saturate the operational environment. Army forces take measures to defeat the enemy's ability to effectively mass effects while creating exploitable advantages to mass effects against enemy capabilities and formations.

G. Combined Arms

Combined arms is the synchronized and simultaneous application of arms to achieve an effect greater than if each element was used separately or sequentially (ADP 3-0). Leaders combine arms in complementary and reinforcing ways to protect capabilities and amplify their effects. Confronted with a constantly changing situation, leaders create new combinations of capabilities, methods, and effects to pose new dilemmas for adversaries. The combined arms approach to operations during competition, crisis, and armed conflict is foundational to exploiting capabilities from all domains and their dimensions.

Complementary capabilities compensate for the vulnerabilities of one system or organization with the capabilities of a different one. Infantry protects tanks from enemy infantry and antitank systems, while tanks provide mobile protected firepower for the infantry. Ground maneuver can make enemy forces displace and become vulnerable to joint fires, while joint fires can disrupt enemy reserves and C2 to enable operations on the ground. Cyberspace and space capabilities and electromagnetic warfare can prevent enemy forces from detecting and communicating the location of friendly land-based fires capabilities, and Army fires capabilities can destroy enemy ground-based cyberspace nodes and electromagnetic warfare platforms to protect friendly communications.

Reinforcing capabilities combine similar systems or capabilities to amplify the overall effects a formation brings to bear in a particular context. During urban operations, for example, infantry, aviation, and armor units working in close coordination reinforce the protection, maneuver, and direct fire capabilities of each unit type while creating

(Operations) I. Foundations 1-11

H. Levels of Warfare

Ref: FM 3-0, Operations (Oct. '22), pp. 1-11 to 1-14.

The levels of warfare are a framework for defining and clarifying the relationship among national objectives, the operational approach, and tactical tasks (ADP 1-01). While the various methods of warfare are ultimately expressed in concrete military action, the four levels of warfare—national strategic, theater strategic, operational, and tactical—link tactical actions to achievement of national objectives as shown in figure 1-2.

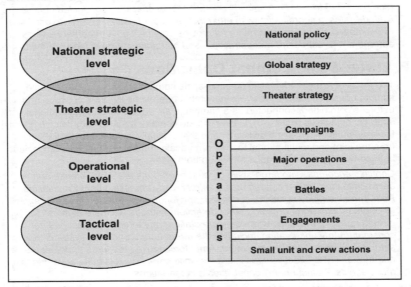

Ref: FM 3-0 (Oct. '22), fig. 1-2. Levels of warfare.

The levels of warfare distinguish four broad overlapping activities—providing national direction and creating national strategy (national strategic), conducting continuous theater campaigning (theater strategic); planning and conducting campaigns and major operations (operational); or planning and executing operations, battles, engagements, and actions (tactical). Some commanders act at more than one level of war. For example, a combatant commander (CCDR) formulates theater strategy and designs the campaign plan. A land component commander assists a CCDR in campaign design and may lead a field army during major operations. The levels of warfare are conceptual, without finite limits or boundaries. They do, however, correlate to specific activities and responsibilities. They help commanders visualize the relationships and actions required to link strategic objectives, military operations at various echelons, and tactical actions. Among the levels of warfare, planning horizons, methods, and products differ greatly. Without this context, tactical operations become disconnected from operational end states and strategic objectives. There are skills and practices related to strategic, theater strategic, operational, and tactical level that differ from each other and are enhanced by specific training and education.

National Strategic Level of Warfare

The national strategic level of warfare is the level of warfare at which the U.S. government formulates policy goals and ways to achieve them by synchronizing action across government and unified action partners and employing the instruments of national power. The instruments of national power are all of the means available to the government in its pursuit of national objectives, expressed as diplomatic, economic, informational, and military. The national strategic level of warfare focuses on developing global strategy and providing global strategic direction. Strategic direction provides context, tasks, and purpose for the employment of the instruments of national power. The specifics of strategic direction address long-term, emerging, and anticipatory issues or concerns that may quickly evolve due to rapidly changing circumstances. Strategic direction is always evolving and adapting.

Theater Strategic Level of Warfare*

The theater strategic level of warfare is the level of warfare at which combatant commanders synchronize with unified action partners and employ all elements of national power to fulfill policy aims within the assigned theater in support of the national strategy. Based on strategic guidance, CCDRs with assigned areas of responsibility and staffs—with input from subordinate commands, including theater armies and supporting commands and agencies—update their strategic estimates and develop theater strategies. A theater strategy is an overarching construct outlining a combatant commander's vision for integrating and synchronizing military activities and operations with the other instruments of national power to achieve national strategic objectives. The theater strategy prioritizes the ends, ways, and means within the limitations established by the budget, global force management processes, and strategic guidance. The theater strategy serves as the basis for development of the combatant command campaign plan (CCP).

The theater strategic level of warfare is introduced as a fourth distinct level from national strategic level in this edition of FM 3-0 (2022).

Operational Level of Warfare

The operational level of warfare is the level of warfare in which campaigns and operations are planned, conducted, and sustained to achieve operational objectives to support achievement of strategic objectives (JP 3-0). The operational level links the employment of tactical forces to the achievement of strategic objectives.

The operational level of warfare generally is the realm of combatant commands and their Service or functional components and subordinate joint task force (JTF) headquarters and their Service or functional components. This includes the theater army headquarters as the Army Service component to a combatant command and any other echelon operating as an ARFOR, JTF headquarters, or land component command. The focus at this level is on operational art—the design of campaigns and operations by integrating ends, ways, and means, while accounting for risk.

Actions at the operational level of warfare usually involve broader aspects of time and space than tactical actions. The theater army's activities continuously support the CCDR in shaping the operational and strategic situation. Operational-level commanders need to understand the complexities of the operational environment and look beyond the immediate situation. Operational-level commanders seek to create the most favorable conditions possible for subordinate commanders by preparing for future events.

Tactical Level of Warfare (See p. 4-11.)

The tactical level of warfare is the level of warfare at which forces plan and execute battles and engagements to achieve military objectives (JP 3-0). Activities at this level focus on tactics—the employment, ordered arrangement, and directed actions of forces in relation to each other (ADP 3-90). Operational-level headquarters determine objectives and provide resources for tactical operations. Tactical-level commanders plan and execute operations to include battles, engagements, and small-unit actions.

Tactical-level combat operations rise to the level of battles or engagements. A battle is a set of related engagements that lasts longer and involves larger forces than an engagement (ADP 3-90). Battles can affect the course of a campaign or major operation, and they are typically conducted by corps and divisions over the course of days or months. An engagement is a tactical conflict, usually between opposing lower echelon maneuver forces (JP 3-0). Engagements are typically conducted at brigade echelons and below. They are usually short, executed in minutes or hours.

(Operations) I. Foundations 1-13

Multidomain Operations

I. Army Strategic Contexts

Ref: FM 3-0, Operations (Oct. '22), pp. 1-14 to 1-16.

Joint doctrine describes the strategic environment in terms of a competition continuum. Rather than a world either at peace or at war, the competition continuum describes three broad categories of strategic relationships—cooperation, competition below armed conflict, and armed conflict. Each relationship is defined as between the United States and another strategic actor relative to a specific set of policy aims. Cooperation, competition, and even armed conflict commonly go on simultaneously in different parts of the world. Because of this, the needs of CCDRs and Army component commanders in one area are affected by the strategic needs of others.

Refer to JP 3-0 for more information about the joint competition continuum. Note. This manual uses "competition" to mean "competition below armed conflict."

Although combatant commands and theater armies campaign across the competition continuum, Army tactical formations typically conduct operations within a context dominated by one strategic relationship at a time. Therefore, Army doctrine describes the strategic situation through three contexts in which Army forces conduct operations:

Army Strategic Contexts

A **Competition (Below Armed Conflict)**

B **Crisis**

C **Armed Conflict**

Ref: FM 3-0 (Oct. '22), fig. 1-3. Army strategic contexts and operational categories.

The Army strategic contexts generally correspond to the joint competition continuum and the requirements of joint campaigns. Because cooperation is generally conducted with an ally or partner to counter an adversary or enemy, Army doctrine considers it part of competition. Army doctrine adds crisis to account for the unique challenges facing ground forces that often characterize transition between competition and armed conflict.

1-14 (Operations) I. Foundations

A. Competition (Below Armed Conflict) *(See pp. 1-63 to 1-76.)*

Competition below armed conflict exists when two or more state or non-state adversaries have incompatible interests, but neither seeks armed conflict. Nation-states compete with each other using all instruments of national power to gain and maintain advantages that help them achieve their goals. Low levels of lethal force can be a part of competition below armed conflict. Adversaries often employ cyberspace capabilities and information warfare to destroy or disrupt infrastructure, interfere with government processes, and conduct activities in a way that does not cause the United States and its allies to respond with force. Competition provides military forces time to prepare for armed conflict, opportunities to assure allies and partners of resolve and commitment, and time and space to set the necessary conditions to prevent crisis or conflict.

B. Crisis *(See pp. 1-77 to 1-86.)*

A crisis is an emerging incident or situation involving a possible threat to the United States, its citizens, military forces, or vital interests that develops rapidly and creates a condition of such diplomatic, economic, or military importance that commitment of military forces and resources is contemplated to achieve national and/or strategic objectives (JP 3-0). Commanders have to consider the possibility that overt military action may escalate a crisis towards armed conflict. The use of space and cyberspace capabilities provides other options that are less likely to cause escalation. The context of crisis is relative to an adversary, which is different from crisis response, which can result from a natural or human disaster. During crisis, armed conflict has not yet occurred, but it is either imminent or a distinct possibility that requires rapid response by forces prepared to fight if deterrence fails.

Note. A crisis can be long in duration, but it can also reflect a near-simultaneous transition to armed conflict. Leaders do not assume that a crisis provides additional time for a transition to armed conflict.

Army forces contribute to joint operations, seeking to deter further provocation and compel an adversary to de-escalate aggression and return to competition under conditions acceptable for the United States and its allies or partners. Through rapid movement and integration with the joint force, Army forces help signal the readiness and willingness to prevail in combat operations. When authorized, Army forces can inform or influence perceptions about an operation's goals and progress to amplify effects on the ground during a crisis; however, commanders ensure their message aligns with reality and that their narratives are truthful and credible. Army forces help the joint force maintain freedom of action and associated positions of relative advantage through the activities they conduct and their presence on the ground. They operate in a way that disrupts adversary risk calculations about the cost of acting contrary to U.S. national interests, compels de-escalation, and fosters a return to competition conditions favorable to the United States. If deterrence fails to end a crisis, Army forces are better postured for operations during armed conflict.

C. Armed Conflict *(See pp. 1-87 to 1-126.)*

Armed conflict occurs when a state or non-state actor uses lethal force as the primary means to satisfy its interests. Armed conflict can range from irregular warfare to conventional warfare and combinations of both. Entering into and terminating armed conflict is a political decision. Army forces may enter conflict with some advanced warning during a prolonged crisis or with little warning during competition. How well Army forces are prepared to enter into an armed conflict ultimately depends upon decisions and preparations made during competition and crisis. At the onset of armed conflict, forward-positioned Army forces may defend key terrain or infrastructure while seeking opportunities to gain the initiative or reposition to more favorable locations with partner forces. Army forces help JFCs gain and maintain the initiative, defeat enemy forces on the ground, control territory and populations, and consolidate gains to establish conditions for a political settlement favorable to U.S. interests. Army forces provide landpower to the joint force and conduct limited contingency or large-scale combat operations to ensure enduring political outcomes favorable to U.S. interests.

(Operations) I. Foundations 1-15

cascading dilemmas for enemy forces. Army artillery can be reinforced by close air support, air interdiction, and naval surface fire support, greatly increasing both the mass and range of fires available to a commander. Space and cyberspace capabilities used to disrupt enemy communications can reinforce a brigade combat team's (BCT's) ground-based jamming effort to increase the disruption to enemy C2. Military information support operations can amplify the effects of physical isolation on an enemy echelon, making it more vulnerable to friendly force exploitation.

The organic composition, training, and task organization of Army units set conditions for effective combined arms. Throughout operations, commanders assess the operational environment and adjust priorities, change task organization, and request capabilities to create exploitable advantages, extend operational reach, preserve combat power, and accomplish missions.

J. Consolidating Gains (Continuously)

Army commanders must exploit successful operations by continuously consolidating gains during competition, crisis, and armed conflict. Consolidate gains are activities to make enduring any initial operational success and to set the conditions for a sustainable security environment, allowing for a transition of control to other legitimate authorities (ADP 3-0). Consolidation of gains is an integral and continuous part of competition, and it is necessary for achieving success across the range of military operations. Successful consolidation of gains requires a realistic and pragmatic assessment of strategic conditions, ally and partner legitimacy, friendly and adversary relative advantages, and the viability of a sustainable political outcome. Operations to inform and influence foreign audiences also play a key role in achieving lasting outcomes.

During competition, Army forces may consolidate gains from previous conflicts for many years as JFCs seek to maintain relative advantages against a specific adversary and sustain enduring political outcomes. U.S. forces in Europe, Japan, the Republic of Korea, and the Middle East remained in place for decades to consolidate gains made in earlier conflicts. Army forces also consolidate gains by continuously developing multinational interoperability and readiness for large-scale combat operations.

During armed conflict, Army forces deliberately plan to consolidate gains throughout an operation as part of defeating the enemy in detail to accomplish overall policy and strategic objectives. Early and effective consolidation activities are a form of exploitation performed while other operations are ongoing, and they enable the achievement of lasting favorable outcomes in the shortest time span. Tactical units consolidating on an objective can be the first step in consolidating gains. In some instances, Army forces will be the lead for integrating forces and synchronizing activities to consolidate gains. In other situations, Army forces will be in support of allies and partners. Army forces may consolidate gains for a sustained period over large land areas. Military governments in occupied territories stabilize civilian populations. Military authorities may temporarily govern areas until populations are stable enough for transition to legitimate civilian authorities. This transition of control to civil authorities reduces demands on combat power.

While Army forces must continuously consolidate gains throughout an operation, consolidating gains becomes the overall focus of Army forces when large-scale combat operations have concluded. During competition, Army forces may consolidate gains from previous conflicts for many years as JFCs seek to maintain relative advantages against a specific adversary. During crisis, Army forces seek to consolidate whatever gains are made relative to a specific adversary so that the crisis does not occur again.

1-16 (Operations) I. Foundations

VI. Understanding an Operational Environment

An operational environment is the aggregate of the conditions, circumstances, and influences that affect the employment of capabilities and bear on the decisions of the commander (JP 3-0). For Army forces, an operational environment includes portions of the land, maritime, air, space, and cyberspace domains understood through three dimensions (human, physical, and information). The land, maritime, air, and space domains are defined by their physical characteristics.

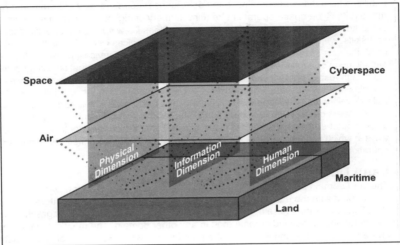

Ref: FM 3-0 (Oct. '22), fig. 1-4. Domains and dimensions of an operational environment. Refer to JP 2-0 and JP 5-0 for more information on describing and analyzing an operational environment from a joint perspective.

An operational environment is the totality of factors that affect what occurs in an assigned area. These factors include actors, events, or actions that occur outside the assigned area. How the many entities behave and interact with each other is difficult to discern. No two operational environments are the same, and all of them continually change. Changes result, in part, from opposing forces and actors interacting, learning, and adapting. The complex and dynamic nature of an operational environment makes determining the relationship between cause and effect challenging, and it contributes to the uncertain nature of war and human competition. This requires that commanders, supported by their staffs, develop and maintain the best possible understanding of their operational environment. Several tools and processes assist commanders and staffs in understanding their operational environment. They include—

- Domains. (*See p. 1-18 to 1-19.*)
- Dimensions. (*See p. 1-20 to 1-21.*)
- Operational and mission variables (*See p. 1-23.*)
- Running estimates (*See p. 3-6 "See Yourself".*)
- Army design methodology (*See p. 3-15.*)
- The military decision-making process (*See p. 3-15.*)
- Building intelligence knowledge (*See chap. 5.*)
- Intelligence preparation of the battlefield (*See pp. 3-20 and 5-6.*)
- Sustainment preparation of the operational environment (*See p. 7-13.*)

(Operations) I. Foundations 1-17

A. (Five) Domains

Ref: FM 3-0, Operations (Oct. '22), pp. 1-18 to 1-21.

Within the context of an operational environment, a domain is a physically defined portion of an operational environment requiring a unique set of warfighting capabilities and skills. Each military Service and branch trains and educates its leaders to be experts about operations in a primary domain, although each Service has some capability in each of the domains, and each develops shared understanding of how to integrate capabilities from different domains. Land operations require mastery of terrain and ground maneuver. Cyberspace operations require mastery of digital information systems and computer code. Space, air, and maritime operations likewise require specific capabilities and skills, which manifest themselves in separate Services within the joint force. Although most domains align with the skills developed in a particular Service, no Service focuses entirely upon or exerts total control of that single domain during operations. Joint commanders assign responsibilities and task-organize based on mission requirements. However, the domains present very different conditions of warfare and require the specialized warfighting skills developed by the different Services and subcomponents within each of the Services. Army leaders do not need to understand all the technical components of what the joint force does in other domains, but they do need to understand the complementary and reinforcing ways in which they can request and employ those capabilities and methods in support of operations on land.

Land Domain *(See also p. 1-34.)*

The land domain is the area of the Earth's surface ending at the high water mark and overlapping with the maritime domain in the landward segment of the littorals (JP 3-31). Variations in climate, terrain, and the diversity of populations have a far greater impact on operations in the land domain than in any other domain. The most distinguishing characteristic of the land domain is the human dimension. Humans transit the maritime, air, and space domains, but they ultimately live, make political decisions, and seek conflict resolution on land.

The nature of combat on land is unique due to the impacts of terrain on all warfighting functions and the application of combat power. For example, terrain provides forces opportunities for evading detection and increasing survivability. It also provides enemy forces the same opportunities. Although technology increases the range of capabilities, complex terrain causes opposing forces to fight at close ranges. Land combatants routinely come face-to-face with one another in large numbers in a wide variety of operational environments containing all types of terrain and potentially nuclear, biological, and chemically degraded environments. When other means fail to drive enemy forces from their positions, Army forces close with and destroy or capture them through close combat. Close combat is warfare carried out on land in a direct-fire fight, supported by direct and indirect fires and other assets (ADP 3-0). The outcome of battles and engagements depends on the ability of Army forces to close with enemy forces and prevail in close combat.

Maritime Domain *(See also pp. 1-35.)*

The maritime domain is the oceans, seas, bays, estuaries, islands, coastal areas, and the airspace above these, including the littorals (JP 3-32). It overlaps with the land domain in the seaward segment of the littoral. Maritime capability may be viewed as global, regional, territorial, coastal, and self-defense forces. Only a few navies are capable of sustained employment far from their countries' shores. However, whether or not their navies are capable of global power projection, most maritime nations also maintain air forces capable of conducting operations over the adjacent maritime domain. This air capability, combined with land-based long-range fires, greatly impacts operations in the maritime domain.

The Navy and its partners employ five functions in a combined arms approach to provide a unique relative advantage for the joint force. These functions are deterrence, operational access, sea control, power projection, and maritime security.

Air Domain *(See also p. 1-34.)*

The air domain is the atmosphere, beginning at the Earth's surface, extending to the altitude where its effects upon operations become negligible (JP 3-30). The speed, range, and payload of aircraft, rockets, missiles, and hypersonic glide vehicles operating in the air domain directly and significantly affect operations on land and sea. Likewise, advances in AMD, electromagnetic warfare, directed energy, and cyberspace capabilities increasingly contest freedom of maneuver in the air.

Control of the air and control of the land are often interdependent requirements for successful campaigns and operations. Control of the air provides a significant advantage when attacking strategically valuable targets at long ranges. However, control of the land is necessary for operating secure airfields and protecting other key terrain that enables air operations. The desired degree of control of the air may vary geographically and over time from no control, to parity, to local air superiority, to air supremacy, all depending upon the situation and the JFC's approved concept of operations.

Space Domain *(See also p. 1-34.)*

The space domain is the area above the altitude where atmospheric effects on airborne objects become negligible. Like the air, land, and maritime domains, space is a physical domain in which military, civil, and commercial activities are conducted. The U.S. Space Command (known as USSPACECOM) has an area of responsibility that surrounds the earth at altitudes equal to, or greater than, 100 kilometers (54 nautical miles) above mean sea level. It has responsibility for planning and execution of global space operations, activities, and missions.

Cyberspace Domain *(See also p. 1-35.)*

For Army forces, the cyberspace domain is the interdependent networks of information technology infrastructures and resident data, including the Internet, telecommunication networks, computer systems, embedded processors and controllers, and relevant portions of the electromagnetic spectrum. Cyberspace is an extensive and complex global network of wired and wireless links connecting nodes that permeate every domain. Cyberspace networks cross geographic and political boundaries to connect individuals, organizations, and systems around the world. Cyberspace allows interactivity among individuals, groups, organizations, and nation-states. Friendly, enemy, adversary, and host-nation networks, communications systems, computers, cellular phone systems, social media, and technical infrastructures are all part of cyberspace. Cyberspace is congested, contested, and critical to successful operations.

Commanders can use cyberspace and electromagnetic warfare capabilities to gain situational awareness and understanding of the enemy through reconnaissance and sensing activities. These reconnaissance and sensing activities augment and enhance the understanding a commander gains from other forms of information collection and intelligence processes. Cyberspace and electromagnetic warfare capabilities enable decision making and protect friendly information. They are a significant means for informing and influencing audiences.

Multidomain Operations

B. (Three) Dimensions

Ref: FM 3-0, Operations (Oct. '22), pp. 1-21 to 1-23.

Understanding the physical, information, and human dimensions of each domain helps commanders and staffs assess and anticipate the impacts of their operations. Operations reflect the reality that war is an act of force (in the physical dimension) to compel (in the information dimension) the decision making and behavior of enemy forces (in the human dimension). Actions in one dimension influence factors in the other dimensions. Understanding the interrelationship enables decision making about how to create and exploit advantages in one dimension and achieve objectives in the others without causing undesirable consequences.

Dimensions

- **Physical Dimension**
- **Information Dimension**
- **Human Dimension**

Physical Dimension

The physical dimension is the material characteristics and capabilities, both natural and manufactured, within an operational environment. While war is a human endeavor, it occurs in a material environment, and it is conducted with physical things. Each of the domains is inherently physical. Terrain, weather, military formations, electromagnetic radiation, weapons systems and their ranges, and many of the things that support or sustain forces are part of the physical dimension. Activities or conditions in the physical dimension create effects in the human and information dimensions.

The electromagnetic spectrum is one of the material characteristics that crosses all the domains. It consists of a range of frequencies of electromagnetic radiation from zero to infinity divided into 26 alphabetically designated bands. The electromagnetic spectrum is relevant in the land, maritime, and air domains because capabilities in those domains depend on electromagnetic spectrum-enabled communications and weapon systems. The electromagnetic spectrum plays a key role in the ability to detect enemy forces that can be identified by their electromagnetic signatures. Conversely, friendly forces must take efforts to mask their electromagnetic signatures to degrade enemy surveillance and reconnaissance efforts.

A physical advantage occurs when a force holds the initiative in terms of a combination of quantitative capabilities, qualitative capabilities, or geographical positioning. Physical advantages are most familiar to tactical forces, and they are typically the immediate goal of most tactical operations. Finding enemy forces, defeating enemy forces, and seizing land areas typically requires the creation and exploitation of multiple physical advantages, including occupation of key terrain, the physical isolation of enemy forces, and the destruction of enemy units. While this dominates tactical operations, leaders understand that physical advantages both complement and are complemented by human and information advantages.

Examples of physical advantage include favorable geography, superior equipment, quantity of resources, and favorable combat power ratios. Superior equipment and favorable geography provide options for seizing the initiative. Superior combat power allows friendly forces to engage enemy forces on favorable terms. The exploitation of physical advantages reduces an enemy force's capacity to fight, creating information and human advantages. Physical advantages implicitly communicate a message that can influence enemy forces' will to fight, sway popular support, and influence enemy risk calculus.

1-20 (Operations) I. Foundations

Information Dimension *(See pp. 2-13 to 2-18.)*

The information dimension is the content, data, and processes that individuals, groups, and information systems use to communicate. Information systems include the technical processes and analytics used to exchange information. The information dimension contains the information itself, including text and images. It also includes the flow or communication pathways of information. Information exchange may be in the form of electromagnetic transmission, print, or speech. The information dimension connects humans to the physical world.

Information transits through all domains in some way or another, whether in electromagnetic transmissions through cyberspace, radar data collected by a destroyer, leaflets dropped from aircraft, social media messaging, books, or satellite photography collected in and transmitted from space. Information, whether true, false, or somewhere in between, is used by friendly, enemy, adversary, and neutral actors to influence the perceptions, decision making, and behavior of individuals and groups. Effective employment of information depends on the audience, message, and method of delivery.

Information is available globally in near-real time. The ability to access information—from anywhere, at any time—broadens and accelerates human interaction, including person to person, person to organization, person to government, and government to government. Social media enables the swift mobilization of people and resources around ideas and causes, even before they are fully understood. Disinformation creates malign narratives that can disseminate quickly and instill an array of emotions and behaviors among groups, ranging from disinterest to violence. From a military standpoint, information enables decision making, leadership, and combat power; it is also a key component of combat power necessary for seizing, retaining, and exploiting the initiative and consolidating gains.

An information advantage is the operational benefit derived when friendly forces understand and exploit the informational considerations of the operational environment to achieve information objectives while denying the threat's ability to do the same. Army forces employ human and physical aspects of the operational environment to gain information advantages. Most types of information advantage result from physical and human factors or activities intrinsic to the operations Army forces conduct. The side possessing better information and using that information more effectively to understand and make decisions has an information advantage. A force that effectively communicates and protects its information while preventing the enemy from doing the same has an advantage. A force that uses information to deceive and confuse an opponent has an advantage. Using information to influence relevant actor behavior more effectively than an adversary or enemy is another information advantage.

Human Dimension

The human dimension encompasses people and the interaction between individuals and groups, how they understand information and events, make decisions, generate will, and act within an operational environment. The will to act and fight emerges from the complex interrelationship of culture, emotion, and behavior. Influencing these factors—by affecting attitudes, beliefs, motivations, and perceptions—underpins the achievement of military objectives.

Commanders and staffs identify relevant actors and anticipate their behavior. Actors are individuals, groups, networks, and populations. Relevant actors are actors who, through their behavior, could substantially impact campaigns, operations, or tactical actions. From this understanding, commanders develop ways to influence relevant actor behavior, decision making, and will through physical and informational means.

A human advantage occurs when a force holds the initiative in terms of training, morale, perception, and will. Human advantages enable friendly morale and will, degrade enemy morale and will, and influence popular support. Examples of human advantages include leader and Soldier competence, morale of troops, and the health and physical fitness of the force. Forces with a cultural affinity to the population in which they operate are also a form of a human advantage. For Army forces, the mission command approach to C2 is a significant human advantage that enhances the friendly decision cycle.

(Operations) I. Foundations 1-21

C. Operational and Mission Variables

The operational and mission variables are tools to assist commanders and staffs in refining their understanding of the domains and dimensions of an operational environment.

See FM 5-0 for more information on operational and mission variables.

Operational Variables - PMESII-PT

Commanders and staffs analyze and describe an operational environment in terms of eight interrelated operational variables: political, military, economic, social, information, infrastructure, physical environment, and time (known as PMESII-PT). The operational variables help leaders understand the land domain and its interrelationships with information, relevant actors, and capabilities in the other domains.

Mission Variables - METT-TC (I)

Commanders analyze information categorized by the operational variables in the context of the missions they are assigned. They use the mission variables, in combination with the operational variables, to refine their understanding of the situation and to visualize, describe, and direct operations. The mission variables are mission, enemy, terrain and weather, troops and support available, time available, and civil considerations, each of which have **informational considerations**. The mission variables are represented as METT-TC (I).

* Informational Considerations *(See pp. 2-13 to 2-18.)*

Informational considerations are those aspects of the human, information, and physical dimensions that affect how humans and automated systems derive meaning from, use, act upon, and are impacted by information.

* METT-TC (I) represents the mission variables leaders use to analyze and understand a situation in relationship to the unit's mission. The first six variables are not new. However, the pervasiveness of information and its applicability in different military contexts requires leaders to continuously assess its various aspects during operations. Because of this, "I" has been added to the METT-TC mnemonic. Information considerations are expressed as a parenthetical variable because they are not an independent consideration, but an important component of each variable of METT-TC that leaders must understand when developing understanding of a situation.

Refer to INFO1: The Information Operations & Capabilities SMARTbook (Guide to Information Operations & the IRCs). INFO1 chapters and topics include information operations (IO defined and described), information in joint operations (joint IO), information-related capabilities (PA, CA, MILDEC, MISO, OPSEC, CO, EW, Space, STO), information planning (information environment analysis, IPB, MDMP, JPP), information preparation, information execution (IO working group, IO weighted efforts and enabling activities, intel support), fires & targeting, and information assessment.

Chap 1
II. Strategic Environment

Ref: FM 3-0, Operations (Oct. '22), chap. 2.

> **Combat Power***
> Combat power is the total means of destructive and disruptive force that a military unit/formation can apply against an enemy at a given time (JP 3-0). It is the ability to fight. The complementary and reinforcing effects that result from synchronized operations yield a powerful blow that overwhelms enemy forces and creates friendly momentum.

* Editor's note: For the purposes of this book, the discussion from FM 3-0 (2022) of combat power is presented in chap. 2, as a means of organizing and introducing chapters 3 through 8 (the six warfighting functions).

I. Strategic Environment

The central challenge to U.S. security is the reemergence of long-term, great power competition with China and Russia as individual actors and as actors working together to achieve common goals.

A. China and Russia *(See pp. 1-90 to 1-93.)*

China uses its rapidly modernizing military, information warfare, and predatory economics to coerce neighboring countries to reorder the Indo-Pacific region to its advantage. Concurrently, Russia seeks veto authority over nations on its periphery in terms of its governmental, economic, and diplomatic decisions, to subvert the North Atlantic Treaty Organization (NATO), and to change European and Middle East security and economic structures to its favor.

B. North Korea and Iran *(See pp. 1-88.)*

In addition to China and Russia, several other states threaten U.S. security. North Korea seeks to guarantee survival of its regime and increase its leverage. It is pursuing a mixture of CBRN, conventional, and unconventional weapons and a growing ballistic missile capability to gain coercive influence over South Korea, Japan, and the United States. Similarly, Iran seeks dominance over its neighbors by asserting an arc of influence and instability while vying for regional hegemony. Iran uses state-sponsored terrorist activities, a network of proxies, and its missile capabilities to achieve its objectives.

C. Non-state Actors (Irregular Warfare) *(See pp. 1-10.)*

While states are the principal actors on the global stage, non-state actors also threaten the strategic environment with increasingly sophisticated capabilities. Terrorists, transnational criminal organizations, threat cyber actors, and other malicious non-state actors have transformed global affairs with increased capabilities of mass disruption. Terrorism remains a persistent tactic driven by ideology and enabled by political and economic structures.

D. Threats / Threat Methods

Ref: FM 3-0, Operations (Oct. '22), pp. 2-7 to 2-10.

Threat

A **threat** is any combination of actors, entities, or forces that have the capability and intent to harm United States forces, United States national interests, or the homeland (ADP 3-0). Threats faced by Army forces are, by nature, hybrid. They include individuals, groups of individuals, paramilitary or military forces, criminal elements, nation-states, or national alliances. In general, a threat can be categorized as an enemy or an adversary:

- An **enemy** is a party identified as hostile against which the use of force is authorized (ADP 3-0). An enemy is also a combatant under the law of war.

- An **adversary** is a party acknowledged as potentially hostile to a friendly party and against which the use of force may be envisaged (JP 3-0). Adversaries pursue interests that compete with those of the United States and are often called competitors.

- **Peer threats** are adversaries or enemies with capabilities and capacity to oppose U.S. forces across multiple domains worldwide or in a specific region where they enjoy a position of relative advantage. Peer threats possess roughly equal combat power to U.S. forces in geographic proximity to a conflict area. Peer threats may also have a cultural affinity with specific regions, providing them relative advantages in the human and information dimensions. *(See discussion below.)*

Peer threats employ strategies that capitalize on their advantages to achieve objectives. When these objectives are at odds with the interests of the United States and its allies, conflict becomes more likely. Peer threats prefer to achieve their goals without directly engaging U.S. forces in combat. They often employ information warfare in combination with conventional and irregular military capabilities to achieve their goals. They exploit friendly sensitivity to world opinion and attempt to exploit American domestic opinion and sensitivity to friendly casualties. Peer threats believe they have a comparative advantage because of their willingness to endure greater hardships, casualties, and negative public opinion. They also believe their ability to pursue long-term goals is greater than that of the United States.

Peer threats employ capabilities from and across multiple domains against Army forces, and they seek to exploit vulnerabilities in all strategic contexts. During conflict, peer threats seek to inflict significant damage across multiple domains in a short amount of time. They seek to delay friendly forces long enough to achieve their goals and end hostilities before friendly forces can decisively respond.

Threat Methods

- **Information Warfare**
- **Systems Warfare**
- **Preclusion**
- **Isolation**
- **Sanctuary**

Peer threats use various methods to render U.S. military power irrelevant whenever possible. Five broad peer threat methods, often used in combination during conventional or irregular conflicts, and below the threshold of conflict, include—

1-24 (Operations) II. Strategic Environment

Information Warfare *(See pp. 2-13 to 2-18.)*

In the context of the threat, information warfare refers to a threat's orchestrated use of information activities (such as cyberspace operations, electronic warfare, and psychological operations) to achieve objectives. Operating under a different set of ethics and laws than the United States, and under the cloak of anonymity, peer threats conduct information warfare aggressively and continuously to influence populations and decision makers. They can also use information warfare to create destructive effects during competition and crisis. During armed conflict, peer threats use information warfare in conjunction with other methods to achieve strategic and operational objectives.

Peer threats use diverse means to conduct information warfare, and these means may include cyberspace operations, perception management, deception, electronic warfare, physical destruction, political warfare, legal warfare, proxies and non-state actors.

Note. Threat forces use the term electronic warfare, which differs from U.S. doctrine's use of electromagnetic warfare.

Threats seek to employ information warfare to attack or disrupt in depth, including within the continental United States, viewing it as a low-cost and low-risk activity. A cyberspace attack may disrupt U.S. infrastructure that impedes deployment of forces, or a disinformation campaign can reduce morale and the will to fight. In some situations, threats use proxies for information warfare to achieve policy aims without having to incur the risks associated with employing military forces or official government entities.

Peer threats typically have fewer policy and legal restrictions than U.S. forces on how they employ information warfare, giving them an initial advantage. They exploit the nature of open societies while restricting their population's access to information. They often obscure their activities to prevent detection or attribution.

Peer threats are free to sow disinformation among U.S. and allied populations while at the same time strictly limiting access to and manipulating the information their own populations receive. They employ all available means to influence a wide range of audiences, including both civilian and military and domestic and international, in support of their goals. Information warfare is a means to exploit shared cultural norms, historical grievances, and a self-serving interpretation of international law to limit U.S. military options and degrade U.S. political will.

Peer threats systematically and continuously combine all of these means to create specific effects within the human, information, and physical dimensions of an operational environment. Peer threats use misinformation, disinformation, propaganda, and information for effect to create doubt, confuse, deceive, and influence U.S. and partner decision makers, forces, and target audiences. They also use information warfare to destroy essential network-based capabilities, such as economic infrastructure, private and government communications, and electrical grids. This use of information warfare is not merely disruptive. It can result in the loss of immense resources and human life, depending on the scale and duration of the attack.

Systems Warfare

Systems warfare is the identification and isolation or destruction of critical subsystems or components to degrade or destroy an opponent's overall system. Peer threats view the battlefield, their own instruments of power, and an opponent's instruments of power as a collection of complex, dynamic, and integrated systems composed of subsystems and components. They use systems warfare to attack critical components of a friendly system while protecting their own system. Simple examples of attacking critical components are adversary use of electronic warfare to disable the links between unmanned aircraft system (UAS) controllers and the aircraft in a specific area, and the emplacement of layered integrated air defense systems from a position of sanctuary to prevent the integration of opposing airpower with ground operations.

Threat Methods (Cont.)

Peer threats believe that a qualitatively or quantitatively weaker force can defeat a superior force, if the weaker force can dictate the terms of combat. Peer threats believe that the systems warfare approach allows them to move away from the conventional approach to combat. Systems warfare makes it unnecessary to match an opponent system-for-system or capability-for-capability. Peer threats seek to locate the critical components of the opposing combat system, determine patterns of interaction and dependencies among components, and identify opportunities to exploit this connectivity.

Systems warfare approaches work in concert with other approaches, and they manifest themselves at the tactical level in terms of integrated fires complexes characterized by surface-to-surface and surface-to-air systems enabled by long-range ISR capabilities. They generally represent one means by which adversaries achieve preclusion at the strategic and operational levels, and they are adversaries' preferred means for destroying friendly forces at the tactical level. An example of systems warfare occurred in Ukraine in 2014.

Preclusion

To preclude is to keep something from happening by taking action in advance. Peer threats use a wide variety of actions, activities, and capabilities to preclude a friendly force's ability to shape an operational environment and mass and sustain combat power. **Antiaccess (A2) and area denial (AD)** are two strategic and operational approaches to preclusion.

- **Antiaccess (A2)** is an action, activity, or capability, usually long-range, designed to prevent an enemy force from entering an operational area (JP 3-0). For example, A2 activities prevent or deny forces the ability to project and sustain forces into a desired area. The employment of A2 capabilities against Army forces begins in the continental United States and extends throughout the strategic support area into a theater. Peer threats have the means to disrupt the United States' force projection capability at home station. These means include ballistic missiles; cruise missiles; and space, cyberspace, and information warfare capabilities.

- **Area denial (AD)** is an action, activity, or capability, usually short-range, designed to limit an enemy force's freedom of action within an operational area (JP 3-0). Usually adversaries do not design area denial to keep friendly forces out, but rather to limit their freedom of action and ability to accomplish their mission within an operational area. Threat forces pursue AD using long-range fires, integrated air defense systems, electronic warfare, CBRN, manmade obstacles, and conventional ground maneuver forces.

Figure 2-1 and figure 2-2 depict employment of A2 and AD approaches in different types of theaters. For illustration purposes, A2 and AD reach are tied to specific capabilities. However, adversary forces can use different actions, activities, or capabilities in an A2 or AD approach.

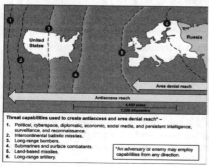

Ref: FM 3-0 (Oct. '22), fig. 2-1.   Ref: FM 3-0 (Oct. '22), fig. 2-2.

1-26 (Operations) II. Strategic Environment

Isolation

Isolation is the containment of a force so that it cannot accomplish its mission. Peer threats will attempt to isolate U.S. forces in several ways. Some examples include—

- Attacking political bonds with allies and partners.
- Preventing or limiting communications to and in an AO.
- Interdicting or severing lines of communication to block support or reinforcement of forward-positioned units.
- Deceiving friendly forces about the current situation and their role in the operational environment.
- Deceiving the public about the current situation to reduce its support of friendly operations that counter threat goals.
- Exploiting inadequate friendly understanding of an operational environment or cultural affinity in an area or region.
- Blocking support or reinforcement of forward-positioned units through direct and indirect fires.
- Using economic coercion.
- Preventing friendly access and overflight.

During competition, peer threats may attempt to isolate friendly forces using disinformation campaigns and the threat of aggression. During crisis, peer threats seek to isolate U.S. forward-positioned forces and prevent their support from the U.S. or elsewhere in theater. During armed conflict, enemy forces identify isolated friendly forces using a variety of capabilities and rapidly attempt to destroy them through long-range, massed, and precision fires.

Sanctuary

Sanctuary is the positioning of threat forces beyond the reach of friendly forces. It is a form of protection derived by some combination of political, legal, and physical boundaries that restricts freedom of action by a friendly force commander. Peer threats will use any means necessary, including sanctuary, to protect key capabilities from destruction, particularly by air and missile capabilities. To create a sanctuary that protects key interests, adversaries employ combinations of both physical and nonphysical means to protect key interests, including—

- International borders.
- Complex terrain.
- Hiding among noncombatants and culturally sensitive structures.
- Counterprecision techniques, including camouflage, concealment, and deception.
- Countermeasures, including decoys, hardened and buried facilities, integrated air defense systems, and long-range fires.
- Information warfare.
- Threatening attacks against the U.S. homeland, possibly using including weapons of mass destruction.
- International law, treaties, and treaty agreements.
- Internal population information control (e.g., denying the internet or jamming external radio and television).

Most means of sanctuary cannot protect an entire enemy force for an extended time. Therefore, a threat will seek to protect selected elements of its forces for enough time to gain the freedom of action necessary to pursue its strategic or diplomatic goals. Threat forces seek to protect their conventional forces, advanced aircraft, and extended-range fires systems. Many peer threats invest in long-range rocket and missile systems capable of counterfire at extreme ranges to allow sanctuary behind international borders. Improved air defense systems, including counter ballistic missile systems, often provide protection for these advanced fires capabilities.

(Operations) II. Strategic Environment 1-27

II. Unified Action and Army Forces

To counter threats and protect national interests worldwide, the Armed Forces of the United States operate as a joint force in unified action. Unified action is the synchronization, coordination, and/or integration of the activities of governmental and nongovernmental entities with military operations to achieve unity of effort (JP 1, Volume 1). Unity of effort is coordination and cooperation toward common objectives, even if the participants are not necessarily part of the same command or organization, which is the product of successful unified action (JP 1, Volume 2). Army forces, as part of unified action, conduct operations in support of the joint force, with multinational allies and partners, and in coordination with other agencies and organizations. The Army's contribution to unified action is multidomain operations which seek to employ all available capabilities in unexpected combinations that create and exploit relative advantages. Leaders must be capable of employing all unified action partners to the greatest extent possible, including conventional forces, special operations forces, allies, partner-nation forces, territorial defense forces, and any other organization or individual whose efforts can legally be harnessed to help achieve objectives.

See pp. 1-30 to 1-31 for an overview and discussion of unified action.

A. Joint Operations and Activities

Single Services may perform tasks and missions to support Department of Defense (DOD) objectives. However, the DOD primarily employs two or more Services (from two military departments) in a single operation from, in, and across multiple domains, particularly in combat, through joint operations. Joint operations are military actions conducted by joint forces and those Service forces employed in specified command relationships with each other, which, of themselves, do not establish joint forces (JP 3-0). A joint force is a force composed of elements, assigned or attached, of two or more military departments operating under a single joint force commander (JP 3-0). Joint operations exploit the advantages of interdependent Service capabilities in multiple domains through unified action.

B. Multinational Operations

Multinational operations is a collective term to describe military actions conducted by forces of two or more nations, usually undertaken within the structure of a coalition or alliance (JP 3-16).

C. Interagency Coordination and Interorganizational Cooperation

Interagency coordination is a key part of unified action. Interagency coordination is the planning and synchronization of efforts that occur between elements of Department of Defense and participating United States Government departments and agencies (JP 3-0). Army forces conduct and participate in interagency coordination using established liaison, personal engagement, and planning processes.

D. Conventional & Special Operations Forces Integration

Army forces integrate conventional and special operations forces to create complementary and reinforcing effects during operations. The mission and operational environment drive the command and support relationships between conventional and special operations forces during an operation. During large-scale combat, conventional forces contribute mass across all warfighting functions required to defeat enemy forces. Special operations forces complement conventional forces by performing their core activities:

- Civil affairs operations.
- Countering weapons of mass destruction. *(continued on p. 1-36.)*

1-28 (Operations) II. Strategic Environment

III. Why the Army Needs Multidomain Operations

Ref: Presentation Slide, FM 3-0 Team Brief, Combined Arms Center (2022).

Multidomain operations **fracture the coherence** of threat operational approaches by repeatedly **disintegrating, dislocating, isolating, and destroying** their interdependent systems and formations and exploiting the opportunities to **defeat enemy forces in detail**. Army forces **combine** maneuver and targeting methods to accrue **relative advantages** over time and **defeat enemy** formations and systems.

Theater Strategic Challenge

Campaign in support of integrated deterrence, counter aggression, and prepare for armed conflict without undesirable escalation.

- Develop multinational interoperability
- Counter threat information warfare
- Control escalation
- Establish forward force posture
- Protect intra-theater LOCs
- Demonstrate credible combat readiness

Operational Challenge

Defeat enemy integrated air defense systems (IADS) and integrated fires control (IFC, see p. 1-100) while preserving combat power necessary to exploit opportunities, extend operational reach, and defeat enemy forces.

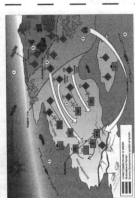

- Employ/protect forward-stationed forces
- Provide air missile defense (AMD)
- Counter enemy AMD
- Apportion joint capabilities to tactical forces
- Provide communications network and sustainment

Tactical Challenge

Defend critical terrain with forward postured forces. Conduct offensive, expeditionary operations against enemy prepared defense within range of enemy ability to mass fires.

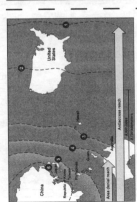

- Employ joint and Army capabilities to achieve convergence
- Protect subordinate echelons
- Maintain tempo and extend operational reach in degraded communications environment

Unified Action

Ref: FM 3-0, Operations (Oct. '22), pp. 2-13 to 2-14 and JP 3-0, Joint Campaigns and Operations (Jun '22), pp. I-4 to I-5.

Joint Operations and Activities

Single Services may perform tasks and missions to support Department of Defense (DOD) objectives. However, the DOD primarily employs two or more Services (from two military departments) in a single operation from, in, and across multiple domains, particularly in combat, through joint operations. Joint operations are military actions conducted by joint forces and those Service forces employed in specified command relationships with each other, which, of themselves, do not establish joint forces (JP 3-0). A joint force is a force composed of elements, assigned or attached, of two or more military departments operating under a single joint force commander (JP 3-0). Joint operations exploit the advantages of interdependent Service capabilities in multiple domains through unified action. Joint planning integrates military power with other instruments of national power (including diplomatic, economic, and informational) to achieve a desired military end state. The end state is the set of required conditions that defines achievement of the commander's objectives (JP 3-0). Joint planning connects the strategic end state to the joint force commander's (JFC's) campaign design and ultimately to tactical missions. JFCs use campaigns and major operations to translate their operational-level actions into strategic results.

Competition Continuum

The competition continuum describes a world of enduring competition conducted through a mixture of cooperation, adversarial competition below armed conflict, and armed conflict. Elements of the Competition Continuum:

- **Cooperation**. Situations in which joint forces take actions with another strategic partner in pursuit of policy objectives.
- **Adversarial Competition**. Joint forces or multinational forces can take actions below armed conflict against a state or non-state adversary in pursuit of policy objectives in response to antagonistic and threatening behavior.
- **Armed Conflict/War**. Armed conflict/war occurs when a state directs its military forces to take actions against an enemy in hostilities or declared war.

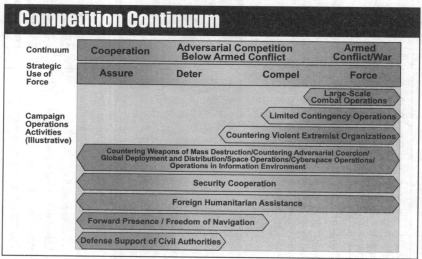

Ref: JP 3-0 (Jun '22), fig. I-1. Competition Continuum

Multidomain Operations

The joint force is organized, trained, and equipped for **sustained large-scale combat** anywhere in the world. The capabilities to conduct large-scale combat operations enable a wide variety of other operations and activities. In particular, opportunities exist prior to large-scale combat to shape an operational environment to prevent, or at least mitigate, the effects of war.

Doctrine categorizes joint operations and activities by their focus. In some cases, the title covers a variety of missions, tasks, and activities. Many activities are accomplished by Army forces and do not constitute joint operations, such as tasks associated with **security cooperation**. Nonetheless, most of these occur under a joint "umbrella," because they contribute to achievement of CCDRs' campaign objectives. Examples include:

- Stability activities
- Defense support of civil authorities
- Foreign humanitarian assistance
- Personnel recovery
- Noncombatant evacuation
- Peace operations
- Countering weapons of mass destruction
- Chemical, biological, radiological, and nuclear response
- Foreign internal defense
- Counterdrug operations
- Combating terrorism
- Counterinsurgency
- Homeland defense
- Mass atrocity response
- Security cooperation
- Military engagement
- Military exercises
- Flexible deterrent options
- Flexible response options

Ref: FM 3-0 (Oct. '22), fig. 2-3. Examples of operations and activities.

Multinational Operations

Multinational operations is a collective term to describe military actions conducted by forces of two or more nations, usually undertaken within the structure of a coalition or alliance (JP 3-16). While each nation has its own interests and often participates within the limitations of national caveats, all nations bring value to an operation. Each nation's force has unique capabilities, and each usually contributes to an operation's legitimacy in terms of international or local acceptability. Army forces should anticipate that most operations will be multinational and plan accordingly.

Refer to FM 3-16 for more information on multinational operations.

Interagency Coordination and Interorganizational Cooperation

Interagency coordination is a key part of unified action. Interagency coordination is the planning and synchronization of efforts that occur between elements of Department of Defense and participating United States Government departments and agencies (JP 3-0). Army forces conduct and participate in interagency coordination using established liaison, personal engagement, and planning processes.

Refer to FM 3-57 for more information on civil-military integration.

Refer to JFODS6: The Joint Forces Operations & Doctrine SMARTbook, 6th Ed. (Guide to Joint, Multinational & Interorganizational Operations). Completely updated for 2023, chapters include joint doctrine fundamentals (JP 1), joint operations (JP 3-0, 2022), joint planning (JP 5-0, 2020), joint logistics (JP 4-0), joint task forces (JP 3-33), joint force operations (JPs 3-30, 3-31, 3-32 & 3-05), multinational operations (JP 3-16),and interorganizational cooperation (JP 3-08).

(Operations) II. Strategic Environment 1-31

G. Army Echelons

Ref: FM 3-0, Operations (Oct. '22), pp. 2-17 to 2-20.

The Army operates through the use of echelons to ensure manageable spans of control for leaders. Echelons generally correspond to a particular level of warfare, but they may contribute to two or more levels depending on the situation.

Generally, higher echelons (for example, divisions and higher) have greater experience in their command teams and staffs. They have the expertise and perspective to coordinate large-scale operations and complex or politically sensitive tasks. They retain control of scarce resources so that they can employ them at the right time and place. This often includes joint air, space, maritime, and cyberspace capabilities. Higher echelons generally employ these critical capabilities to set conditions for lower echelon success and to weight the main effort appropriately. Higher echelons maneuver subordinate formations and use capabilities from all domains to shape the environment and create and exploit relative advantages. Generally speaking, the joint force command degrades enemy strategic capabilities to enable forcible entry and sustained operations. The land component command sets the theater, defeats enemy long-and mid-range fires, provides operational-level sustainment, and apportions joint capabilities to corps. Corps, operating as tactical formations, defeat enemy mid-range fires, employ joint capabilities to set conditions for divisions to maneuver, and maintain the tempo of operations through sustainment and other rear operations. Divisions defeat enemy short-range fires, mass effects on enemy forward echelons, and synchronize BCT maneuver in close combat with enemy forces. BCTs conduct close operations to defeat and destroy enemy forces during battles and engagements.

Theater Army

The theater army's mission is the most diverse and complex of any Army echelon. The theater army headquarters is tailored to a specific CCDR with the ability to conduct both operational and administrative C2 over Army forces theater wide. It provides enabling capabilities appropriate to theater conditions, such as theater intelligence, theater sustainment, theater signal, theater fires, theater information activities, civil affairs, engineer, and theater medical. In theaters without assigned field armies, corps, or divisions, the theater army assumes direct responsibility across warfighting functions for its tactical commands. The theater army is the Army Service component command to a geographic combatant command.

Refer to FM 3-94 and ATP 3-93 for additional information on theater army administrative and operational requirements.

Field Army

A field army is constituted to meet specific requirements. A field army may consist of a headquarters battalion with subordinate companies and special troops, a variable number of attached corps, an attached expeditionary sustainment command, a variable number of divisions normally attached to corps, and other attached functional and multifunctional brigades. When required, a field army is an operational headquarters that provides C2 over multiple corps. During operations, forces are assigned or attached to the field army. Field armies are most likely to be employed in theaters where peer adversaries have the capability of conducting large-scale combat. These regions include the U.S. European Command and U.S. Indo-Pacific Command.

Corps

The corps is the most versatile echelon above brigade due to its ability to operate at both the tactical and operational levels. While it is organized, staffed, trained, and equipped to fight as a tactical formation, the corps may be called upon to become a joint and multinational headquarters for conducting operations. When operating as the senior Army headquarters under a joint task force (JTF), the corps will serve as the ARFOR. The corps can also serve as the coalition forces land component commander (CFLCC) when

1-32 (Operations) II. Strategic Environment

properly augmented with joint and multinational personnel. If the corps is uncommitted to specific CCDR requirements, it focuses on building and sustaining readiness to prevail in large-scale combat operations. The roles of the corps include acting as the—

- Senior Army tactical formation in large-scale combat, commanding two to five Army divisions together with supporting brigades and commands.
- ARFOR (with augmentation) within a joint force for campaigns and major operations when a field army is not present.
- JTF headquarters (with significant augmentation) for crisis response and limited contingency operations.
- CFLCC (with significant augmentation) commanding Army, Marine Corps, and multinational divisions together with supporting brigades and commands when a field army is not present.

During large-scale combat operations, a corps headquarters normally functions as a tactical headquarters under a joint or multinational land component. The corps is the echelon best positioned and resourced to achieve convergence with Army and joint capabilities.

Refer to FM 3-94 and ATP 3-92 for more information about Army corps.

Division

The division is the Army's principal tactical warfighting formation during large-scale combat operations. Its primary role is to serve as a tactical headquarters commanding brigades. A division conducts operations in an AO assigned by its higher headquarters—normally a corps. It task-organizes its subordinate forces according to the mission variables to accomplish its mission. A division typically commands between two and five BCTs, a mix of functional and multifunctional brigades, and a variety of smaller enabling units. The division is typically the lowest tactical echelon that employs capabilities from multiple domains to achieve convergence during large-scale combat operations. Winning battles and engagements remains the division's primary purpose. During limited contingencies, it can organize itself to serve in multiple roles. The roles of the division include acting as a—

- Tactical headquarters.
- ARFOR headquarters (with significant augmentation).
- CFLCC (with significant augmentation).
- JTF headquarters (with significant augmentation).

Refer to FM 3-94 and ATP 3-91 for more information about Army divisions.

Brigade Combat Teams (BCT)

A BCT is the Army's primary combined arms, close-combat maneuver force. BCTs maneuver against, close with, and destroy enemy forces. BCTs seize and retain key terrain, exert constant pressure, and break the enemy's will to fight. They are the principal ground maneuver units of a division or a JTF. Divisions seek to employ BCTs in mutually supporting ways to the greatest extent possible. However, BCTs must be capable of fighting isolated from higher echelon headquarters and adjacent units during periods of degraded communication and when operations are widely distributed.

Refer to FM 3-96 for more information on Army BCTs.

Multifunctional and Functional Brigades

Theater armies, corps, and divisions are task-organized with an assortment of multifunctional and functional brigades to support their operations. These brigades add capabilities such as intelligence, attack and reconnaissance aviation, fires, protection, contracting support, or sustainment. The theater army may tailor subordinate corps and divisions with combinations of multifunctional brigades. Multifunctional brigades provide a variety of functions in support of operations. A functional brigade provides a single function or capability.

(Operations) II. Strategic Environment 1-33

H. Domain Interdependence

Ref: FM 3-0, Operations (Oct. '22), pp. 2-15 to 2-17.

The Army provides forces and capabilities from all domains to the joint force. Army forces employ joint capabilities from all domains to complement and reinforce their own capabilities. Understanding domain interdependences helps leaders better mitigate friendly vulnerabilities while creating and exploiting relative advantages. Successful operations in an environment where the enemy can contest every domain requires continuous joint integration down to the lowest tactical echelons.

See pp.1-18 to 1-19 for an overview and description of the five domains.

Land Capabilities enable...

Air Operations *(See pp. 6-6 to 6-11.)*

Land capabilities enable air operations in multiple ways. Some of these ways include—

- Fixing enemy ground forces for destruction from the air.
- Providing air-delivered fires through rotary-wing and UAS platforms.
- Controlling, securing, and defending airports and airfields.
- Securing land-based C2 nodes for air operations.
- Destroying enemy surface-to-air systems.
- Employing surface-to-air fires.
- Integrating all-source intelligence to identify threats to friendly air capabilities.
- Providing logistics support to other Service components.

Air capabilities enable land operations in multiple ways. Some of these ways include—

- Providing air-to-ground fires.
- Providing offensive and defensive depth through air interdiction and strategic attack.
- Protecting ground forces from air attack.
- Employing airborne platforms for information collection.
- Providing aerial movement of personnel, equipment, and supplies.
- Employing airborne electromagnetic warfare platforms.

Space Operations *(See p. 6-12.)*

Land capabilities enable space operations in multiple ways. Some of these include—

- Destroying enemy space ground stations, ground links, and launch sites with surface-to-surface fires.
- Securing ground links and launch sites.
- Securing bases and C2 nodes for units controlling space capabilities.
- Securing bases and C2 nodes from which to launch attacks against enemy space capabilities.

Space capabilities enable land operations in multiple ways. Some of these include—

- Enabling geolocation and timing-dependent technology, including global positioning systems and precise and accurate fires.
- Enabling a global C2 network through satellite communications.
- Enhancing situational understanding by providing meteorological, oceanographic, and space environmental factors and detailed imagery of land areas and enemy dispositions on land.

1-34 (Operations) II. Strategic Environment

- Deceiving, disrupting, degrading, denying, or destroying enemy space systems.
- Conducting navigation warfare to disrupt enemy use of positioning, navigation, and timing-enabled devices.
- Enabling theater missile warning and other warning intelligence.

Cyberspace Operations *(See pp. 2-16 to 2-17 and 6-13.)*
Land capabilities enable cyberspace operations in multiple ways. These include—

- Securing critical cyberspace infrastructure including data storage facilities, wired network transport, ground-based repeaters, and terminals.
- Conducting information activities that protect and defend joint communications networks and data.
- Conducting physical attacks against enemy cyberspace-based capabilities and infrastructure on land.
- Defeating enemy forces collecting information through cyberspace.

Cyberspace capabilities enable land operations in multiple ways. Some of these ways include—

- Enabling secure global communications and a shared COP.
- Supporting decision making and logistics.
- Facilitating high-volume data storage and knowledge management.
- Networking sensors and fires platforms.
- Attacking enemy networks including C2, integrated air defense systems, and integrated long-range fires systems.
- Enabling rapid communication to audiences through social media and other applications.
- Enabling targeted influence operations.

Maritime Operations *(See pp. 1-27 to 1-138.)*
Land capabilities enable maritime operations in multiple ways. Some of these ways include—

- Attacking land-based threats to maritime capabilities, including enemy air bases, surface-to-surface fires, and sensors.
- Protecting ports and defending land areas that control maritime choke points.
- Denying maritime areas with surface-to-surface fires and surface-to-air fires.
- Integrating joint all-source intelligence to identify threats to maritime capabilities.
- Providing directed logistics support to maritime oriented forces operating from land.

Maritime capabilities enable land operations in multiple ways. Some of these ways include—

- Increasing operational reach and lethality through long-range fires systems and information collection.
- Providing access to otherwise inaccessible land areas.
- Providing and protecting transportation of units, equipment, and supplies on a large scale, over strategic distances.
- Integrating with all-source intelligence.
- Preventing enemy forces from using sea lines of communications and supply routes.
- Attacking enemy maritime threats to land forces.

(Operations) II. Strategic Environment 1-35

- Counterinsurgency.
- Counterterrorism.
- Direct action.
- Foreign humanitarian assistance.
- Foreign internal defense.
- Hostage rescue and recovery.
- Military information support operations.
- Security force assistance
- Special reconnaissance.
- Unconventional warfare.
- SOF contributions during deep and extended deep operations are often critical to setting conditions for conventional close and rear operations.
- SOF contributions during deep and extended deep operations are often critical to setting conditions for conventional close and rear operations.

E. Joint Interdependence

Joint interdependence is the purposeful reliance by one Service on another Service's capabilities to maximize complementary and reinforcing effects of both. The degree of interdependence varies with specific circumstances. The Army depends on the other Services for strategic and operational mobility, joint fires, and other key enabling capabilities. The Army supports the other Services, combatant commands, and unified action partners with ground-based indirect fires and air and missile defense (AMD), defensive cyberspace operations, electromagnetic warfare, communications, intelligence, rotary-wing aircraft, logistics, and engineering. The Army's ability to set and sustain a theater of operations is essential to allowing the joint force freedom of action. The Army establishes, maintains, and defends vital infrastructure. It also provides the JFC with unique capabilities, such as port and airfield opening, logistics, CBRN defense, and reception, staging, onward movement, and integration (RSOI).

F. Army Force Posture

The Army postures forces in a way that balances the need for sustainable readiness with the need for responsiveness. Forward-stationed rotational forces provide CCDRs with support to operations during competition and rapid response during crisis. These forces are usually small in number and may be vulnerable if the situation rapidly escalates to armed conflict. Forces based in strategic support areas allow for unit training and a sustainable readiness cycle. These forces are part of the Army's global response capability, or they are in support of regional contingency plans that typically have deployment timelines that occur over months.

Army Reserve Components support a wide variety of domestic and global Army operations. Although they constitute about half of the Army's organized units, they provide about 80 percent of the Army's sustainment units, over 70 percent of maneuver support units, a fourth of the Army's mobilization base expansion capability, and most of its civil affairs capacity. The Army Reserve Components are also the Army's major source of trained individual Soldiers for strengthening headquarters and filling vacancies in the Regular Army during a crisis. Reserve Components provide a key resource for reconstitution operations during armed conflict. It is critical for planners to understand that Reserve Components forces have mobilization requirements that take time and typically have deployment time limits that must factor into force management and contingency plans.

Refer to ADP 1 for more information on Army reserve forces. See Chapter 6 for more information on reconstitution. Refer to FM 3-0, chap. 5 for information on reserve mobilization.

1-36 (Operations) II. Strategic Environment

Chap 1

Multidomain Operations

III. Fundamentals of Operations

Ref: FM 3-0, Operations (Oct. '22), chap. 3.

I. Multidomain Operations: the Army's Operational Concept

The Army's operational concept is multidomain operations. Multidomain operations are how Army forces contribute to and operate as part of the joint force. Army forces, enabled by joint capabilities provide the lethal and resilient landpower necessary to defeat threat standoff approaches and achieve joint force objectives.

> **Multidomain operations** are the **combined arms** employment of **joint and Army** capabilities to create and exploit **relative advantages** that achieve objectives, defeat enemy forces, and consolidate gains on behalf of joint force commanders. *(See pp. 1-2 to 1-3.)*

The employment of joint and Army capabilities, integrated across echelons and synchronized in a combined arms approach, is essential to defeating threats able to contest the joint force in all domains. Army forces integrate land, maritime, air, space, and cyberspace capabilities that facilitate maneuver to create physical, information, and human advantages joint force commanders exploit across the competition continuum. Commanders and staffs require the knowledge, skills, and attributes to integrate capabilities rapidly and at the necessary scale appropriate to each echelon.

During competition, theater armies strengthen landpower networks, set the theater, and demonstrate readiness for armed conflict through the command and control (C2) of Army forces supporting the CCP. During crisis, theater armies provide options to combatant commanders (CCDRs) as they facilitate the flow and organization of land forces moving into theater. During armed conflict, theater armies enable and support joint force land component commander (JFLCC) employment of land forces. The JFLCC provides C2 of land forces and allocates joint capabilities to its corps and other subordinate tactical formations. Corps integrate joint and Army capabilities at the right tactical echelons and employ divisions to achieve JFLCC objectives. Divisions, enabled and supported by the corps, defeat enemy forces, control land areas, and consolidate gains for the joint force. Defeating or destroying enemy capabilities that facilitate the enemy's preferred layered stand-off approaches are central to success. Ultimately, operations by Army forces both enable and are enabled by the joint force.

Because uncertainty, degraded communications, and fleeting windows of opportunity characterize operational environments during combat, multidomain operations require disciplined initiative cultivated through a mission command culture. Leaders must have a bias for action and accept that some level of uncertainty is always present. Commanders who empower leaders to make rapid decisions and to accept risk within the commander's intent enable formations at echelon to adapt rapidly while maintaining unity of effort.

Tenets and Imperatives

There are no absolute rules for warfare. However, given analysis of the current strategic environment and assessments of the best ways to employ Army forces, doctrine emphasizes tenets and imperatives for operations that improve their prospects of success without dictating how exactly to solve a tactical or operational problem.

(Operations) III. Fundamentals 1-37

II. Tenets

The tenets of operations are desirable attributes that should be built into all plans and operations, and they are directly related to how the Army's operational concept should be employed. Commanders use the tenets of operations to inform and assess courses of action throughout the operations process. The degree to which an operation exhibits the tenets provides insight into the probability for success. The tenets of operations are—

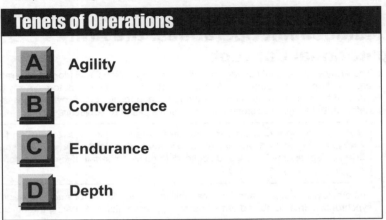

The Army provides forces capable of transitioning to combat operations, fighting for information, producing intelligence, adapting to unforeseen circumstances, and defeating enemy forces. Army forces employ capabilities from multiple domains in a **combined arms approach** that creates **complementary and reinforcing effects** through **multiple domains**, while preserving **combat power** to maintain options for the joint force commander (JFC). **Creating and exploiting relative advantages** require Army forces to operate with endurance and in depth. Endurance enables the ability to absorb the enemy's attacks and press the fight over the time and space necessary to accomplish the mission. Depth applies combat power throughout the enemy's formations and the operational environment, securing successive operational objectives and consolidating gains for the joint force.

A. Agility

The ability to act faster than the enemy is critical for success. Agility is the ability to move forces and adjust their dispositions and activities more rapidly than the enemy. Agility requires sound judgment and rapid decision to gain a relative advantage and control making, often gained through the creation and exploitation of the terms and information advantages. Agility requires leaders to anticipate needs or opportunities, and it requires trained formations able to change direction, tasks, or focus as quickly as the situation requires. Change may come in the form of a transition between phases of an operation or the requirement to adapt to a new opportunity or hazard.

The time available to create and exploit opportunities against adaptive threats is usually limited. Agile units rapidly recognize an opportunity and take action to exploit it. Speed of recognition, decision making, movement, and battle drills enable agility. During armed conflict, this often requires units to change their location and disposition rapidly. Units must be able to employ capabilities and then rapidly task-organize them again for movement or new tasks while remaining dispersed for survivability. C2 and sustainment nodes must maintain a level of functionality on the move and be able to rapidly emplace and displace in order to reduce the probability of enemy detection. Nodes that are critical to success and susceptible to enemy detection and destruction are most vulnerable, and they must be the most agile.

Agility helps leaders influence tempo. Tempo is the relative speed and rhythm of military operations over time with respect to the enemy (ADP 3-0). It implies the ability to understand, decide, act, assess, and adapt. During competition, commanders act quickly to control events and deny enemy forces relative advantages. By acting faster than the situation deteriorates, commanders can change the dynamics of a crisis and restore favorable conditions. During armed conflict, commanders normally seek to maintain a higher tempo than enemy forces do. A rapid tempo can overwhelm an enemy force's ability to counter friendly actions, and it can enable friendly forces to exploit a short window of opportunity.

B. Convergence

Convergence is an outcome created by the concerted employment of capabilities from multiple domains and echelons against combinations of decisive points in any domain to create effects against a system, formation, decision maker, or in a specific geographic area. Its utility derives from understanding the interdependent relationships among capabilities from different domains and combining those capabilities in surprising, effective tactics that accrue advantages over time. When combined, the complementary and reinforcing nature of each friendly capability presents multiple dilemmas for enemy forces and produces an overall effect that is greater than the sum of each individual effect. The greater degree to which forces achieve convergence and sustain it over time the more favorable the outcome.

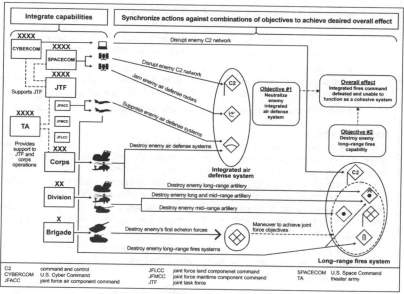

Ref: FM 3-0 (Oct. '22), fig. 3-1. Convergence.

Convergence occurs when a higher echelon and its subordinate echelons create effects from and in multiple domains in ways that defeat or disrupt enemy forces long enough for friendly forces to effectively exploit the opportunity. Convergence broadens the scope of mass, synchronization, and combined arms, by applying combat power to combinations of decisive points, instead of just one, across time, space, and domains. Convergence is a way to balance the principles of mass, objective, and economy of force, massing combat power on some parts of the enemy force while employing different techniques against other decisive points to create cumulative effects the enemy cannot overcome. Convergence requires the synchronization of specific targets and broad objectives by the senior tactical echelon below the land component command.

(Operations) III. Fundamentals 1-39

Multidomain Operations

Convergence
Ref: FM 3-0, Operations (Oct. '22), pp. 3-3 to 3-6.

Convergence
Convergence is an **outcome** created by the **concerted employment** of capabilities from multiple domains and echelons **against combinations of decisive points** in any domain to create effects against a system, formation, decision maker, or in a specific geographic area.

Convergence requires the synchronization of specific targets and broad objectives by the senior tactical echelon below the land component command. The degree to which a formation achieves convergence in an operation depends on how well leaders are able to—

- Develop an understanding of the enemy system, its capabilities, requirements, decision processes, and vulnerabilities through effective surveillance that provides mixed, redundant, and overlapping coverage.
- Determine the desired overall effect or opportunity and the individual effects and objectives that precipitate the opportunity.
- Integrate Army and joint capabilities at the echelons where they are most effective.
- Consider all domains and redundant methods of attack to increase the probability of success.
- Synchronize the employment of each capability and echelon to generate simultaneous, sequential, and enduring effects against the enemy system.
- Assess the individual effects and the probability that the desired overall effect has been achieved. Commanders prepare to re-attack or adapt a course of action if the desired effect is not achieved, or if other opportunities emerge.
- Assume risk and rapidly exploit the opportunities convergence provides.

Integration
Convergence requires the integration of the capabilities at the echelons where their employment is most effective. Integration is the arrangement of military forces and their actions to create a force that operates by engaging as a whole (JP 1, Vol. 1). Commanders generally integrate Army capabilities through task organization and support relationships. Commanders allocate the employment of joint capabilities to subordinate echelons; integrating these capabilities requires an understanding of joint processes. The degree to which commanders effectively integrate joint and Army capabilities at all echelons directly influences success during operations.

- Military forces comprise a wide variety of components that leaders must arrange into a coherent and effective whole. Army leaders integrate—
- Joint capabilities.
- Multinational, interagency, and interorganizational capabilities.
- Echelons and staffs.
- Different types of units to achieve a combined arms approach.

Almost every leader activity, in some way, orients on integrating parts of the force to achieve unity of purpose and unity of effort. There are many intellectual tools leaders use to facilitate integration. Common ones include—

- The joint and Army targeting processes.
- Mission analysis to integrate the activities of multiple staff proponents.
- The nesting concept advocated for in the mission command approach to C2.

1-40 (Operations) III. Fundamentals

- Reception, staging, onward movement, and integration (RSOI) for new forces entering an operation.
- Engagement area development to integrate all weapons systems into a defense.

Synchronization

Once leaders have integrated the right capabilities, they must synchronize their employment and effects. Synchronization is the arrangement of military actions in time, space, and purpose to produce maximum relative combat power at a decisive place and time (JP 2-0).

Understanding the following factors enables leaders to determine when to initiate employment of a capability and how to adapt to changes in the operational environment during execution:

- The desired overall effect over time.
- How the individual effects complement each other over time.
- The time it takes each capability or formation to generate its individual effects from the start of employment.
- Whether each individual effect is enduring, simultaneous, or sequenced with the other effects.
- The consequences of an individual effect not occurring at the planned time.

Individual effects can be enduring, simultaneous, or sequential. Enduring effects provide a continuous impact on the threat until they are no longer necessary. Enduring effects can have a debilitating effect on enemy forces, but they may require significant resources to sustain. Simultaneity is the execution of related and mutually supporting tasks at the same time across multiple locations and domains (ADP 3-0). Simultaneous effects, the result of attacking enemy forces in multiple domains at the same time and across the depth of the enemy's echelons, can have a paralyzing effect on enemy decision making and the effectiveness of the enemy's most critical systems for a limited period of time.

Leaders synchronize actions and effects through C2 and the operations process. The mission, commander's intent, and concept of operations form the basis for detailed synchronization. Commanders determine the degree of control necessary to synchronize operations. They balance synchronization with agility and initiative, but they never surrender the initiative for the sake of synchronization. Excessive synchronization can lead to too much control, which limits the initiative of subordinates and undermines mission command.

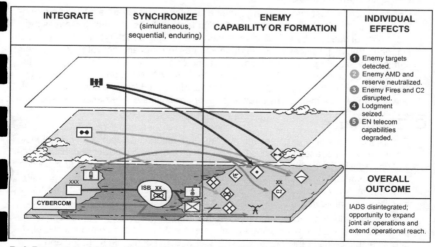

Ref: Presentation Slide, FM 3-0 Team Brief, Combined Arms Center (2022).

Achieving Convergence

Achieving convergence requires detailed, centralized planning and mission orders that enable decentralized execution. Redundant and resilient communications enable synchronized action. However, leaders must anticipate degraded communications and be prepared to rely on mission orders, accept risk, and make decisions to accomplish the mission. During execution, leaders seek to maintain the conditions of convergence through rapid transitions, adjusting priorities, shifting the main effort, or adapting to maintain momentum. Longer periods of convergence allow for greater opportunities to expand advantages and achieve objectives.

Leaders must understand the various processes for requesting joint capabilities and integrating them with ground maneuver. Air, space, and cyberspace tasking cycles operate on different time horizons and have different requirements for requesting effects. These cycles may vary depending on the theater and the situation. Whenever possible, leaders anticipate requirements for these effects during planning and provide ample time for the joint force to generate them. Leaders may request effects on shorter timelines, but they should not make them essential to mission success.

During competition, the theater army establishes conditions for convergence that enable deterrence, provide options during crisis, and enable success at the outset of armed conflict. Intelligence, sustainment, positioning of forces, and other activities to set the theater facilitate situational understanding, decision making, integration, and synchronization during armed conflict. The theater army requests cyberspace and space effects through the combatant command to ensure there is enough time to integrate and synchronize these capabilities. The theater army balances the use of these capabilities during competition with the need to preserve them for use by Army formations during crisis or armed conflict. When armed conflict occurs in a theater, the theater army continues to facilitate convergence by providing capabilities to the land component command and shaping the operational environment outside the joint operational area.

During armed conflict, the land component command apportions joint capabilities to subordinate echelons. Corps integrate joint capabilities with ground maneuver at the appropriate echelon where forces employ them to achieve convergence and achieve objectives. The advantages provided by maritime, air, space, and cyberspace capabilities will not be available all the time, so tactical echelons must be ready to exploit their effects when generated.

Convergence is most effective when its effects accrue and create a cycle of expanding opportunity. Employing multiple and redundant methods of attack increases the probability of success by avoiding dependence on a single method of detecting, tracking, and attacking. Success causes enemy forces to react and activate more of their capabilities, creating another opportunity in one or more domains. The corps and its subordinate echelons align their operations on land with the opportunities created by the effects generated by the other components of the joint force, preserving combat power to maximize their ability to exploit the opportunities convergence presents.

C. Endurance

Endurance is the ability to persevere over time throughout the depth of an operational environment. Endurance enhances the ability to project combat power and extends operational reach. Endurance is about resilience and preserving combat power while continuing operations for as long as is necessary to achieve the desired outcome. During competition, Army forces improve endurance by setting the theater across all warfighting functions and improving interoperability with allies and other unified action partners.

Endurance reflects the ability to employ combat power anywhere for protracted periods in all conditions, including environments with degraded communications, chemical, biological, radiological, and nuclear (CBRN) contamination, the and high casualties. Endurance stems from the ability to time and distance necessary

1-42 (Operations) III. Fundamentals

throughout organize, protect, and sustain a force, regardless of the distance from its support area and the austerity of the environment. Endurance involves anticipating requirements and making the most effective and efficient use of resources.

As forces fight through successive engagements, maintaining mutual support among units helps prevent them becoming isolated, being defeated in detail, and culminating early. Protection prevents or mitigates enemy effects and preserves combat power, postponing culmination and prolonging effective operations. One way Army forces preserve combat power is by maintaining dispersion to the greatest degree possible. Leaders can mass combat power from dispersed positions and generate the desired effects without concentrating forces any more than is necessary. During operations, commanders and staffs integrate, synchronize, and simultaneously apply protection capabilities.

Leadership and tactics contribute to endurance. Plans that allow for different units to be the main effort using follow and support or follow and assume techniques prevent early culmination in the units first committed to close combat. Realistically determining what tempo friendly forces can maintain given enemy resistance, weather, and physical distances and the impact they have on Soldiers, leaders, and equipment increases endurance over time.

Sustainment operations are essential to endurance. Using all methods for continuously delivering sustainment through land, maritime, and air capabilities improves endurance. Lleaders must anticipate degraded communications and combine analog systems for communication with predictive analysis and disciplined initiative to ensure commanders can maintain acceptable tempo for as long as necessary.

D. Depth

Depth is the extension of operations in time, space, or purpose to achieve definitive results (ADP 3-0). While the focus of endurance is on friendly combat power, the focus of depth is on enemy locations and dispositions across all domains. Commanders achieve depth by understanding the strengths and vulnerabilities of the enemy's echeloned capabilities, then attacking them throughout their dispositions in simultaneous and sequential fashion. Although simultaneous attacks through all domains in depth are not possible in every situation, leaders seek to expand their advantages and limit enemy opportunities for sanctuary and regeneration. Leaders describe the depth they can achieve in terms of operational reach.

Operational reach is the distance and duration across which a force can successfully employ military capabilities (JP 3-0). Staffs assess operational reach based on available sustainment, the range of capabilities and formations, and courses of action compared with the intelligence estimates of enemy capabilities and courses of action. This analysis helps the commander understand the limits on friendly operations, the risks inherent in the mission, and likely points in time and space for transitions.

Below the threshold of armed conflict, the theater army creates depth by improving the infrastructure for force projection and by improving interoperability with multinational forces to the degree required by operation plans (OPLANs) and contingency operations. It also adds depth to its operations by expanding influence with allies and partners, populations, and other relevant actors through joint exercises, sustained forward positioning of advisor teams, and forward basing of combat formations.

During armed conflict, the JFLCC creates depth by facilitating access to Army and other joint capabilities, especially space and cyberspace capabilities that improve the protection of tactical formations and degrade enemy integrated air defense systems. The JFLCC also requests that the JFC influence the extended deep area in support of land operations. The corps directs fires into its deep area to defeat enemy long-range fires, disrupt enemy sustainment and C2, separate maneuver echelons, and shape the success of future close operations. Special operations forces operating in the extended deep area can detect targets and enable the employment of joint fires to support conventional operations.

(Operations) III. Fundamentals 1-43

III. Imperatives

Ref: FM 3-0, Operations (Oct. '22), pp. 3-8 to 3-18.

Imperatives are actions Army forces must take to defeat enemy forces and achieve objectives at acceptable cost. They are informed by the operational environment and the characteristics of the most capable threats Army forces can encounter. Imperatives include—

See Yourself, See the Enemy, and Understand the Operational Environment *(See pp. 3-16 to 3-17.)*

Commanders visualize operational environments in terms of the factors that are relevant to decision making. Operational environments are dynamic and contain vast amounts of information that can overload C2 systems and impede decision making. Commanders simplify information collection, analysis, and decision making by focusing on how they see themselves, see the enemy, and understand the operational environment. These three categories of factors are interrelated, and leaders must understand how each one relates to the others in the current context.

As part of the operations process, Army leaders use different methodologies to understand and weigh options. These methodologies include the Army design methodology, the military decision-making process, and the rapid decision-making and synchronization process. Each methodology provides a process that allows commanders and staffs to see themselves, see the enemy, and understand the operational environment.

Account for Being Under Constant Observation and All Forms of Enemy Contact *(See pp. 4-2 to 4-3.)*

Air, space, and cyberspace capabilities increase the likelihood that threat forces can gain and maintain continuous visual and electromagnetic contact with Army forces. Enemy forces possess a wide range of space-, air-, maritime-, and land-based ISR capabilities that can detect U.S. forces. Leaders must assume they are under constant observation from one or more domains and continuously ensure they are not providing lucrative targets for the enemy to attack. Leaders consider nine forms of contact in multiple domains.

Create and Exploit Relative Advantages (Physical, Information, and Human) in Pursuit of Decision Dominance *(See p. 4-4.)*

The employment of lethal force is based on the premise that destruction and other physical consequences compels enemy forces to change their decision making and behavior, ultimately accepting defeat. The type, amount, and ways in which lethal force compels enemy forces varies, and this depends heavily on enemy forces, their capabilities, goals, and the will of relevant populations. Understanding the relationship between physical, information, and human factors enables leaders to take advantage of every opportunity and limit the negative effects of undesirable and unintended consequences.

Make Initial Contact with Smallest Element Possible *(See p. 4-4.)*

Army forces are extremely vulnerable when they do not sufficiently understand the disposition of enemy forces and become decisively engaged on terms favorable to enemy forces. To avoid being surprised and incurring heavy losses, leaders must set conditions for making enemy contact on terms favorable to the friendly force. They anticipate when and where to make enemy contact, the probability and impact of making enemy contact, and actions to take on contact. Quickly applying multiple capabilities against enemy forces while preventing the bulk of the friendly force from being engaged itself requires an understanding of the forms of contract.

1-44 (Operations) III. Fundamentals

Multidomain Operations

Impose Multiple Dilemmas on the Enemy *(See p. 4-5.)*

Imposing multiple dilemmas on enemy forces complicates their decision making and forces them to prioritize among competing options. It is a way of seizing the initiative and making enemy forces react to friendly operations. Simultaneous operations encompassing multiple domains—conducted in depth and supported by deception—present enemy forces with multiple dilemmas. Employing capabilities from multiple domains degrades enemy freedom of action, reduces enemy flexibility and endurance, and disrupts enemy plans and coordination. The application of capabilities in complementary and reinforcing ways creates more problems than an enemy commander can solve, which erodes both enemy effectiveness and the will to fight.

Anticipate, Plan, and Execute Transitions *(See p. 3-18.)*

Transitions mark a change of focus in an operation. Leaders plan transitions as part of the initial plan or parts of a branch or sequel. They can be unplanned and cause the force to react to unforeseen circumstances. Transitions can be part of progress towards mission accomplishment, or they can reflect a temporary setback.

Designate, Weight, and Sustain the Main Effort *(See p. 3-19.)*

Commanders frequently face competing demands for limited resources. They resolve these competing demands by establishing priorities. One way in which commanders establish priorities is by designating, weighting, and sustaining the main effort. The main effort is a designated subordinate unit whose mission at a given point in time is most critical to overall mission success (ADP 3-0). Commanders provide the main effort with the appropriate resources and support necessary for its success. When designating a main effort, commanders consider augmenting a unit's task organization and giving it priority of resources and support.

Consolidate Gains Continuously *(See p. 3-19.)*

Leaders add depth to their operations in terms of time and purpose when they consolidate gains. Commanders consolidate gains at the operational and tactical levels as a strategically informed approach to current operations with the desired political outcome of the conflict in mind. During competition and crisis, commanders expand opportunities created from previous conflicts and activities to sustain enduring U.S. interests, while improving the credibility, readiness, and deterrent effect of Army forces. During large-scale combat operations, commanders consolidate gains continuously or as soon as possible, deciding whether to accept risk with a more moderate tempo during the present mission or in the future as large-scale combat operations conclude.

Consolidating gains at every echelon leads to better transitions out of armed conflict and into post-conflict competition. It serves as a preventative against the rise of an insurgency by those wishing to prolong the conflict.

Understand and Manage the Effects of Operations on Units and Soldiers *(See p. 2-9.)*

Continuous operations rapidly degrade the performance of people and the equipment they employ, particularly during combat. In battle, Soldiers and units are more likely to fail catastrophically than gradually. Commanders and staffs must be alert to small indicators of fatigue, fear, indiscipline, and reduced morale, and they must take measures to deal with these before their cumulative effects drive a unit to the threshold of collapse. Staffs and commanders at higher echelons must take into account the impact of prolonged combat on subordinate units, which causes efficiency to drop, even when physical losses are not great. Leaders consider the isolation Soldiers experience when not being able to remain connected with family and friends via social media and other platforms for extended periods. Well-trained, physically fit Soldiers in cohesive units retain the qualities of tenacity and aggressiveness longer than those who are not.

(Operations) III. Fundamentals 1-45

Operational Approach and Operational Framework

The **operational approach** provides the logic for how tactical tasks ultimately achieve the desired end state. It provides a unifying purpose and focus to all operations. Sound operational approaches balance risk and uncertainty with friction and chance. The operational approach provides the basis for detailed planning, allows leaders to establish a logical operational framework, and helps produce an executable order. *(See discussion below.)*

An **operational framework** organizes an area of geographic and operational responsibility for the commander and provides a way to describe the employment of forces. The framework illustrates the relationship between close operations, operations in depth, and other operations in time and space across domains. As a visualization tool, the operational framework bridges the gap between a unit's conceptual understanding of the environment and its need to generate detailed orders that direct operations. *(See pp. 1-54 to 1-62.)*

IV. Operational Approach

Through operational art, commanders develop their operational approach—a broad description of the mission, operational concepts, tasks, and actions required to accomplish the mission (JP 5-0). An operational approach is the result of the commander's visualization of what needs to be done in broad terms to solve identified problems. It is the main idea that informs detailed planning. When describing an operational approach, commanders—

Operational Approach

- **Consider ways to defeat enemy forces in detail and potential decisive points.**
- **Employ combinations of defeat mechanisms to isolate and defeat enemy forces, functions, and capabilities.**
- **Assess options for assuming risk.**

A. Defeating Enemy Forces in Detail

Armed conflict implies the need to defeat enemy forces. **Defeat** is to render a force incapable of achieving its objectives (ADP 3-0). When used as a task or effect in operations, defeat provides maximum flexibility to the commander in how to accomplish the mission. Senior leaders assign defeat as a task when the situation is still developing, or when the commander on the ground, by virtue of experience and proximity to the problem, is uniquely capable of deciding how to employ lethal force to accomplish objectives. As a task, defeat is appropriate for theater strategic and operational-level echelons, but it is often too vague for tactical echelons below corps level, where more specific outcomes or a higher level of destruction might be necessary to ensure the overall defeat of enemy forces. As a purpose or an effect, defeat is often used to describe the ultimate outcome of an operation.

Defeat inevitably leads to transition. Strategic defeat occurs when an enemy's political leadership and national will acquiesce to the friendly political will, and the situation transitions to a more desirable form of competition below armed conflict. Operational defeat occurs when enemy forces no longer have the will or ability to pursue military objectives, and the friendly force has achieved most or all of its objectives. At the tactical level, an attacking force defeats an enemy defense when it causes enemy forces to transition to a retrograde and cease defending friendly objectives. A defending force defeats an enemy attack when it causes enemy forces to culminate and transition to the defense before achieving their objectives.

1-46 (Operations) III. Fundamentals

When U.S. forces possess overwhelming advantages across all domains, the JFC is able to attack all elements of the enemy force with a high degree of simultaneity. Simultaneity disrupts the enemy's C2 system and rapidly disintegrates each component of the threat warfighting system at the same time. However, peer threats, by definition, possess a scale and quality of warfighting capability that is too extensive to attack at once. When fighting a peer threat, commanders identify weaknesses between enemy units or in enemy formations and warfighting systems that provide opportunities to defeat them in detail.

Defeat in detail is concentrating overwhelming combat power against separate parts of a force rather than defeating the entire force at once (ADP 3-90). Traditionally, commanders of a smaller force use this technique to achieve success against a larger enemy force. However, defeat in detail also applies to operations that focus effort on a specific enemy function, capability, echelon, domain, or dimension.

Defeat in detail requires leaders to evaluate enemy forces in the context of all the relevant domains and dimensions of an operational environment. Commanders must understand the various parts of an enemy force and its vulnerabilities, and then discern the best ways to project combat power against those vulnerabilities. By comparing enemy weaknesses to friendly advantages, leaders begin to see opportunities and formulate options. Sometimes enemy vulnerabilities and friendly advantages intersect at a single place and time in a way that is decisive to mission accomplishment. That single place and time is a decisive point— key terrain, key event, critical factor, or function that, when acted upon, enables commanders to gain a marked advantage over an enemy or contribute materially to achieving success (JP 5-0). Decisive points help commanders select clear, conclusive, attainable objectives that directly contribute to achieving an end state through convergence or other means.

B. Defeat and Stability Mechanisms

Defeat Mechanism

A defeat mechanism is a method through which friendly forces accomplish their mission against enemy opposition (ADP 3-0). Army forces at all echelons commonly use combinations of four defeat mechanisms: destroy, dislocate, disintegrate, and isolate. Applying more than one defeat mechanism simultaneously creates multiple dilemmas for enemy forces and complementary and reinforcing effects not attainable with a single mechanism. Commanders may have an overarching defeat mechanism or combination of mechanisms that accomplish the mission, with supporting defeat mechanisms for components of an enemy formation or warfighting system. Defeat mechanisms can guide subordinate development of tactical tasks, purposes, and effects in their operations, facilitating control and initiative.

During competition, commanders take actions that set conditions for the future application of defeat mechanisms and demonstrate the capability to impose the defeat mechanisms on enemy forces. These activities include posturing forces, penetrating enemy networks, and conducting exercises with allies and partners.

Commanders determine the speed and degree to which a defeat mechanism must impact an enemy force or warfighting system. Although rapid defeat is typically desirable, it may be more feasible or acceptable to take a gradual approach to completing a defeat. Rendering an enemy incapable of achieving its objectives does not usually require total annihilation. To determine the degree of impact on the enemy force, commanders consider causing only minor degradation to a threat warfighting system or unit when it is sufficient to prevent the enemy from achieving its objective. This preserves friendly combat power and applies the economy of force principle. In other cases, especially main efforts against determined peer threat forces, commanders typically require a significant portion of an enemy's force be destroyed.

Army forces at all echelons commonly use combinations of four defeat mechanisms: destroy, dislocate, disintegrate, and isolate. See following page for an overview and discussion.

(Operations) III. Fundamentals 1-47

Multidomain Operations

Defeat Mechanisms

Ref: FM 3-0, Operations (Oct. '22), pp. 3-19 to 3-20.

Army forces at all echelons commonly use combinations of four defeat mechanisms: destroy, dislocate, disintegrate, and isolate. Applying more than one defeat mechanism simultaneously creates multiple dilemmas for enemy forces and complementary and reinforcing effects not attainable with a single mechanism.

Destroy. When commanders destroy, they apply lethal force against an enemy capability so that it can no longer perform its function. Destroy is a tactical mission task that physically renders an enemy force combat-ineffective until it is reconstituted. Alternatively, to destroy a combat system is to damage it so badly that it cannot perform any function or be restored to a usable condition without being entirely rebuilt (FM 3-90-1). Destruction and the threat of destruction lie at the core of all the defeat mechanisms and make them more compelling. The other mechanisms work when friendly action has caused enemy forces to face a grim reality: their ability to fight and relative advantages are degraded, and their options are to surrender, withdraw, or be destroyed.

Dislocate. Dislocate is to employ forces to obtain significant positional advantage in one or more domains, rendering the enemy's dispositions less valuable, perhaps even irrelevant. Typically, the impact of dislocation increases when the friendly force exploits advantages in multiple domains. Commanders often achieve dislocation through deception and by placing forces in locations where enemy forces do not expect them. Achieving dislocation requires an understanding how enemy forces are oriented and how quickly they can shift. Envelopments and turning movements enable physical dislocation. Deception can create and enhance psychological effects of dislocation.

Isolate. Isolate means to separate a force from its sources of support in order to reduce its effectiveness and increase its vulnerability to defeat (ADP 3-0). Isolation can encompass multiple domains and can have both physical and psychological effects detrimental to accomplishing a mission. Isolating an enemy force from the electromagnetic spectrum increases the effects of physical isolation by reducing its ability to communicate and degrading its situational awareness. The ability of an isolated unit to perform its intended mission generally degrades over time, decreasing its ability to interfere with an opposing force's course of action. When commanders isolate, they deny enemy forces access to capabilities that enable them to maneuver at will in time and space.

Disintegrate. Disintegrate means to disrupt the enemy's command and control, degrading the synchronization and cohesion of its operations. Disintegration prevents enemy unity of effort and leads to a degradation of the enemy's capabilities or will to fight. It attacks the cohesion of enemy formations and their ability to employ combined arms approaches and work effectively together. Commanders can achieve disintegration by targeting enemy functions essential to the threat's ability to act as a whole. They often achieve disintegration by specifically targeting an enemy's command structure, communications systems, the linkages between them, and the capabilities they control. Disintegration can be achieved through the employment of the other three defeat mechanisms in combination, particularly when directed toward systems like integrated fires commands and integrated air defense systems heavily dependent upon C2 and sensor nodes.

Cyberspace, space, and electromagnetic warfare capabilities can help disintegrate enemy formations by degrading communications and disrupting the quality of enemy information and decisions. Separating enemy reserves and follow-on echelons from the main body with maneuver forces or fires is a physical way to isolate echelons, achieve favorable force ratios, and destroy those echelons. This in turn disintegrates the coherence of an enemy's attack or defense. Destroying enemy sustainment capability separates enemy fires and maneuver from fuel and ammunition and delays resupply operations.

1-48 (Operations) III. Fundamentals

Stability Mechanism

A stability mechanism is the primary method through which friendly forces affect civilians in order to attain conditions that support establishing a lasting, stable peace (ADP 3-0). As with defeat mechanisms, combinations of stability mechanisms produce complementary and reinforcing effects that accomplish the mission more effectively and efficiently than single mechanisms do alone. The four stability mechanisms are compel, control, influence, and support:

Compel means to use, or threaten to use, lethal force to establish control and dominance, affect behavioral change, or enforce compliance with mandates, agreements, or civil authority.

Control involves imposing civil order.

Influence means to alter the opinions, attitudes, and ultimately behavior of foreign, friendly, neutral, and threat audiences through messages, presence, and actions.

Support establishes, reinforces, or sets conditions necessary for the instruments of national power to function effectively.

C. Risk

Commanders accept risk on their own terms to create opportunities and apply judgment to manage those hazards they do not control. Risk is an inherent part of every operation and cannot be avoided. Commanders analyze risk in collaboration with subordinates to help determine what level and type of risk exists and how to mitigate it. When considering how much risk to accept with a course of action, commanders consider risk to the force against the probability of mission success during current and future operations. They assess options in terms of weighting the main effort, economy of force, and physical loss in the context of what they have been tasked to do.

Leaders consider risk across all domains. Accepting risk in one domain may create opportunities in other domains. For example, the risk of seizing an airfield puts ground forces at risk, but it creates an opportunity to receive reinforcements and supplies that extend operational reach. During combat against an enemy with capabilities comparable to that of the United States, the greatest opportunity may come from the course of action with the most risk. An example of this is committing significant forces to a potentially costly frontal attack that fixes the bulk of enemy forces in place to set the conditions for their envelopment by other forces. Another is taking a difficult but unexpected route to achieve surprise. Accepting significant risk is necessary when seeking to create an advantage where none exists otherwise.

The unrealistic expectation of avoiding all risk is detrimental to mission accomplishment. While each situation is different, commanders avoid undue caution or commitment of resources to guard against every perceived threat. Waiting for perfect intelligence and synchronization may increase risk or close a window of opportunity. Mission command requires that commanders and subordinates accept risk, exercise initiative, and act decisively, particularly when the outcome is uncertain.

Commanders determine how to impose risk on enemy forces. Viewing the situation through the enemy's perspective, commanders seek to create multiple dilemmas and increase the number and severity of hazards with which enemy forces must contend. Leaders consider the human and information factors that govern the manner in which enemy forces assess costs and benefits and calculate risk. Commanders disrupt this risk calculation when they increase perceived costs to enemy forces and reduce the perception of potential benefits. Commanders do this by imposing dilemmas on enemy forces, not based on what a U.S. or allied leader views as a problem, but on what an enemy commander views as detrimental. Some dilemmas are universally accepted as costly, but others are cultural or personal. Commanders rely on military intelligence and experience to develop this level of situational understanding.

(Operations) III. Fundamentals 1-49

V. The Elements of Operational Art

Ref: ADP 3-0, Operations (Jul '19), chap. 2.

In applying operational art, Army commanders and their staffs use intellectual tools to help understand an operational environment and visualize and describe their approach to conducting an operation. Collectively, these tools are the elements of operational art. They help commanders understand, visualize, and describe the integration and synchronization of the elements of combat power and their commander's intent and guidance. Commanders selectively use these tools in any operation. Their broadest application applies to long-term operations.

Not all elements of operational art apply at all levels of warfare. A company commander concerned about the tempo of an upcoming operation is probably not concerned with an enemies' center of gravity. A corps commander may consider all elements of operational art in developing a plan to support the JFC.

As some elements of operational design apply only to JFCs, the Army modifies the elements of operational design into elements of operational art by adding Army-specific elements. They adjust current and future operations and plans as the operation unfolds, and they reframe as necessary.

Elements of Operational Art

Operational art consists of these elements:

• End state and conditions.	• Phasing and transitions.
• Center of gravity.*	• Culmination.*
• Decisive points.*	• Operational reach.*
• Lines of operations and lines of effort.*	• Basing.*
• Tempo.	• Risk.

*Common to elements of operational design.

Ref: ADP 3-0 (2016), table 2-2. Elements of operational art.

End State and Conditions. The end state is a set of desired future conditions the commander wants to exist when an operation ends. Commanders include the end state in their planning guidance. A clearly defined end state promotes unity of effort; facilitates integration, synchronization, and disciplined initiative; and helps mitigate risk. An end state should anticipate future operations and set conditions for transitions.

Center Of Gravity (COG). A center of gravity is the source of power that provides moral or physical strength, freedom of action, or will to act (JP 5-0). The loss of a center of gravity can ultimately result in defeat. A center of gravity is an analytical tool for planning operations. It provides a focal point and identifies sources of strength and weakness. However, the concept of center of gravity is only meaningful when considered in relation to the objectives of the mission. Because most enemies represent adaptive, complex systems, they are likely to have multiple centers of gravity. Destroying or capturing one is unlikely to win a campaign or resolve most conflicts.

Decisive Points. A decisive point is a geographic place, specific key event, critical factor, or function that, when acted upon, allows commanders to gain a marked advantage over an enemy or contribute materially to achieving success (JP 5-0). Decisive points help commanders select clear, conclusive, attainable objectives that directly contribute to achieving an end state. Geographic decisive points can include port facilities, distribution networks and nodes, and bases of operation. Specific events

1-50 (Operations) III. Fundamentals

and elements of an enemy force can be decisive points. Examples of such events include commitment of the enemy operational reserve and reopening a major oil refinery. Space and cyberspace-enabled capabilities may also represent decisive points.

Lines of Operations and Lines of Effort.
Lines of operations and lines of effort link objectives to the end state physically and conceptually. Commanders may describe an operation along lines of operations, lines of effort, or a combination of both. The combination of them may change based on the conditions within an area of operations. Commanders synchronize and sequence actions, deliberately creating complementary and reinforcing effects. The lines then converge on the well-defined, commonly understood end state outlined in the commander's intent.

Tempo.
Tempo is the relative speed and rhythm of military operations over time with respect to the enemy. It reflects the rate of military action. Controlling tempo helps commanders keep operational initiative during combat operations or rapidly establish a sense of normalcy during humanitarian crises. During combat operations, commanders normally seek to maintain a higher tempo than enemy forces do. A rapid tempo can overwhelm an enemy force's ability to counter friendly actions. During other operations, commanders act quickly to control events and deny enemy forces positions of advantage. By acting faster than the situation deteriorates, commanders can change the dynamics of a crisis and restore favorable conditions.

Phasing and Transitions.
A **phase** is a planning and execution tool used to divide an operation in duration or activity. A change in phase usually involves a change of mission, task organization, or rules of engagement. **Transitions** mark a change of focus between phases or between the ongoing operation and execution of a branch or sequel.

Culmination.
The culminating point is a point at which a force no longer has the capability to continue its form of operations, offense or defense (JP 5-0). Culmination represents a crucial shift in relative combat power. It is relevant to both attackers and defenders at each level of warfare. While conducting offensive operations, the culminating point occurs when a force cannot continue the attack and must assume a defensive posture or execute an operational pause. While conducting a defense, it occurs when a force can no longer defend itself and must withdraw or risk destruction. The culminating point is more difficult to identify when Army forces perform stability tasks. Two conditions can result in culmination while performing stability tasks: units being too dispersed to achieve security and units lacking required resources to achieve the end state.

Operational Reach.
Operational reach reflects the ability to achieve success through a well-conceived operational approach, and it is applicable to Army forces operating as part of the joint force. Operational reach is a tether; it is a function of intelligence, protection, sustainment, endurance, and combat power relative to enemy forces. The limit of a unit's operational reach is its culminating point.

Basing.
Army basing overseas typically falls into two general categories: permanent (bases or installations) and nonpermanent (base camps). A base is a locality from which operations are projected or supported (JP 4-0). A base camp is an evolving military facility that supports the military operations of a deployed unit and provides the necessary support and services for sustained operations (ATP 3-37.10). Base camps are nonpermanent by design and designated as bases when the intention is to make them permanent.

Risk.
Risk is the probability and severity of loss linked to hazards. Risk, uncertainty, and chance are inherent in all military operations. When commanders accept risk, they create opportunities to seize, retain, and exploit operational initiative and achieve decisive results. The willingness to incur risk is often the key to exposing enemy weaknesses that an enemy considers beyond friendly reach. Understanding risk requires accurate running estimates and valid assumptions. Embracing risk as opportunity requires situational awareness and imagination, as well as audacity.

(Operations) III. Fundamentals 1-51

VI. Strategic Framework

Ref: FM 3-0, Operations (Oct. '22), pp. 3-21 to 3-23.

The strategic framework accounts for factors in the strategic environment and the connection of strategic capabilities to operational- and tactical-level operations. The strategic framework includes four areas: strategic support area, joint security area, extended deep area, and assigned operational area.

See pp. 1-56 to 1-57 for discussion of assigned operational areas.

Strategic Support Area

The strategic support area describes the area extending from a theater of operations to a base in the United States or another CCDR's area of responsibility. It contains those organizations, lines of communications, and other agencies required to support deployed forces. It also includes the airports and seaports supporting the flow of forces and sustainment into a theater. Finally, a strategic support area may contain key operational capabilities, such as cyberspace assets, that are employed from outside an operational area but create effects inside it. Most friendly nuclear, space, and cyberspace capabilities and important network infrastructure are controlled and located in the strategic support area.

Joint Security Area

A joint security area is a specific area to facilitate protection of joint bases and their connecting lines of communications that support joint operations (JP 3-10). The joint security area (JSA) is inside, or immediately adjacent to, an operational area where significant forces and sustainment from two or more services are positioned to conduct or support operations. Joint security on land includes bases, mission-essential assets, lines of communications, and convoy security. A senior Army commander is often designated with responsibility for joint security operations on land. *Refer to JP 3-10 for more information on JSAs.*

Extended Deep Area

The extended deep area is comprised of operational and strategic deep areas. These areas typically do not fall within the land component command's AO, but they are part of its area of interest because enemy capabilities and vulnerabilities in the extended deep area can have significant impacts on the outcomes of operations. Extended deep areas are typically the purview of the joint force headquarters or another combatant command. Typically, the joint force air component command (known as the JFACC) is the supported command in extended deep areas. Army forces may be tasked to support it with long-range precision fires.

Operational deep areas are generally inside the area of interest and immediately beyond the land component's initially assigned AO. These areas may or may not be within the boundaries of a joint operations area (JOA) or a theater of operations. Operational deep areas are often beyond the feasible movement of conventional forces without significant support from the joint force. Operational deep areas contain enemy supporting formations and capabilities for their main forces. Enemy forces can generate significant combat power from these areas, and the capabilities that reside there are often vital to their conduct of operations. In most campaign designs, operational objectives for friendly forces reside initially in the operational deep area.

Strategic deep areas are beyond the feasible range of movement for conventional ground forces or policy prohibits their operations. These areas are where the CCDR, other combatant commands, and national agencies can employ strategic intelligence capabilities, joint fires, special operations forces, and space and cyberspace capabilities. Many enemy space, cyberspace, and information warfare capabilities reside in strategic deep areas across international boundaries and outside the JOA, and they often comprise multiple areas of influence.

1-52 (Operations) III. Fundamentals

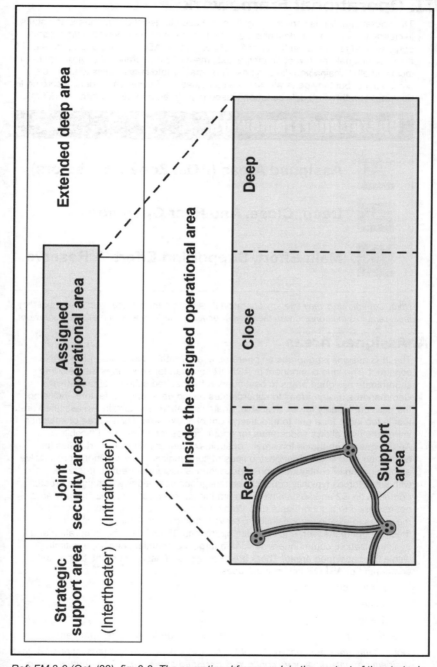

Ref: FM 3-0 (Oct. '22), fig. 3-2. The operational framework in the context of the strategic framework.

VII. Operational Framework

The operational framework is a cognitive tool used to assist commanders and staffs in clearly visualizing and describing the application of combat power in time, space, purpose, and resources in the concept of operations (ADP 1-01). Commanders build their operational framework on their assessment of the operational environment, including all domains and dimensions. They may create new models to fit the circumstances, but they generally apply a combination of common models according to doctrine. The three models commonly used to build an operational framework are—

Operational Framework

 Assigned Areas (AOs, Zones, & Sectors)

 Deep, Close, And Rear Operations

 Main Effort, Supporting Effort, & Reserve

Note. Commanders may use any operational framework models they find useful, but they must remain synchronized with their higher echelon headquarters' operational framework.

A. Assigned Areas

The JFC assigns land forces an operational area within a joint organizational construct. The land component or ARFOR commander subdivides their AO into subordinate assigned areas to best support the desired scheme of maneuver. Commanders assign areas to subordinates based on a range of factors, including the mission, friendly forces available, enemy situation, and terrain. An assigned area that is too large for a unit to effectively control or exceeds a unit's area of influence increases risk, allows sanctuaries for enemy forces, and limits joint flexibility. An assigned area that is too small constrains maneuver, limits opportunities for dispersion, and creates congested lines of communication. Most operations involve a combination of contiguous and noncontiguous assigned areas. Large areas with small forces typically conduct noncontiguous operations which place greater demands on C2 and sustainment. Commanders retain responsibility for any area not assigned to a subordinate unit. Within their assigned area, units use control measures to assign responsibilities, prevent fratricide, facilitate C2, coordinate fires, control maneuver, and organize operations. To facilitate this integration and synchronization, commanders designate targeting priorities, effects, and timing within their assigned areas. There are three types of assigned areas that a land component or ARFOR commander uses:

- Area of operations
- Zone
- Sector

See following pages (pp. 1-56 to 1-57) for discussion of these assigned areas.

Note. Contiguous boundaries do not imply units are capable of mutual support or that their subordinate units have contiguous assigned areas. Therefore, mutual support between adjacent units and subordinate units must be part of commander dialogue to ensure the formation is assuming risk deliberately and at the right echelon.

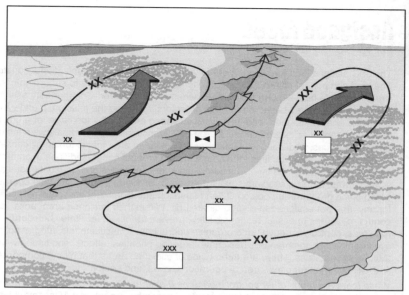

Ref: FM 3-0 (Oct. '22), fig. 3-3. Notional corps area of operations with noncontiguous divisions.

Operational environments are larger than the associated assigned area. They influence and are influenced by factors outside unit boundaries. To account for these factors commanders typically consider areas of influence and areas of interest relative through all domains and dimensions.

Area of Influence

An area of influence is an area inclusive of and extending beyond an operational area wherein a commander is capable of direct influence by maneuver, fire support, and information normally under the commander's command or control (JP 3-0). The ranges of a unit's maneuver and fires capabilities typically define its area of influence; however, commanders consider all forms of contact they can make with enemy forces when visualizing their area of influence. A unit's area of influence contracts or expands based on the capabilities allocated by the higher headquarters and adjusts as the unit repositions its capabilities on the battlefield. An area of influence is normally larger than its associated assigned area, but it is smaller than its area of interest. Units typically have areas of influence that overlap with adjacent unit assigned areas. A unit might desire to collect information on or strike enemy forces traversing through an adjacent unit assigned area. This situation requires control measures to enable friendly forces to maintain pressure on enemy forces while mitigating the risk of fratricide. Understanding an area of influence helps commanders and staffs plan branches and sequels to the current operation in preparation for operations outside of the current assigned area.

Area of Interest

An area of interest is that area of concern to the commander, including the area of influence, areas adjacent to it, and extending into enemy territory (JP 3-0). This visualization tool enables commanders and staffs to understand the impact of threats outside their assigned area and how their operation is progressing along with their adjacent and higher units. An area of interest includes those aspects of the domains from which enemy forces can employ capabilities that jeopardize the accomplishment of the mission. The area of interest can shift according to the situation. For

(Operations) III. Fundamentals 1-55

Multidomain Operations

Assigned Areas

Ref: FM 3-0, Operations (Oct. '22), pp. 3-23 to 3-25.

The JFC assigns land forces an operational area within a joint organizational construct. The land component or ARFOR commander subdivides their AO into subordinate assigned areas to best support the desired scheme of maneuver. Commanders assign areas to subordinates based on a range of factors, including the mission, friendly forces available, enemy situation, and terrain. An assigned area that is too large for a unit to effectively control or exceeds a unit's area of influence increases risk, allows sanctuaries for enemy forces, and limits joint flexibility. An assigned area that is too small constrains maneuver, limits opportunities for dispersion, and creates congested lines of communication. Most operations involve a combination of contiguous and noncontiguous assigned areas. Large areas with small forces typically conduct noncontiguous operations which place greater demands on C2 and sustainment. Commanders retain responsibility for any area not assigned to a subordinate unit. Within their assigned area, units use control measures to assign responsibilities, prevent fratricide, facilitate C2, coordinate fires, control maneuver, and organize operations. To facilitate this integration and synchronization, commanders designate targeting priorities, effects, and timing within their assigned areas. There are three types of assigned areas that a land component or ARFOR commander uses: area of operations, zone, and sector.

While there are many other control measures that enable terrain management (for example, position areas for artillery or tactical assembly areas) only AOs, zones, and sectors are part of the assigned area model. Commanders and staffs use AOs when the operation requires a higher level of control. Zones are best for front line units conducting high tempo offensive operations characterized by direct fire contact with the enemy and a fluid forward line of troops. Sectors are best for front line units conducting a defense, making it easier for a higher headquarters to conduct deep operations and for subordinate units to have mutually supporting fires.

Area of Operations (AO)

An area of operations is an operational area defined by a commander for the land or maritime force commander to accomplish their missions and protect their forces (JP 3-0). An area of operations is defined by its boundaries. Within their AO, units integrate assigned and supporting capabilities, synchronize warfighting functions, and generate combat power against enemy forces to accomplish the mission. Responsibilities for an AO include—

- Terrain management.
- Information collection, integration, and synchronization.
- Civil affairs operations.
- Movement control.
- Clearance of fires.
- Security.
- Personnel recovery.
- Airspace management.
- Minimum-essential stability operations tasks which are—
- Establish civil security.
- Provide immediate needs (access to food, water, shelter, and medical treatment).

Note. Commanders can add, remove, or adjust AO responsibilities based on the situation and mission variables. A land AO by definition does not include a volume of airspace to control. Airspace control authorities delegate airspace control to Army commanders based on the situation. All commanders must be prepared to enable or coordinate airspace management.

1-56 (Operations) III. Fundamentals

Zone

A zone is an operational area assigned to a unit in the offense that only has rear and lateral boundaries. The non-bounded side of a zone is open towards enemy forces. A higher echelon headquarters uses fire support coordination and maneuver control measures such as a limit of advance and a coordinated fire line to synchronize its deep operations with those of a subordinate unit. Zones allow higher headquarters to adjust deep operations without having to change unit boundaries. This gives greater flexibility to the higher headquarters for controlling deep operations, allowing subordinate units to focus on close and rear operations. Units treat everything behind the forward line of troops as an AO with its associated responsibilities. Units can subdivide a zone into subordinate AOs, zones, or sectors.

Sector

A sector is an operational area assigned to a unit in the defense that has rear and lateral boundaries and interlocking fires. The non-bounded side of a sector is open towards the enemy. A higher echelon headquarters uses fire support coordination and maneuver control measures such as battle positions and trigger lines to synchronize subordinate units. Higher headquarters are responsible to synchronize employment of combat power forward of the main battle and security areas or coordinated fire line. Higher headquarters use sectors to synchronize and coordinate subordinate force engagement areas and allow for mutually supporting fields of fire, which do not require deconfliction between adjacent units. Units treat everything behind the forward line of troops as an AO with its associated responsibilities. Units can subdivide a sector into subordinate areas of operations, zones, or sectors.

Mutual Support

Commanders and staffs consider mutual support when considering how large an area to assign subordinates. Mutual Support is that support which units render each other against an enemy, because of their assigned tasks, their position relative to each other and to the enemy, and their inherent capabilities (JP 3-31). In Army doctrine, mutual support is a planning consideration related to force disposition, not a command relationship. Mutual support has two aspects: supporting range and supporting distance.

When two units are mutually supporting, their assigned areas are generally contiguous with each other. Units with non-contiguous areas are generally not mutually supporting. Supporting range is the distance one unit may be geographically separated from a second unit yet remain within the maximum range of the second unit's weapons systems (ADP 3-0). Supporting range depends on available weapons systems, and it is normally the maximum range of the supporting unit's indirect fire weapons, although certain capabilities employed via space and cyberspace can be used at much longer ranges. Terrain, visibility, and weather may limit the supporting range. If one unit cannot effectively or safely fire in support of another, the first may not be in supporting range even though its weapons have the required range. At higher echelons, communications are also a consideration. If two units cannot effectively coordinate the use of indirect fire, then they may not be considered in supporting range of each other.

Supporting distance is the distance between two units that can be traveled in time for one to come to the aid of the other and prevent its defeat by an enemy or ensure it regains control of a civil situation (ADP 3-0). These factors affect supporting distance:

- Terrain and mobility.
- Distance.
- Enemy capabilities, including those employed from the space, cyberspace, air, and maritime domains.
- Friendly capabilities, including those employed from the space, cyberspace, air, and maritime domains.

(Operations) III. Fundamentals 1-57

example, Army forces track the location of enemy AMD, artillery, and armored formations outside the assigned area whose movement or employment may impact current and future operations. The area of interest also includes adjacent and other friendly forces whose actions or inactions of friendly forces could affect operations.

Commanders consider all forms of contact possible with enemy forces and update their area of interest as the situation develops, including the effects of enemy influence and disinformation. An area of interest surrounds an assigned area, extending forward and to the flanks of the assigned area, and overlapping with adjacent unit assigned areas. Depending on the operation, it may also extend rearwards, especially when enemy forces contest operational and strategic lines of communications. A unit's area of interest is especially important for helping leaders assess risks and maintain situational awareness of factors that will become important as operations progress. It is particularly important that the area of interest account for human, information, and physical dimension factors across all domains.

B. Deep, Close, and Rear Operations

Within assigned areas, commanders organize their operations in terms of time, space, and purpose by synchronizing deep, close, and rear operations. An echelon's focus in time, space, and purpose—not necessarily their the physical location—determines whether they are deep, close, or rear operations. This model assists commanders and staffs in synchronizing capabilities that reside outside of their unit's assigned area, (for example, from air, space, and cyberspace) with operations inside their assigned areas. The degree of convergence that a corps can achieve to set conditions for its subordinate divisions depends on its ability to synchronize close, deep, and rear operations among its subordinate echelons and with the joint force.

Divisions and higher echelons typically align their deep, close, and rear operations to corresponding areas, due to the scale of forces and physical considerations involved. This facilitates their C2 of forces spread over wide distances whose physical locations do not correspond to the location and purpose of their effects. Typically, divisions and corps assign command posts to enable control of these areas. For example, a division may position an artillery battery in a position area for artillery located in the rear area but employ its fires in support of close operations. In this case, the rear command post might control the battery's sustainment and protection, but the division main command post will control its priorities for providing indirect fire support. At brigade echelons and below, differentiating between close, deep, and rear may have less utility during large-scale combat operations because of the high tempo, narrow focus, and short planning horizons. At every echelon, however, commanders must understand the relationship among these operations and their combined impact on mission accomplishment.

See following pages for an overview and further discussion. Refer to FM 3-94 for more information about deep, close, and rear areas.

Support Area Operations *(See pp. 7-22 to 7-26.)*

Support area operations are a critical part of rear operations. Support area operations are the tactical actions securing lines of communications, bases, and base clusters that enable an echelon's sustainment and command and control.

A support area is where units position, employ, and protect base sustainment assets and lines of communications required to sustain, enable, and control operations. Support area operations include sustainment for the echelon and relevant security operations. Support area operations enable the tempo of deep and close operations. Support area operations require detailed planning to coordinate among the various units providing sustainment, protection, and security. A maneuver enhancement brigade or BCT typically provides C2 for support area operations for a division or corps due to the level of security, planning, and integration required.

1-58 (Operations) III. Fundamentals

Notional Corps Deep, Close, & Rear Areas

Ref: FM 3-0, Operations (Oct. '22), p. 3-28.

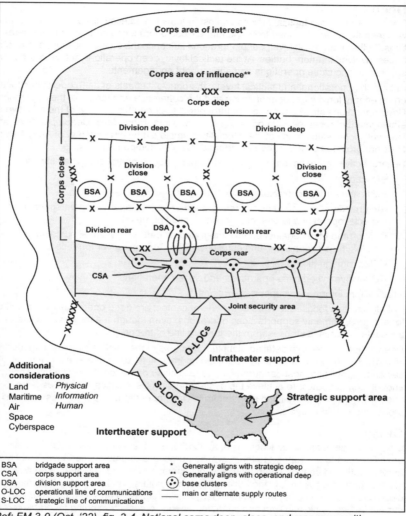

BSA	bridgade support area
CSA	corps support area
DSA	division support area
O-LOC	operational line of communications
S-LOC	strategic line of communications

* Generally aligns with strategic deep
** Generally aligns with operational deep
⊙ base clusters
—— main or alternate supply routes

Ref: FM 3-0 (Oct. '22), fig. 3-4. Notional corps deep, close, and rear areas with contiguous divisions.

Note. The symmetry of figure 3-4 provides the simplest way for understanding the operational framework which is a mental model. Applying the model to real situations results in significant variations.

Multidomain Operations

Deep, Close, and Rear Operations

Ref: FM 3-0, Operations (Oct. '22), pp. 3-29 to 3-31.

Deep

Deep operations are tactical actions against enemy forces, typically out of direct contact with friendly forces, intended to shape future close operations and protect rear operations. At the operational level, deep operations influence the timing, location, and enemy forces involved in future battles. At the tactical level, deep operations set conditions for success during close operations and subsequent engagements.

At both the operational and tactical levels, the principal effects of deep operations focus on an enemy force's freedom of action and the coherence and tempo of their operations. Deep operations strike enemy forces throughout their depth and prevent the effective employment of reserves, C2 nodes, logistics, and long-range fires. Deep operations are inherently joint, since many of the capabilities employed by or in support of Army formations are provided by a joint headquarters or Service component.

Several activities are typically conducted as part of deep operations. They include—

- Deception.
- ISR and target acquisition.
- Interdiction (by ground or air fires, ground or aerial maneuver, cyberspace forces, special operations forces, or any combination of these).
- Long-range fires against enemy integrated air defense systems, sustainment nodes, fires capabilities, and echeloned follow-on maneuver formations.
- Electromagnetic warfare.
- Offensive cyberspace operations and space operations.
- Military information support operations.

Not all activities focused forward of the line of contact are deep operations. Counterfire, for example, primarily supports close operations, even though the targets attacked may be located at great distances from the forward line of troops.

Deep operations require detailed planning. Because of the relative scarcity of resources with which to perform these activities, deep operations focus on the enemy vulnerabilities and capabilities most dangerous to the next close operation. Attacks must employ enough combat power to achieve the desired result. This is critical when—as is frequently the case—maintaining momentum in close operations depends on successful prosecution of deep operations.

Close

Close operations are tactical actions of subordinate maneuver forces and the forces providing immediate support to them, whose purpose is to employ maneuver and fires to close with and destroy enemy forces. Until enemy forces are defeated or destroyed in close operations, they retain the ability to fight and hold ground. At the operational level, close operations comprise the efforts of large tactical units—corps and divisions—to win current battles by closing with and defeating enemy forces after setting favorable terms to do so. At the tactical level, close operations comprise the efforts of smaller tactical units to win current engagements through movement combined with direct and indirect fires while physically in contact with the enemy forces they intend to destroy and defeat. Close operations concentrate overwhelming combat power at the right time and place to create and then exploit windows of opportunity to achieve assigned objectives.

Close operations include the deep, close, and rear operations of their subordinate maneuver formations. For example, divisions and separate brigades conduct corps close operations. Brigade combat teams (BCTs) are the primary forces conducting division close operations.

1-60 (Operations) III. Fundamentals

The positioning of assets and capabilities does not determine whether they are part of the close operation. For example, some reconnaissance and target acquisition units, while located forward near the line of contact, may have a purpose that supports deep operations.

Close operations are inherently lethal because they involve direct fire engagements at relatively short ranges with enemy forces seeking to mass direct, indirect, and aerial fires against friendly forces. Deep and rear operations set conditions for the success of close operations. The measure of success of deep and rear operations is their positive impact on increasing the effectiveness and reducing the cost of close operations.

Activities are part of close operations if their purpose contributes to defeating committed enemy forces that are or will be in direct physical contact with friendly forces. The activities that comprise close operations include—

- Maneuver of subordinate formations (including counterattacks).
- Close combat (including offensive and defensive operations).
- Indirect fire support (including counterfire, close air support, electromagnetic attack, and offensive space and cyber operations against enemy forces).
- Information collection.
- Sustainment support of committed units.

Rear

Rear operations are tactical actions behind major subordinate maneuver forces that facilitate movement, extend operational reach, and maintain desired tempo.

This includes continuity of sustainment and C2. Rear operations support close and deep operations. At the operational level, rear operations sustain current operations and prepare for the next phase of the campaign or major operation. These operations are distributed, complex, and continuous. At the tactical level, rear operations enable the desired tempo of combat, assuring that friendly forces have the agility to exploit any opportunity.

Rear operations typically include five broad activities: positioning and moving reserves; positioning and repositioning aviation, fire support, and AMD units; conducting support area operations; securing sustainment and C2 nodes; and controlling tactical unit movement between the division or corps rear boundary and units conducting close operations. Rear operations typically include efforts that consolidate gains to make conditions created by deep and close operations more permanent. All of these activities compete for limited terrain and lines of communications. Division and corps rear command posts are generally responsible for rear operations. There are several considerations for conducting rear operations. They include—

- C2.
- Information collection activities to detect enemy forces.
- Establishment and maintenance of routes.
- Terrain management.
- Movement control.
- Protection of critical friendly capabilities.
- Information activities.
- Infrastructure repair and improvement.
- Defeating bypassed forces and continuing to consolidate gains.
- Minimum-essential stability tasks.

Enemy deep operations often target friendly rear operations because they are often both vulnerable and essential to friendly mission success. Commanders commit combat power to protect rear operations, but they balance those requirements against those necessary for successful close and deep operations. Units involved in rear operations must protect themselves using both passive and active measures. Commanders and staffs must continuously reevaluate the possibility of more serious threats to rear operations and develop plans to meet them with minimum disruptions to ongoing close operations.

(Operations) III. Fundamentals 1-61

C. Main Effort, Supporting Effort, & Reserve

Ref: FM 3-0, Operations (Oct. '22), p. 3-32.

Main Effort

Commanders designate a subordinate unit as a main effort when its mission at a given point in time is most critical to overall mission success. Commanders weight the main effort with additional combat power. Typically, commanders shift the main effort one or more times during execution. When commanders designate a unit as the main effort, it receives priority of support and resources to maximize combat power. Commanders establish clear priorities of support, and they shift resources and priorities to the main effort as circumstances and the commander's intent require. The unit that directly accomplishes the mission is usually the main effort when it conducts its mission. Commanders typically designate priority for sustainment to units that they anticipate to be the main effort. This helps maximize the combat power of a unit before it becomes the main effort. Shifting a priority of sustainment to the current main effort might be too late to be effective.

Supporting Effort

A supporting effort is a designated subordinate unit with a mission that supports the success of the main effort (ADP 3-0). Commanders resource supporting efforts with the minimum assets necessary to accomplish the mission. Forces often realize success of the main effort through success of supporting efforts. A main effort in an earlier phase can be a supporting effort for a main effort in a later phase.

Reserve

A reserve is that portion of a body of troops that is withheld from action at the beginning of an engagement to be available for a decisive movement (ADP 3-90). A reserve is an uncommitted force, and it does not normally have a full suite of combat multipliers available to it until it is committed. It is the echelon's main effort once it is committed. Commanders constitute a reserve and base the size of the reserve on the level of uncertainty in the current tactical situation. Commanders consider survivability, mobility, and the most likely mission when positioning their reserve. While commanders can assign their reserve a wide variety of tasks, through planning priorities, to perform on commitment, a reserve remains prepared to accomplish other missions. The primary purposes for a reserve are to—

- Exploit success.
- Counter tactical reverses that threaten the integrity of the friendly force's operations.
- Retain the initiative.

Once a reserve is committed, units reconstitute a new one whenever possible. When a commander assigns a unit the mission of being the reserve, the commander gives the unit a list of planning priorities. Typically a reserve has no more than three planning priorities because of the time it takes to adequately prepare for each priority.

See p. 3-19 for discussion of main effort, supporting effort, & reserve as an imperative.

Chap 1

Multidomain Operations

IV(A). Competition

Ref: FM 3-0, Operations (Oct. '22), chap. 4.

This chapter begins with an overview of how the Army contributes to **competition below the threshold of armed conflict** as part of the joint force. It describes methods employed by adversaries and how Army forces contest adversary activities by supporting combatant command campaign plans and preparing for large-scale combat operations with unified action partners. The chapter concludes by discussing how Army forces consolidate gains and transition to crisis or armed conflict as branches to joint campaigns.

I. Overview of Operations During Competition

Competition below armed conflict occurs when an adversary's national interests are incompatible with U.S. interests, and that adversary is willing to actively pursue them short of open armed conflict. While neither side desires, at least initially, to use military force as the primary method to achieve its goal, the adversary is willing to employ national instruments of power, including military force, below the threshold of actual armed conflict to achieve its aims. The resulting tension between the two sides creates potential for violent escalation when one side challenges the status quo.

> Army forces are successful during competition when they deter adversary malign action, enable the attainment of other national objectives, and maintain the ability to swiftly and effectively transition to armed conflict when deterrence fails.

Operations during competition are intended to deter malign adversary action, set conditions for armed conflict on favorable terms when deterrence fails, and shape an operational environment with allies and partners in ways that support U.S. strategic interests and policy aims. Theater armies support combatant commanders (CCDRs) as they conduct operations to deter adversaries and achieve national objectives. Their operations, conducted as part of a combatant command campaign plan, are conducted over time and across broad areas without armed conflict. This may include cooperative training, support to local institutions, construction projects, and a range of other activities. In many cases, enduring engagement is necessary, especially given the tendency of adversaries to pursue strategic objectives over long periods of time that do not comport with the shorter political-strategic cycles found in the U.S. or among many of its allies and partners.

Army forces contribute to conventional deterrence during competition by preparing for armed conflict, including large-scale combat operations. This includes assisting allies and partners to improve their military capabilities and capacity. Preparation for combat operations and demonstrating the interoperability of the U.S. joint force with allies and partners presents the strongest deterrence to adversaries. Deliberate messaging that communicates the will and capability to conduct combat operations can amplify the deterrent effect of physical actions on the ground. Interoperability, coupled with the demonstrated capabilities and capacity of Army forces, reinforces a unified approach to defending mutual interests. Even a small contingent of forward-stationed U.S. Army forces are a challenge to defeat when operating with allies and partners. A force ready for large-scale combat operations contributes to the potency and integration of the other instruments of national power, provides CCDRs capabilities for graduated responses, and enables the Army to help the joint force achieve national strategic objectives through competition rather than armed conflict.

(Operations) IV(a). Competition 1-63

Multidomain Operations

II. Adversary Methods During Competition

Ref: FM 3-0, Operations (Oct. '22), pp. 4-2 to 4-3.

To effectively plan, prepare, execute, and assess operations during competition requires a broad understanding of the strategic environment and common adversary methods and objectives. Adversaries use a range of techniques to hinder the United States from achieving its objectives during competition and further their own interests. Forward-positioned Army forces may be able to detect and assess such adversary activities. By understanding and effectively countering adversary techniques, Army forces can help the joint force and interagency partners achieve their objectives.

Activities to Achieve Strategic Goals

Adversaries employ all of their instruments of national power in a combination of ways to pursue strategic interests without direct military confrontation with the United States. For example, Russia applies its elements of national power through an approach called "New-Type War" (also labeled "Russian New Generation Warfare"). This approach allows Russia to achieve many of its strategic goals below the level of armed conflict and with limited employment of military forces. If coercion through diplomatic, information, and economic instruments fails, Russia is prepared to employ its conventional military power and proxy forces as needed. China also relies on a comprehensive use of its instruments of national power. Like Russia, China seeks to achieve many of its strategic objectives with nonmilitary instruments of national power and keep military forces in a supporting role that reinforces facts established on the ground with other than overt military action.

By using all instruments of national power, an adversary is able to further its interests through a range of nonmilitary and military activities that may provide advantages over U.S. forces. Examples of nonmilitary activities include Russia and China's diplomatic efforts to establish security cooperation agreements with neighboring countries as a way to expand regional influence. Another example is China's use of infrastructure projects, as part of "The Belt and Road Initiative", to grow its economic influence. In both examples, adversaries primarily use nonmilitary means to achieve strategic objectives while weakening U.S. influence and undermining political-military partnerships between the United States and other countries within these same regions.

Adversaries can pursue more aggressive options through military activities that safeguard their interests abroad, maintain regional stability, and exert influence regionally and globally. These activities may include controlling or reducing access to certain areas of the global commons, challenging the established borders of other nations, or using the threat of force to influence the decisions of neighboring countries. Adversaries may pursue these activities overtly with the use of conventional military forces or covertly through a combination of proxy forces, unconventional warfare, and information warfare.

Proxy forces are generally non-state actors aligned with respective state actors, and they perform activities on behalf of or in accordance with the state actor's strategic objectives. Examples of proxies include paramilitary groups, criminal organizations, private civilian organizations, private companies, special interest groups, and religious groups. Covert methods, such as the use of proxy forces, provide adversaries with plausible deniability and cost savings in achieving strategic objectives.

1-64 (Operations) IV(a). Competition

Activities to Counter a United States Response

While adversaries desire to further their interests and achieve their goals without U.S. involvement, they will be prepared to counter a response from the U.S. military. To do this, an adversary may attempt to prevent or constrain the United States' ability to project forces to the region and limit U.S. response options by using the following methods:

- Conduct information warfare activities to manipulate the acquisition, transmission, and presentation of information in such a way that legitimizes the adversary's actions and portrays the United States as the aggressor.
- Conduct preclusion activities through nonlethal means to undermine relationships, raise political stakes, manipulate public opinion, and erode resolve to constrain or eliminate basing rights, overflight corridors, logistics support, and concerted allied action.
- Isolate the United States from allies and partners by fostering instability in critical areas and among relevant actors to increase U.S. operational requirements.
- Create sanctuary from U.S. and partner forces through international law and treaty agreements, monitoring and attacking partner forces from across international borders, and using proxy forces.
- Conduct systems warfare by executing cyberspace attacks against critical force projection and sustainment infrastructure nodes to delay or disrupt the United States' ability to deploy forces. Systems warfare approaches include nonattributable attacks on domestic infrastructure and the employment of networked military capabilities that support isolation and preclusion efforts.

Activities to Preclude United States Access to a Region

Adversaries seek to establish conditions that limit or prevent U.S. access to a region, typically in locations close to their borders. This includes forward positioning of layered and integrated air defenses, early warning surveillance radars, rocket artillery, electronic warfare capabilities, and counter-space capabilities. Additionally, adversaries may seek to position intermediate-range ballistic missiles, cruise missiles, fixed-wing aircraft, unmanned aircraft systems, and naval surface and subsurface forces to shape an operational environment in their favor. Positioning systems that support an antiaccess (A2) strategy allows adversaries to deny or disrupt U.S. access to a region in the event of hostilities while providing leverage against friendly partner nations with the potential use of force. Furthermore, the positioning of systems capable of delivering conventional and nuclear munitions creates additional challenges for the United States. An adversary's ability to establish, maintain, and demonstrate robust A2 systems bolsters its domestic narratives while eroding partner nation trust and confidence.

Friendly forces must assume they are always under observation because of all the means available to a peer adversary, particularly those available in space and cyberspace. In addition to forward positioning capabilities that support A2 and area denial (AD) approaches, these adversaries seek understanding of the disposition, readiness, and activities of U.S. forces within a contested region. Adversary activities include reconnaissance of U.S. military installations, unit movements, ports of embarkation and debarkation, and staging areas to identify potential targets for ballistic missiles and long-range fires. Adversaries employ cyberspace tools to conduct reconnaissance of friendly networks to identify vulnerabilities for possible exploitation. An adversary may conduct probing actions in the air and maritime domains to test responses by U.S. and other friendly forces. The intelligence gained through these activities will prepare an adversary for hostilities in the event a situation escalates to armed conflict.

Refer to ATP 7-100 series for a detailed discussion on specific threat capabilities and employment strategies. See Chapters 6 and 7 for specific examples of how adversaries are likely to employ A2 and AD capabilities in the beginning of a conflict.

(Operations) IV(a). Competition 1-65

III. Preparation for Large-Scale Combat Operations

Army forces that cannot credibly execute operations during armed conflict neither deter adversaries nor assure allies and other unified action partners. Preparation for large-scale combat operations is therefore the primary focus of Army conventional forces during competition. While there are multiple forms of armed conflict, large-scale combat among state actors is the most complex and lethal form of armed conflict, and it demands significant focus along multiple lines of effort to prepare for it. Some of the activities Army forces execute to prepare for armed conflict include—

- Setting the theater.
- Building allied and partner capabilities and capacity.
- Improving joint and multinational interoperability.
- Protecting forward-stationed forces.
- Preparing to transition and execute operation plans (OPLANs).
- Training and developing leaders for operations in specific theaters.

A. Set the Theater (See pp. 7-6 to 7-14.)

Setting the theater is the broad range of activities continuously conducted to establish conditions for the successful execution of operations in a theater. Setting the theater never ends. It is conducted to enhance an operational environment in ways favorable to friendly forces and occurs during competition, crisis, and armed conflict. While setting the theater occurs across each strategic context, its importance is greatest during competition because that is when the most time is available. Army forces must set the theater during competition to enable quick transitions during crisis and conflict, when time favors the aggressor. Army forces use military engagements, security cooperation, and other activities to assess and understand the current conditions within the theater and execute specific theater setting activities to enable joint forces and other unified action partners.

Setting the theater requires a comprehensive approach among unified action partners and bilateral or multilateral diplomatic agreements that allow U.S. forces access to ports, terminals, airfields, and bases in the area of responsibility (AOR) to support future operations. This includes but is not limited to theater opening; reception, staging, onward movement, and integration (RSOI); establishing networks; classifying routes; and other operational activities that set the conditions for operations in the AOR. Information activities are a significant part of setting the theater. They enable decision making, protect friendly information, inform domestic and international audiences, and influence foreign audiences, while helping to counter adversary information warfare.

Setting the theater is a continuous activity for all staff sections and warfighting functions. It involves significant sustainment, air and missile defense (AMD), engineering, information collection, intelligence, and communications focused on setting conditions to counter known or potential threats to U.S. interests across the AOR. All warfighting functions, functional areas, and branches that comprise staffs and commands conduct preparation of the operational environment to address unique considerations for setting the theater within their respective areas of expertise (for example, civil preparation of the environment and joint intelligence preparation of the operational environment).

For more information about the land component's roles and responsibilities for setting the theater in conflict, refer to JP 3-31 and JP 3-35. For additional information about the subordinate Army tasks and activities associated with setting the theater, refer to ATP 3-93 and FM 4-0.

1-66 (Operations) IV(a). Competition

B. Build Allied and Partner Capabilities and Capacity

Army forces fight as part of a joint and multinational force. The United States cannot achieve its security interests without the cooperation of treaty allies, partner nations, and other unified action partners. Helping partner nations build, rebuild, or maintain their national security institutions is a critical step in maintaining regional stability, and it is ultimately less expensive than requiring U.S. forces to do so. Additionally, by maintaining partner-nation security institutions, the Army helps add to the aggregate force that is available to potentially deter adversary forces or counter them if they choose to pursue their goals through armed conflict. Forward-stationed U.S. Army forces, by themselves, generally do not enjoy favorable combat power ratios with peer adversaries. Allies and other partners provide the bulk of forces initially able to conduct operations during armed conflict. This combined force capability enhances deterrence for both the partner nation and the United States.

Combined training and exercises with partners play a key role in building allied and partner capabilities and shaping an operational environment. Such events are the most overt and visible means of demonstrating friendly capabilities, interoperability, and will. Exercises also help set the theater. Multinational forces that maintain high levels of combat readiness provide the credibility essential to assure partners and deter adversaries. Combined exercises build relationships and mutual respect among allies and other multinational partners, identify systems and processes to employ partner capabilities effectively, and reveal shortfalls to be improved upon. Training exercises occur at all echelons of command, from tactical units to large, combined task forces. The application of lessons learned during these exercises is key to improving multinational interoperability during competition. An example of a failure to prepare for large-scale combat operations occurred in the Philippines in 1941.

C. Interoperability

The ability of Army forces to fight as a cohesive whole, integrated with the joint force, allies, and partners, is vital to maximizing combat power and creating a deterrent effect in a theater. Interoperability is the ability to act together coherently, effectively, and efficiently to achieve tactical, operational, and strategic objectives (JP 3-0). An Army formation that is interoperable with joint and multinational partners is substantially more capable than one that is not. Interoperability with any unified action partner is essential to effective operations. Interoperability standards and procedures must be trained, tested, and refined during competition; it is too late to seek interoperability once a crisis or armed conflict begins.

Interoperability starts with mutual understanding across echelons throughout a multinational force. Effective interoperability includes understanding technical challenges and developing methods to bridge gaps, understanding the tactical capabilities of each member in the multinational force, and integrating partners into a unified operational approach. During competition, the theater army or a delegated command is responsible for building the infrastructure that enables this. Communication is primarily achieved through liaison teams, understanding staff processes, and ensuring adequate access to partner nation command and control (C2) systems (within the limits of national caveats). Understanding foundational interoperability requirements like NATO doctrine; American, British, Canadian, Australian, and New Zealand (known as ABCANZ) Armies Program interoperability standards; and Combined Forces Command (in the Republic of Korea) processes is critical to communicate and interoperate with allied forces. It is essential that these requirements and standards are incorporated into routine training and exercise planning at all echelons to build the required interoperable readiness needed in a time of crisis or conflict.

Refer to FM 3-16 for more information on multinational operations and interoperability.

(Operations) IV(a). Competition 1-67

Multidomain Operations

D. Protect Forward-Stationed Forces

Peer threats possess reconnaissance and surveillance, fires; special operations forces; and other capabilities that can range forward-stationed Army forces and place them at risk. The protection of Army forces forward, under the assumption that deterrence is not guaranteed, is essential. Army forces implement procedures and conduct necessary activities to ensure they, and the elements of the joint or partner force they protect, can endure an initial attack with little early warning. This includes preparation for threat attacks from any domain that is informed by understanding what holds friendly forces at risk and how a threat may attack. An adversary could attack using capabilities from domains other than land, either to set conditions on the ground or as a means of escalation designed to limit friendly options. Preparation by Army forces therefore includes planning and integration with other elements of the joint force. Coordination for Army and joint capabilities that are able to protect friendly forces during armed conflict and enable them to endure until they can be supported is critical to establishing deterrence.

The demonstrated ability of Army forces to withstand an adversary's initial attack adds to the integrated deterrence effect on adversaries and may dissuade them from escalation. Host-nation capabilities may constitute a significant part of force protection and forward defense, so they must be integrated into theater protective efforts. Forward deployed forces that cannot be adequately protected or quickly repositioned during adversary escalation to armed conflict should be relocated to more defensible locations.

E. Prepare to Transition and Execute Operation Plans

Army forces at every echelon prepare to execute OPLANs that they are expected or likely to support. The foundation for this is active, continuous information collection and intelligence analysis. Higher echelons, such as the theater army and corps, identify initial targets and the required Army and joint capabilities needed to attack those targets in the initial stages of an armed conflict. They likewise consider whether general defense plans that provide guidance for subordinate unit immediate action during the early stages of a conflict initiated with few indications or warnings are necessary or prudent.

Preparation to execute OPLANs must extend to all echelons and partners. Lower tactical echelons train tactical tasks related to the parts of an OPLAN they support or execute. Units conduct deployment rehearsals and emergency deployment readiness exercises to improve response times and validate plans. Rehearsals with unified action partners build mutual understanding and improve interoperability. Units conduct thorough reconnaissance of all lines of communications, infrastructure, avenues of approach, assembly areas, and potential firing points or battle positions. Leaders and Soldiers should walk the actual terrain that engagements and battles could be fought on and, when possible, they should use this terrain for rehearsals. A shared understanding of OPLANs, terrain, and adversaries down to the lowest tactical echelon will allow an effective transition to armed conflict.

F. Train and Develop Leaders

Leaders prepare themselves, their subordinates, and their organizations for operations in specific combatant command AORs. When developing expertise in specific regions, units become familiar with applicable OPLANS and coordinate with the theater army, the assigned military intelligence brigade-theater (MIB-T), and other theater army-assigned units as appropriate. This regionally specific readiness augments ongoing training and leader development activities conducted across the force.

See pp. 2-7 to 2-12 for discussion on the role of leadership during operations.

1-68 (Operations) IV(a). Competition

IV. Interagency Coordination

Ref: FM 3-0, Operations (Oct. '22), pp. 4-8 to 4-9.

Military engagement, security cooperation, and deterrence activities usually involve a combination of military forces and capabilities separate from, but integrated with, the efforts of interagency participants. These actions are coordinated by diplomatic chiefs of mission and country teams. Understanding their roles and relationships is critically important. The Department of State is responsible for the diplomatic instrument of national power. Chiefs of mission are the final approval authorities for all U.S. military activities that occur in the nation they are responsible for, and they have the authority to modify the execution of planned activities during competition.

Refer to JP 5-0 for more information on country-specific plans.

Activities that occur during competition encompass a wide range of actions where the military instrument of national power supports and is subordinate to the other instruments of national power. Competition overseas generally requires cooperation with international organizations (for example, the United Nations) and government entities in other countries to protect and enhance mutual national security interests, deter conflict, and set conditions for future contingency operations.

United States Diplomatic Mission

U.S. diplomatic missions include representatives of all U.S. departments and agencies physically present in the country. Chiefs of missions are the principal officers in charge of diplomatic missions. They are often, but not always, ambassadors. They oversee all U.S. government programs and interactions with and in a host nation. The chief of mission is the personal representative of the President and reports through the Secretary of State, ensuring all in-country activities serve U.S. interests and regional and international objectives.

The United States maintains different types of diplomatic missions in different countries. Some countries have only a consulate, many have only an embassy, and others have an embassy and a number of consulates. Typically, Army elements conducting security cooperation activities coordinate with diplomatic mission officials, even in nations with only a consulate. Relationships with consular offices are determined on a case-by-case basis. The same entities and offices existing in an embassy are present or liaised at consulates.

Refer to FM 3-22 for a detailed explanation of this role in relation to Army operations.

Country Team

The country team is the point of coordination within the host country for the diplomatic mission. The members of the country team vary depending on the levels of coordination needed and the conditions within that country. The country team is made up of the senior member of each represented U.S. department or agency, as directed by the chief of mission. The team may include the senior defense official or defense attaché, the political and economic officers, and any other embassy personnel desired by the ambassador.

Refer to JFODS6: The Joint Forces Operations & Doctrine SMARTbook, 6th Ed. (Guide to Joint, Multinational & Interorganizational Operations). Completely updated for 2023, chapters include joint doctrine fundamentals (JP 1), joint operations (JP 3-0, 2022), joint planning (JP 5-0, 2020), joint logistics (JP 4-0), joint task forces (JP 3-33), joint force operations (JPs 3-30, 3-31, 3-32 & 3-05), multinational operations (JP 3-16), and interorganizational cooperation (JP 3-08).

V. Competition Activities

Ref: FM 3-0, Operations (Oct. '22), pp. 4-9 to 4-12.

Competition involves activities conducted under numerous programs within a combatant command. The CCDR uses these activities to improve security within partner nations, enhance international legitimacy, gain multinational cooperation, and influence adversary decision making. Competition activities include obtaining access for U.S. forces, maintaining sufficient forward-based presence within a theater to influence conditions in the strategic environment, and mitigating conditions that could lead to a crisis or armed conflict. At any time during competition, but especially during times of heightened tension, leaders must take great care to ensure Army forces avoid activities that accidently provoke crisis or armed conflict. Army forces, as directed by the theater army, must stay within an activity level that meets the CCDR's intent for readiness without unintentionally increasing tensions.

Activities that occur during competition are directly tied to authorities provided in various titles of the United States Code and approved programs, and they are integrated and synchronized with the Department of State, other government agencies, country teams, and ambassadors' plans and objectives. The Department of State and the United States Agency for International Development (USAID) help produce the joint regional strategy to address regional goals, management, and operational considerations. Each country team develops both an integrated country strategy and a country development cooperation strategy to address joint mission goals and coordinated strategies for development, cooperation, security, and diplomatic activities. Working with the Department of State and various country teams, the CCDR develops country-specific security cooperation plans, which are codified in the country-specific security cooperation section of the combatant command campaign plan (CCP). Some CCPs include regional country plans, posture plans, and theater distribution plans that facilitate synchronization of resources, authorities, processes, and timelines to favorably affect conditions within the CCDRs' AORs.

Army forces execute activities during competition that support joint force campaigning goals, satisfy interagency requirements, and set the necessary conditions to employ Army combat power during crisis and armed conflict. The theater army works with the CCDR to develop objectives for the employment of Army forces in theater and develops support plans to address Army-specific activities. Army forces provide security cooperation capabilities across any given theater of operations by conducting military engagement, security cooperation, nuclear deterrence, counter-weapons of mass destruction activities, and humanitarian assistance.

Military Engagement

Military engagement is contact and interaction between individuals or elements of the Armed Forces of the United States and those of another nation's armed forces, or foreign and domestic civilian authorities or agencies, to build trust and confidence, share information, coordinate mutual activities, and maintain influence (JP 3-0). Military engagement occurs as part of security cooperation, but it also extends to interaction with domestic civilian authorities. Army forces will also routinely communicate with nongovernmental organizations, either directly or indirectly, to ensure expectations and roles are understood.

CCDRs and Army senior leaders seek out partners and communicate with adversaries to discover areas of common interest and tension. This increases the knowledge base for subsequent decisions and resource allocation. Such military engagements can reduce tensions and may prevent conflict, or, if conflict is unavoidable, they may allow the United States to enter into conflict with greater access and stronger alliances or coalitions. Army forces support military engagement through deliberate interactions with unified action partners at the junior Soldier through senior leader levels. The State Partnership Program provides a good example of how powerful military engagement can be.

1-70 (Operations) IV(a). Competition

Security Cooperation

Security cooperation is all Department of Defense interactions with foreign security establishments to build security relationships that promote specific United States security interests, develop allied and partner nation military and security capabilities for self-defense and multinational operations, and provide United States forces with peacetime and contingency access to allied and partner nations (JP 3-20). These efforts may include Army forces participating in joint and multinational exercises and employing regionally aligned forces. Conducting security cooperation is one of the Army's primary stability tasks.

- **Security assistance** is a group of programs the U.S. Government uses to provide defense articles, military training, and other defense-related services by grant, lease, loan, credit, or cash sales.
- **Security force assistance** is the Department of Defense activities that support the development of the capacity and capability of foreign security forces and their supporting institutions (JP 3-20).
- **Foreign internal defense** is participation by civilian agencies and military forces of a government or international organizations in any of the programs and activities undertaken by a host nation government to free and protect its society from subversion, lawlessness, insurgency, terrorism, and other threats to its security (JP 3-22).
- **Security sector reform** is a comprehensive set of programs and activities undertaken by a host nation to improve the way it provides safety, security, and justice (JP 3-07).

Nuclear Deterrence and Countering Weapons of Mass Destruction

U.S. nuclear capabilities are foundational to the deterrence of adversary weapons of mass destruction use. To ensure the credibility of this deterrent, joint and Army forces must integrate the planning and operations of nuclear and conventional forces. Further, Army forces must plan, train, and exercise to conduct operations under the adversary threat or use of weapons of mass destruction in order to deny the adversary any perceived advantage that might result from employing weapons of mass destruction. To do so, commanders and staffs must continuously assess, protect, and mitigate the effects of adversary chemical, biological, radiological, and nuclear (CBRN) weapons use and contamination hazards. They must train under simulated weapons of mass destruction conditions. When under threat of nuclear attack, commanders must balance the risk of dispersing forces to mitigate the impact of nuclear effects across their AO against the ability to concentrate sufficient combat power to achieve objectives. In a chemically contaminated environment, a commander's decision-making ability is complicated by the effects on Soldier stamina, reaction times, and sustainment. Each of these environments requires unique actions to ensure a formation's ability to maneuver, fight, and sustain operations.

Humanitarian Assistance

USAID is the lead U.S. government agency, responsible to the Secretary of State, for administering civilian foreign aid and providing humanitarian assistance and disaster relief. USAID often works in concert with Army forces when Soldiers are tasked to provide assistance. It can supplement forces conducting civil affairs operations that the DOD conducts to build relationships and win the trust, confidence, and support of local populations.

Refer to TAA2: Military Engagement, Security Cooperation & Stability SMARTbook (Foreign Train, Advise, & Assist) for further discussion. Topics include the Range of Military Operations (JP 3-0), Security Cooperation & Security Assistance (Train, Advise, & Assist), Stability Operations (ADRP 3-07), Peace Operations (JP 3-07.3), Counterinsurgency Operations (JP & FM 3-24), Civil-Military Operations (JP 3-57), Multinational Operations (JP 3-16), Interorganizational Cooperation (JP 3-08), and more.

(Operations) IV(a). Competition 1-71

VI. Relative Advantages During Competition

During competition, Army forces seek relative advantages at the theater strategic, operational, and tactical levels. Relative advantages are advantages that Army forces provide the joint force commander (JFC) in relation to a specific adversary, and they are always contextual. They are necessary to deter adversaries, assist the joint force in promoting U.S. interests, and set conditions to conduct operations during crisis and armed conflict. These advantages augment unified action partner activities, and they address Service-specific issues identified during combatant command campaign development. Identifying, achieving, and maintaining these advantages helps the Army employ combat power effectively during crisis and armed conflict. A relative advantage is temporary. Adversaries quickly adapt to counter advantages (especially technological ones) once they are created or employed, and they seek to reduce or eliminate their effectiveness.

Understanding advantage relative to an adversary requires understanding the adversary's capabilities and will, friendly capabilities and will, and the operational environment within the theater. It further requires understanding of the interrelated influences of each dimension in an operational environment, including how physical, human, and information factors affect each other in a specific context. Changes in one dimension often have outcomes in the other two and in more than one physical domain.

A. Physical Advantages During Competition

Due to the expected tempo of operations, a sufficient number of Army forces comprised of the right capabilities must be forward stationed to provide CCDRs with a credible deterrent force and the ability to respond, when necessary, to adversary actions. Physical advantage encompasses combat power and the correlation of forces: the ability to deliver effects, superior range, and the ability to concentrate superior capabilities at the right places and times. Examples of activities that create physical advantages during competition include—

- Working with allies to conduct a deployment exercise of a theater-tailored unit to improve its OPLAN integration and interoperability.
- Surveying a potential assembly area with a forward engineer support team to determine if the area is of sufficient size to accommodate a properly dispersed Army formation.
- Hardening facilities against attack and rehearsing drills in response to potential adversary courses of action.
- Maintaining stocks of key supplies and equipment (Army pre-positioned stocks [APS]) in or near areas of concern to accelerate deployment of forces during crises or armed conflict.

B. Information Advantages During Competition

Information activities play a key role during competition. They include Army support to the combatant command and unified action partner strategic messaging. Coordinating with interagency and other unified action partners helps to develop and deliver coherent messages that counter adversary disinformation. Army forces reinforce strategic messaging by maintaining and demonstrating U.S. Army readiness for operations. Examples of relative information advantages are—

- Identifying targets and conducting target development on threat capabilities.
- Setting the conditions for convergence by developing methods to penetrate adversary computer networks.
- Discrediting adversary disinformation by helping the JFC inform domestic and international audiences through Army and joint information activities.

1-72 (Operations) IV(a). Competition

- Promoting the purpose and outcomes of multinational exercises and training events.
- Continuously monitoring the operational environment to detect changes to adversary methods or narratives.

C. Human Advantages During Competition

The institutional depth and professionalism of U.S. Army personnel contribute to the morale and will of partner security forces as Army forces interact across all ranks and echelons. Army formations serve as a professional force operating under the rule of law as guests in a specific region to facilitate the accomplishment of mutual military training goals. This can be a powerful advantage over adversaries who seek to extract concessions, including financial and informational gains, from other countries or groups. This bond of trust forms the foundation of the U.S. alliance system, and it is the primary means to ensure the security of the United States and its partners. Examples of activities that help achieve human advantages include—

- Training U.S. and partner nation forces in multinational exercises at combat training centers.
- Routine interaction with allies and other unified action partners that builds and maintains human, technical, and procedural interoperability through agreed-to standards.
- Hosting international officers at U.S. professional military education programs and sending U.S. officers to international military schools.
- Sustained presence by theater-aligned advisor teams that builds relationships and promotes interoperability over time.

VII. Consolidating Gains During Competition

Army forces continuously consolidate gains to maintain an operational environment that is advantageous to U.S. strategic interests. Experience proves that what Army forces do during competition helps ensure stability and reduces the potential for manmade crises or armed conflict throughout a region, even in locations where no previous combat has occurred. Examples of consolidating gains during competition range from transportation system improvements (including port, airfield, and rail lines of communications), increasing theater supply stocks, intelligence cooperation, and providing Army medical personnel to support a combatant command's humanitarian and civic assistance activities. Army forces contributing to humanitarian relief efforts with allies and partners help cement existing international relationships or set conditions for new ones in other places.

Army forces consolidate gains most effectively by maintaining a persistent or permanent presence in a theater of operations. This presence enables the cultivation of relationships on a predictable and reliable basis and provides Army forces a high degree of regular access to allies and partners. The enduring results of these activities help ambassadors, country teams, and JFCs gain a greater degree of influence with allies and partners as they pursue mutually beneficial objectives. In addition to this increased influence, Army consolidating gains activities contribute to joint efforts to support deterrence.

Consolidation of gains during competition following armed conflict or crisis is significantly different than during steady-state competition. In areas that have not seen recent armed conflict or a disruptive crisis, Army forces consolidate gains by reinforcing the success of steady-state competition activities. They do this by following through on what was begun earlier in consistent ways that provide predictability to allies and partners. In most cases, these activities will be indistinguishable from other competition activities designed to build partner capabilities and improve other advantages relative to threat forces.

(Operations) IV(a). Competition 1-73

Consolidating gains following armed conflict requires significant operations that, if not properly conducted, could result in a return to crisis or conflict. These efforts include information collection and intelligence analysis to understand threats, their support from the population, and what options are available to defeat them. Consolidating gains also includes stability tasks related to providing security, food, water, shelter, and medical treatment to the population. When appropriate, Army forces then work to restore or rebuild civil institutions and to transition security and stability tasks to those institutions.

Refer to FM 3-07 and FM 3-57 for additional details on stability operations and governance.

When immediate concerns are addressed after a crisis or armed conflict, the theater army and supporting forces focus most of their efforts on theater strategic consolidation of gains. They work with the theater's other components, the combatant command, interagency partners (primarily the Department of State), partner nations, and other unified action partners to develop and achieve long-term objectives. In general, these consolidation of gains activities are less intense and occur over longer periods of time. Army forces build on the success of past conflicts by conducting targeted engagements with unified action partners. Examples of this include routine engagements with Republic of Korea, Japanese, and NATO forces by forward-positioned and rotational units.

See ATP 3-93 for a detailed overview of routine theater army activities to consolidate gains during competition.

Peace operations are a means of consolidating gains. Peace operations are multiagency and multinational crisis response and limited contingency operations involving all instruments of national power with military missions to contain conflict, redress the peace, and shape the environment to support reconciliation and rebuilding and facilitate the transition to legitimate governance (JP 3-07.3). They usually occur under agreements brokered through organizations like the United Nations or through regional bodies like the African Union.

Refer to JP 3-07.3 and ATP 3-07.31 for additional details on peace operations.

VIII. Transition to Crisis and Armed Conflict

Transitions are inherently complex and unpredictable because anticipated environmental conditions can quickly change and alter the perception of strategic leaders who do not have all the information necessary for clear understanding. A response by one side can result in the perception of escalation by the other, leading to increased tensions. A crisis requiring a response can also occur because of unforeseen environmental changes. Transition from competition to crisis or armed conflict is often based on four types of decisions, resulting actions, and the follow-on associated effects from the initial action. Examples include—

- A decision by national command authorities to escalate or initiate armed conflict. Examples include the 2003 invasion of Iraq, the 2011 Libya strike, and the 2020 strike against Iranian General Qasem Soleimani.

- A decision by adversaries to escalate or initiate armed conflict. Examples include Hezbollah's rocket strikes against Israel in 2006, the Russo-Georgia War in 2008, and the Russian invasions of Ukraine in 2014 and 2022.

- A decision by allied nations to escalate or initiate an armed conflict. An example of this is the 1967 Six-Day War that occurred between Egypt, Syria, Jordan, Iraq, and Israel.

- Decisions made in response to a rapid environmental change that neither side planned for but causes tensions to rise. An example is the change brought to the strategic environment by the 2020 pandemic.

1-74 (Operations) IV(a). Competition

Army forces and leaders anticipate the potential for conflict in their operational environment. This is informed by input from the intelligence community and direction from strategic-level leaders. SFABs and other regionally aligned units have access to sensitive areas and ally and partner leaders. Their access and robust communications enable them to gain insight on actual conditions on the ground and provide real-time updates to decision makers during fluid situations. Decisions made before and during the initial stages of a crisis or armed conflict have significant impact on the decisions made by adversaries and the ultimate outcome of a particular situation.

Once a crisis or armed conflict starts, adversaries use all capabilities at their disposal to disrupt the deployment of Army forces. They will attempt to prevent the Army and the joint force from obtaining the needed time to deploy and build combat power. This chaos, with its resultant frictions, is the environment into which Army forces will respond.

A. Conflict Type Determination

Army forces and leaders need to anticipate the type of conflict the Nation will fight. This is informed by input from the intelligence community and direction from strategic-level leaders. Decisions made before and during the initial stages of a crisis or armed conflict have significant impact on the decisions made by adversaries and the ultimate outcome of a particular situation. The initial decisions or recommendations by Army strategic leaders impact the ability of the Army to project force in a timely manner.

B. Force Protection

Indications and warnings in a theater may prompt a decision to mobilize and deploy Army forces in anticipation of a crisis or armed conflict. Army forces anticipate and react to adversary actions targeting them where they are located during the initial stages of an operation, whether in the United States or forward deployed. Adversaries seek to degrade and disrupt the ability of Army forces to deploy. Adversaries may employ cyberspace attacks to inflict power outages at home station, target transportation networks to delay shipment of unit equipment, conduct social media attacks on Service or family members, and instigate protests that lower popular support for Army forces. Insider threats and proxies can conduct acts of terror, sabotage, subterfuge, and other activities against U.S. forces stationed in the United States and abroad. Adversaries may immediately employ lethal capabilities against Army forces using their air-, sea-, cyber-, and space-based capabilities to exploit surprise. Forward stationed forces should be prepared to deploy from garrison to dispersed locations to prepare a defense against an enemy attack. Force protection during transition will include physical security measures, operations security, and active information efforts to counter adversary efforts to misinform and otherwise influence Soldiers, Family members, and supporting organizations and communities.

C. Noncombatant Evacuation Operations (NEO)

A transition to crisis or armed conflict may require a noncombatant evacuation operation (NEO). Army forces conduct NEOs under a wide range of conditions. They may be conducted under relatively stable conditions or under unstable conditions that involve enemy combatants. Ideally, leaders anticipate a NEO requirement and are able to execute it prior to crisis or armed conflict. Uncertain adversary intentions and the threat of violence often create desperation amongst evacuees and local populations and increase the complexity and risk for forces conducting NEOs.

Once a NEO is requested, approved, and directed, the CCDR directs forces to conduct evacuation operations in support of the Department of State and Chief of Mission. NEOs, especially those of significant scale, will require Army forces that would otherwise be dedicated to other missions.

Refer to JP 3-68 for more information on NEOs.

(Operations) IV(a). Competition 1-75

D. Initial Employment Of Forward-Stationed Forces

A key strategic decision during competition or crisis is whether forward-stationed units will defend forward to hold terrain or displace to more advantageous positions. This decision should occur during competition unless a crisis unfolds in unanticipated ways and forces a decision point. During transition, forward-stationed Army forces have three courses of action. The JFC can integrate Army forces with host-nation land component forces as part of a mobile or area defense, assign U.S. Army forces a theater reserve role, or implement a plan that combines both courses of action.

A theater reserve role allows the JFC to preserve Army combat power for future offensive operations. Army forces may be required to defend key terrain and infrastructure to allow for receiving deploying forces. Army leaders advise the JFC on the best ways to use forward-stationed forces based on current conditions in the AO.

1-76 (Operations) IV(a). Competition

Chap 1

Multidomain Operations

IV(B). Crisis

Ref: FM 3-0, Operations (Oct. '22), chap. 5.

I. Overview of Operations During Crisis

A crisis is an incident or situation involving a threat to the United States, its citizens, military forces, or vital interests that develops rapidly and creates a condition of such diplomatic, economic, or military importance that commitment of military forces and resources is contemplated to achieve national objectives (JP 3-0). A crisis may be the result of adversary actions or indicators of imminent action, or it may be the result of natural or human disasters. During a crisis, opponents are not yet using lethal force as the primary means for achieving their objectives, but the situation potentially requires a rapid response by forces prepared to fight to deter further aggression. When directed, the Army provides a JFC with capabilities to help deter further provocation and sufficient combat power to maintain or reestablish conventional deterrence. The introduction of significant land forces demonstrates the will to impose costs, provides options to joint force and national leaders, and signals a high level of national commitment. The effects of a persistent presence on the ground among allied or partner forces cannot be easily replicated with air or maritime power alone.

> Success during a crisis is a return to a state of competition in which the United States, its allies, and its partners are in positions of increased advantage relative to the adversary. Should deterrence fail, Army forces are better positioned to defeat enemy forces during conflict.

Crisis response operations are characterized by high degrees of volatility and uncertainty. A crisis may erupt with no warning, or it may be well anticipated. Its duration is unpredictable. Additionally, adversaries may perceive themselves in a different context or state of conflict than U.S., allied, and partner forces. What is seen by one side as a crisis might be perceived by the other as armed conflict or competition. Army leaders must demonstrate flexibility anticipate changes in an operational environment, and provide JFCs with credible, effective options. This requires trained forces agile enough to adapt quickly to new situations and commanders and staffs adept at linking tactical actions to attaining policy objectives.

Regardless of the capabilities employed, there are generally two broad outcomes from a crisis. Either deterrence is maintained, and de-escalation occurs, or armed conflict begins. While this requires that Army forces be prepared for either type of transition, forces deploying during crisis always assume they are deploying to fight. While Army forces prepare for armed conflict, they avoid sending signals that armed conflict is inevitable, regardless of what the adversary does, to avoid inadvertent escalation. Generally, senior leaders at the corps and higher echelons influence those perceptions through public communications in support of the JFC and national leaders.

Note. Army forces also respond to crises related to disaster response, humanitarian assistance, and defense support to civil authorities when tasked. These crisis contexts and response options are covered in separate doctrinal publications. Refer to JP 3-28, JP 3-29, ADP 3-07, ADP 3-28, FM 3-07, and ATP 3-57.20 for more information on these types of crises and associated response options.

(Operations) IV(b). Crisis 1-77

II. Adversary Methods During Crisis

A crisis is frequently caused by an adversary acting aggressively to coerce and intimidate its opponents with the threat of force. Once an adversary crosses a U.S., allied, or partner crisis threshold, it attempts to shape and control the crisis to limit or prevent a U.S. military response. An adversary's attempts to control the situation involves escalating or de-escalating its activities based on an assessment of the situation, which includes a calculation of risk. Adversary forces conduct a detailed analysis of their available capabilities, capacity, and operational reach within a theater relative to friendly forces; their overall desired end state; and their willingness to achieve that end state before deciding to escalate. Even after careful analysis, the way a situation develops can be unpredictable. Some peer adversaries view conflict as a continuous condition in which heightened or reduced periods of violence occur and recur. Changing the intensity of their actions, even when that reduces tension, does not end their campaign to oppose U.S. interests.

A. Adversary Activities to Shape a Crisis

As a crisis develops, peer adversaries will attempt to shape the situation to their benefit through information warfare and preclusion focused on the U.S. joint force. They may use diplomatic, economic, and information means to divide an opponent's political leadership from its civilian population. They create separation by introducing or exacerbating distrust and division between different groups to weaken an opponent's political leadership and to create dissatisfaction among an opponent's civilian population. Adversaries position military forces in ways to increase uncertainty for opponents and to complicate their decision making. These activities create conditions for the adversary to exploit situations with minimal interference from the U.S., allied, or partner military forces.

Adversaries may use proxy forces to conduct information warfare, unconventional warfare, and criminal activities, although the balance and utility of these forces in crisis differs from their use during competition. Proxy forces, whether they are a militant separatist group, private military company, or criminal network, bring different capabilities to a situation, and the employment of their capabilities shifts as the strategic context changes. For example, while criminal networks can still accomplish useful tasks in environments marked by increased levels of violence, they do not have the same level of utility that they did during competition. Similarly, separatist groups cannot typically operate without significant support from their sponsor's military or security services, and that support is likely to be focused elsewhere at the beginning of a crisis. Despite their limitations, proxy forces provide adversaries with another tool to shape a crisis situation.

B. Activities to Control Escalation

Peer adversaries may attempt to control the escalation of a crisis to avoid armed conflict with the United States by initiating actions to prevent or counter a U.S. response. These actions may focus on allies or partners using diplomatic, information, military, and economic instruments. Adversary measures include setting fait accompli conditions on the ground designed to make military responses either too expensive to employ or too late to affect the political situation. An adversary also has other options to control escalation, which include accelerating its operational timelines, employing information warfare, increasing support to proxy forces, and increasing the number of forward deployed units in the region. Adversary forces may also initiate crises in other theaters to distract U.S. forces and diffuse their response in the area of greatest interest. In extreme cases, an adversary may conduct a limited attack in response to U.S. reactions to the activities that precipitated the original crisis.

1-78 (Operations) IV(b). Crisis

Multidomain Operations

C. Activities to Mitigate United States Deterrence

As adversary forces plan for operations during a crisis, they consider several key actions to mitigate U.S. deterrence efforts and to ensure these operations do not interfere significantly with their interests. These actions may include—

- Conducting limited attacks to expose friendly force vulnerabilities. These attacks may also degrade the deterrence value of deployed forces and destroy credibility among current and potential partners.
- Disrupting or delaying the deployment of Army and joint forces through cyberspace attacks and denial of space capabilities.
- Exploiting gaps in national interests among the United States, partner nations, and potential partners by attacking weaker countries with whom the United States has no treaty obligations to defend.
- Conducting deception operations to conceal their real intent.
- Increasing the use of proxy forces to coopt, coerce, or influence the local population, organizations, and governments within a crisis region.
- Creating multiple dilemmas for the United States by attacking or threatening the use of force against potential partner nations in regions outside of the crisis region.
- Impacting the will of the public through information warfare, including cyberspace attacks.
- Threatening the use of nuclear weapons to prevent intervention by the United States, allies, and partners.

III. Operations Security

Operations security is vital to the success of operations during crisis. Continuously employing the operations security process generates measures and countermeasures to limit an adversary's ability to discern friendly intent, knowing that friendly forces are always under observation and at risk of detection. Operations security is a function of how tasks and activities are conducted and how individual Soldiers and units are successful in meeting the directed standards. Army units in a joint operations area (JOA) exercise strict operations security to protect friendly information and protect the network against cyberspace attacks. They do this by ensuring no use of personal electronic devices, minimizing electromagnetic emissions, and limiting communications on command and control (C2) information systems to the maximum possible extent. This protects Soldiers from social media and other information-related attacks and limits the information available to adversaries that can be used to target family members. It also makes it more difficult for adversaries to identify units and their locations and reduces the incentive for adversary forces to strike targets they view as lucrative enough to risk conflict to destroy. Stress caused by adversary social media attacks during crisis is potentially circumvented by avoiding social media altogether, since the combined effects of a disinformation campaign could degrade Soldier performance and morale far more than not having access to personal devices and media accounts. Operations security is a continuous activity at every echelon down to the individual Soldier level.

IV. Relative Advantages During Crisis

During crisis, Army forces capitalize on the knowledge and experiences gained and use the systems, processes, and infrastructure developed while setting the theater to respond to adversary aggressions or threats. While this preparation and experience provide Army forces, allies, and other coalition forces with a good starting point in mature theaters, the transition into crisis will most likely be chaotic, and it will present Army leaders with unforeseen challenges that require rapid response. Army forces therefore build upon the information, human, and physical advantages gained during competition to mitigate friction, deter adversaries, and when necessary, transition into armed conflict.

(Operations) IV(b). Crisis 1-79

V. Army Support to the Joint Force during Crisis

Ref: FM 3-0, Operations (Oct. '22), pp. 5-4 to 5-7.

The military supports unified action partners during crisis by providing flexible deterrent and response options.

A **flexible deterrent option (FDO)** is a planning construct intended to facilitate early decision making by developing a wide range of interrelated responses that begin with deterrent-oriented actions carefully tailored to create a desired effect.

A **flexible response option (FRO)** is a military capability specifically task-organized for effective reaction to an enemy threat or attack and adaptable to the existing circumstances of a crisis.

FDOs and FROs occur across the diplomatic, informational, military, and economic (known as DIME) instruments of national power, and they are not just confined to the military. They are most effective when integrated and implemented in a nearly simultaneous manner

Determining what threat and enemy forces perceive as important will inform U.S. understanding of their desired end state, associated courses of action, and employment of forces. This allows strategic leaders to determine the appropriate amount of military force to apply in concert with diplomatic, information, and economic activities to prevent adversaries from achieving their objectives.

Diplomatic
- Support international organization declarations.
- Increase diplomatic engagement to show resolve.
- Restrict travel of U.S. citizens.

Informational
- Increase public awareness of problems and potential for conflict.
- Publicize violations of international law.
- Counteradversary disinformation.

Economic
- Enact trade sanctions.
- Cancel or restrict U.S. funded programs.
- Reduce security assistance programs.

Military
- Increase readiness posture of in-place forces.
- Deploy additional forces into theater.
- Increase defense support to public diplomacy.

Ref: FM 3-0 (Oct. '22), fig. 5-1. Simultaneous flexible deterrent and response option examples.

While FDOs are primarily intended to prevent a crisis from developing or worsening, FROs are designed to preempt or respond to attacks against U.S. interests. FDOs are preplanned, deterrence-oriented actions carefully tailored to bring an issue to early resolution without armed conflict, and they can be initiated before or after unambiguous warning of threat action. In comparison, FROs can be employed in response to aggression by adversaries, and they are intended to facilitate early decision making by developing a wide range of actions carefully tailored to produce desired effects. FDOs and FROs must be deliberately tailored in terms of timing, efficiency, and effectiveness to avoid undesired effects, such as eliciting an armed response should adversary leaders perceive that friendly FDOs or FROs are being used as preparation for a preemptive attack.

1-80 (Operations) IV(b). Crisis

FDOs and FROs serve three basic purposes. First, they provide a visible and credible message to adversaries about U.S. will and capability to resist aggression. Second, they position U.S. forces in a manner that facilitates implementation of the operations or contingency plan should armed conflict occur. Third, they provide options for joint and national senior leaders. They allow for measured increases in pressure to avoid unintentionally provoking combat operations, and they enable decision makers to develop the situation to gain a better understanding of adversary capabilities and intentions. FDOs and FROs are elements of contingency plans executed to increase deterrence in addition to, but outside the scope of, the ongoing joint operations. The key goals of FDOs and FROs are—

- Communicate the strength of U.S. commitment to treaty obligations and regional peace and stability.
- Confront adversaries with unacceptable costs for their possible aggression.
- Isolate adversaries from regional neighbors and attempt to split adversary coalitions.
- Rapidly improve the military balance of power in the theater of operations without precipitating an armed response from adversaries.
- Develop the situation and better understand adversary capabilities and intentions.

Leaders exercise restraint and carefully calculate risk before recommending an increase in Army forces to address a crisis. Peer adversaries have global capabilities, and they can create multiple dilemmas for U.S. forces by escalating a crisis horizontally in a different theater. Surging forces in one region may address a crisis, but it potentially creates opportunities for adversaries or enemies in another region. Leaders must anticipate second- and third-order effects on other combatant commands and the risk to the homeland when forces are committed to address a specific crisis.

Army contribution examples to joint flexible deterrent options
Command and control headquarters—establishment of a field army or deployment of a corps or division.
Air defense to protect key infrastructure and population centers from theater ballistic missiles.
Additional personnel to expand the capability of theater-assigned headquarters.
Intelligence assets to support situational understanding, targeting, and information activities.
Deploying a security force assistance brigade to establish liaison capability or conduct security force assistance.
Building or expanding infrastructure and increasing sustainment capacity to facilitate reception, staging, onward movement, and integration.
Army contribution examples to joint flexible response options
Airborne or air assault units positioned to conduct joint forcible entry.
A brigade combat team drawing Army pre-positioned stocks.
Port opening to receive the joint force.
Multi-domain task force to respond to adversary antiaccess and area denial activities.
Special operations forces to conduct foreign internal defense, direct action, or special reconnaissance.
Civil affairs to enable civil-military operations and interorganizational cooperation.
Chemical, biological, radiological, and nuclear units for response to weapons of mass destruction employment.

Ref: FM 3-0 (Oct. '22), table 5-1. Potential Army contributions to joint flexible deterrent and response options.

Refer to JFODS6: The Joint Forces Operations & Doctrine SMARTbook, 6th Ed. (Guide to Joint, Multinational & Interorganizational Operations). Completely updated for 2023, chapters include joint doctrine fundamentals (JP 1), joint operations (JP 3-0, 2022), joint planning (JP 5-0, 2020), joint logistics (JP 4-0), joint task forces (JP 3-33), joint force operations (JPs 3-30, 3-31, 3-32 & 3-05), multinational operations (JP 3-16), and interorganizational cooperation (JP 3-08).

A. Physical Advantages During Crisis

Achieving physical advantage during crisis consists of working with host-nation forces to form a credible defense and ensuring the survivability of allied forces in theater. If there is key or decisive terrain, Army forces and host-nation partners may seek to deter adversaries by setting a defense of that ground early in a crisis. Army forces in theater assume conflict is imminent and take all available measures to protect against attack in every domain where an attack could occur.

During crisis, Army combat power will likely be limited initially to a small number of forward-stationed forces, those forces that can draw Army pre-positioned stocks (APS) rapidly, and forces used to threaten adversary forces with forcible entry into their area of operations (AO). This combat power will most likely be used in a defensive posture until the JFC receives enough land forces to make offensive operations feasible. The intent should be to increase the combat power of Army forces to a point where they can credibly threaten adversary forces with offensive operations. Ideally, this will deter further enemy action. If, however, deterrence fails, this force facilitates armed conflict that will terminate on terms favorable to U.S. interests. In well-developed theaters, Army combat power will likely be forward stationed and integrated with partner forces as a key part of their defensive plans. This credible land force, capable of disrupting or significantly degrading an adversary's initial attack, maximizes the deterrent potential of Army forces.

B. Information Advantages During Crisis

Two key information activities are protecting friendly information and degrading the threat's ability to communicate, sense, make effective decisions, and maintain influence with relevant actors and populations. An example is the use of strategic messaging to undermine the credibility of an adversary by exposing violations of international law and showing that adversary narratives are false. Achieving information advantages is a commander-driven, combined arms activity that employs capabilities from every warfighting function. During crisis, commanders lead their staffs to refine information activities based upon plans and processes developed during competition. Examples include commanders and staffs focusing on the challenges and tasks of establishing a mission-partner environment, building or modifying an intelligence architecture, and creating or refining common operating procedures with allies and other partners.

C. Human Advantages During Crisis

While enduring relationships with alliance and coalition partners may be in place at the theater strategic level as a crisis develops, at the operational and tactical levels it is likely that units have less experience operating with one another. Forces deploying into a theater may have experience working with the security forces of partner nations if they were regionally aligned or worked together in a professional military education or training setting, but most will not have such experience. This requires leaders who have worked with joint and multinational partners to focus their staffs on the most critical interoperability tasks necessary for effective coalition operations. It also requires awareness of the difficulty in fully understanding situations when dealing with other cultures. Employing the liaison networks built by the theater army during competition will enable simultaneous in-theater training exercises with the deployment of Army forces. This facilitates early shared understanding, helping leaders and subordinate units integrate with allied and partner forces in the most expeditious and efficient manner possible while also signaling determination to adversaries. Demonstrated readiness for combat operations and interoperability among U.S., allied, and partner forces helps to upset adversary risk calculations and deter further aggression.

1-82 (Operations) IV(b). Crisis

VI. Force Projection (See p. 4-10.)

The demonstrated ability to project Army forces into an operational area is an essential element of conventional deterrence. Army forces depend almost entirely upon joint lift capabilities for deployment. Force projection is the ability to project the military instrument of national power from the United States, or another theater in response to requirements for military operations (JP 3-0).

During a crisis, ground forces provide a JFC with more enduring options than forces primarily concentrated in or transiting other domains. Army forces are capable of occupying ground indefinitely. They must be sustained just like other Services, but the Army's ability to maintain persistent presence is far greater because of the nature of operations on land. The potentially close physical proximity of ground forces to adversary forces provides the JFC with greater understanding and can help the JFC dictate the tempo of operations.

Army forces achieve persistent presence by deploying forces into a theater to support forward-stationed U.S. forces, or those of allies or partners. These forces are likely already executing operations directed in response to provocations, indications, or warnings that hostile activities may commence. At the direction of the JFC, Army forces execute tasks, activities, and operations designed to deter further malign activity and set conditions for success should deterrence fail. The forward presence or projection of Army formations into a theater provides capabilities that create tactical and operational dilemmas for threat forces, enabling the JFC to seize and retain the initiative. Prompt deployment of land forces in the initial phase of a crisis can preclude the need to deploy larger forces later, and it assures allies and other partners. Effective early intervention can also deny adversaries the time necessary for them to set conditions in their favor.

Deployment alone does not guarantee success. Achieving successful deterrence involves convincing adversaries that the deployed force is able to markedly reduce the adversary force's chance of success during armed conflict. Adversaries measure the ability of Army forces to conduct operations during armed conflict through careful observation of how well those forces are prepared to conduct large-scale combat operations and the capabilities those forces introduce into a specific context as part of the overall joint response.

As Army forces prepare to respond to a crisis, the JFC conducts a final review of deploying forces, ensuring they are deployed in the proper sequence and are able to be task-organized effectively for the anticipated mission. Threat forces are likely to detect force projection activities using space and cyberspace capabilities, human intelligence, and open-source collection efforts. Planners should anticipate adversary forces using all available means to contest the deployment of forces, beginning from home station, during transit, and upon arrival in theater. Therefore, operations security, dispersion of forces, deception operations, and physical security are critical planning considerations. Senior commanders and planners must understand the risks and shape deployments to satisfy both speed and operational readiness.

Force projection is particularly important during crisis, as Army forces have an unknown amount of time to shape a developing situation. It can occur, however, in any context. Forces projected forward during competition to conduct exercises, bolster allies and partners, and conduct other activities are under observation. Adversaries assess the speed and efficiency of these routine deployments, which can have a deterrent effect. Given the fluid nature of a crisis, force projection may continue well after a crisis has transitioned to armed conflict.

See pp. 1-139 to 1-144 for further discussion on contested deployments.

(Operations) IV(b). Crisis 1-83

Force Projection Planning

Ref: FM 3-0, Operations (Oct. '22), pp. 5-8 to 5-13. See also p. 4-10.

Sound force projection planning encompasses—

Opening the Theater *(See pp. 7-8 to 7-9.)*

During the transition to crisis or armed conflict, Army forces open the theater to receive deploying forces. Army forces execute existing plans to establish and open air, sea, and rail terminals. Distribution systems and intermediate staging bases may be established where required. Higher echelon (including theater, corps, and division enablers) and rapidly deployable C2 elements begin to integrate with host-nation forces as quickly as possible to set the conditions for RSOI of follow-on tactical forces. This includes coordination with the forces of other supporting nations to assure effective distribution of services, facilities, and supplies to all deploying units across the alliance or coalition. During theater opening, designated arriving forces draw available APS. This provides the JFC with increased capacity and capability during the initial stages of a crisis or armed conflict. Army forces must be prepared for combat while conducting theater opening operations. The first deploying units require the capability to defend themselves while they provide reaction time and maneuver space for follow-on forces.

Mobilization

Mobilization is the process by which the Armed Forces of the United States, or part of them, are brought to a state of readiness for war or other national emergency (JP 4-05). During mobilization, the Army focuses its efforts on filling joint manning documents to augment combined and joint task force (JTF) headquarters, land component headquarters, and Army units designated for deployment. During crisis, strategic leaders may decide to mobilize select portions of the U.S. Army National Guard and U.S. Army Reserve to provide key capabilities to JFCs. During armed conflict, it is likely that strategic leaders will remove some or all mobilization limitations, enhancing the Army's ability to respond to an aggressive act by an enemy with the necessary capabilities. An example of a limitation that is lifted for armed conflict would be ordering a full mobilization of the Army National Guard or Army Reserve in lieu of a selected reserve call up or partial mobilization.

Deployment

Deployment is the movement of forces into and out of an operational area (JP 3-35). Proper planning establishes what, where, and when forces are needed to achieve objectives. How the JFC intends to employ forces is the foundation of the deployment structure and timing. For example, a JFC may deploy a combat-ready brigade combat team (BCT) or division early in a crisis to stabilize a situation or secure ports for follow-on forces, accepting risks to the movement efficiency of follow-on forces. Corps and division staffs examine all deployment possibilities and conduct parallel planning.

Most Army equipment travels via strategic sealift. It will take weeks or months for the equipment to arrive in theater. Commanders and planners must not underestimate the joint deployment challenges of operating against peer adversary forces with robust air, maritime, space, and cyberspace capabilities.

Protection During Transit

Threats may attempt to impede or prevent unit deployments. This creates a requirement for coordination of the physical security of deploying unit personnel and equipment as they move to ports of embarkation. Physical security is required for personnel and equipment while awaiting transport at ports of embarkation, during movement, and after arrival at ports of debarkation. Planning for physical security remains a focus in unit staging areas, along routes upon which units and supplies move, and for tactical assembly areas prior to onward movement into AOs.

1-84 (Operations) IV(b). Crisis

Reception, Staging, Onward Movement, and Integration (RSOI)

RSOI is the process that delivers combat power to the JFC in a theater of operations or a JOA. RSOI is the responsibility of the theater army and its associated theater sustainment command (TSC). During crises involving a peer adversary, RSOI must occur rapidly in as many dispersed locations as possible to complicate adversary targeting. It is a theater-level process, with careful coordination required between units, theater sustainment personnel, host-nation support, and commercial entities. Effective RSOI matches personnel with their equipment, minimizes staging and sustainment requirements while transiting these ports of debarkation, and begins onward movement as quickly as possible. Deploying units need to understand and implement previously developed plans to accomplish integration and maintain combat readiness upon their arrival.

Initial Employment of Forces

The initial employment of Army forces during a crisis will most likely be as part of FDOs or FROs. This employment may represent the opening stages of a joint operation or a show of force demonstration. The objective of this early employment is to deter an adversary from further aggression, expand the theater to receive follow-on Army and joint forces, and form a credible defense with host-nation forces to prevent adversary gains. Without a robust theater infrastructure, a large number of forward-stationed forces, or a robust APS inventory that enables rapid deployment, Army forces can only provide limited support to partner forces. While immediate support may be limited, even limited support could prove decisive when it is obvious that additional capabilities are quickly moving into the theater to address initial shortfalls.

Sustainment

Sustainment is central to force projection, and sustainment preparation of an operational environment is the basis for sustainment planning. Corps, division, and brigade planners focus on identifying the resources available in an operational area for use by friendly forces and ensuring access to them. The theater army is a key partner in providing this information to deploying units. A detailed estimate of requirements allows planners to advise the commander of the most effective method of providing adequate and responsive support, while minimizing the vulnerable sustainment footprint. There is no fundamental difference in sustainment preparation of an operational environment during competition, crisis, or armed conflict, except that sustainment activities intensify as Army forces respond to crisis and prepare for armed conflict, since time available decreases and requirements from risks to units on the ground increase exponentially. Proper sustainment permits the Army to project force over time and through the necessary depth of an AO.

Redeployment

Redeployment is the transfer or rotation of forces and materiel to support another commander's operational requirements, or to return personnel, equipment, and materiel to the home and/or demobilization stations for reintegration and/or out-processing (JP 3-35). National strategic leaders determine the appropriate time for the redeployment of Army forces. Usually, redeployment of Army forces does not occur until tensions reduce and conditions permit the transition of security and stability responsibilities to other legitimate authorities.

Refer to SMFLS5:The Sustainment & Multifunctional Logistics SMARTbook (Guide to Operational & Tactical Level Sustainment), chapter nine for 44 pages on deployment operations from ATP 3-35 and JP 3-35. Topics include Predeployment, Movement; (RSO&I) Reception, Staging, Onward movement, Integration; and redeployment.

VII. Consolidating Gains

During and after crisis response, Army forces consolidate gains to deny adversary forces the means to extend the crisis or create a similar crisis in the future. This will often entail maintaining an enhanced force posture in a JOA for a period of time to demonstrate U.S. willingness to defend allies and partners. Army forces continue to support improvements to host-nation capabilities through a security cooperation plan designed to make them less vulnerable to future crisis. If an adversary directly targets partner forces, or acts through a proxy, the United States must be prepared to reconstitute the partner's forces as quickly as possible.

The ability of Army forces to reconstitute partner nation forces is especially important to JFCs since, in many areas, only the Army has the capacity to conduct a comprehensive security cooperation program. Many allies and partners rely primarily on their armies and do not have robust navies or air forces. Consolidating gains during and after crisis response creates enduring change that reinforces deterrence against adversaries and improves relative advantages for U.S., allied, and partner forces.

VIII. Transition to Competition or Armed Conflict

There are two outcomes of a crisis—a de-escalation to competition or an escalation to armed conflict. Transitions are typically points of friction. Commanders emphasize information collection prior to and during transitions to maintain a detailed understanding of the threat and continuously assess the situation in order to position their forces to retain the initiative.

Transition Back to Competition

During a crisis, partner nation security forces and government institutions may suffer losses that reduce capability and capacity due to the actions of adversary or proxy forces. Army forces may be tasked to execute security cooperation programs to help restore or maintain partner nation capabilities and capacity as a means to consolidate gains. Army forces seek to restore partner security forces and government institutions as quickly as possible to maintain popular support. Doing so reduces the need for large numbers of U.S. forces to deploy in the future or be maintained in theater to support or enable a partner nation's security. A quick recovery also highlights the strength of the alliance or bilateral relationship of a partner nation with the United States.

Army forces use products developed from the civil preparation of the environment to help rebuild partner-nation security forces. Army forces work with partner nations to do this and do not act unilaterally. The security force assistance brigade (SFAB) and civil affairs units are the ideal core for this effort, but all types of Army units may contribute.

Army forces may help the JFC exploit favorably resolved crises to establish new patterns of behavior for a theater. While crises are generally uncertain and volatile, their resolution and resulting transition back to competition provide opportunities to capitalize on changes in an operational environment. Army forces support joint efforts to create and reinforce changes that benefit the United States and its allies and partners in competition and provide improved relative advantages that assist in deterring future adversary malign behavior.

Transition to Armed Conflict

Army forces responding to a crisis are prepared for and expect to fight. This saves time during the transition and requires an understanding of the OPLAN or likely concept of operations as early as possible. Forward-positioned forces reposition into battle positions or tactical assembly areas and take all available measures to protect themselves from attack in every domain as they prepare for combat. When located with allied or partner units, Army forces synchronize their activities to ensure unity of purpose and mutual support. Depending upon the enemy and distance from the United States, Army forces should expect to receive little support in the opening stages of a conflict and plan accordingly.

1-86 (Operations) IV(b). Crisis

Chap 1

Multidomain Operations

IV(C). Armed Conflict

Ref: FM 3-0, Operations (Oct. '22), chap. 6.

Section I of this chapter introduces large-scale combat operations and the ways in which they vary. It addresses topics applicable to both offensive and defensive operations, including enemy methods, relative advantages, integrating with the joint force, defeat mechanisms, and enabling operations. Section II describes defensive operations. Section III describes offensive operations. Section IV describes transition to post-conflict competition and stability operations.

I. Armed Conflict & Large-Scale Combat

Armed Conflict

Armed conflict encompasses the conditions of a strategic relationship in which opponents use lethal force as the primary means for achieving objectives and imposing their will on the other. The employment of lethal force is the defining characteristic of armed conflict, and it is the primary function of the Army. Lethality's immediate effect is in the physical dimension—reducing the enemy's capability and capacity to fight. However, the utility of lethal force extends into the information and human dimensions where it, along with the other instruments of national power, influence enemy behavior, decision making, and will to fight.

During armed conflict, operations usually reflect combinations of conventional and irregular warfare approaches. Leaders apply doctrine for large-scale combat operations during limited contingencies that require conventional warfare approaches. Irregular warfare includes counterinsurgency and unconventional warfare, which other publications specifically address. The initial actions of large-scale combat operations will likely overlap with actions initiated during competition and crisis. For example, while some units are engaged in offensive or defensive operations, other units may be completing non-combat evacuations while in contact with enemy forces.

Large-Scale Combat Operations *(See p. 1-11.)*

Large-scale combat operations are extensive joint combat operations in terms of scope and size of forces committed, conducted as campaigns aimed at achieving operational and strategic objectives through the application of force. Large-scale combat on land occurs within the framework of a larger joint campaign, usually with an Army headquarters forming the base of a joint force headquarters. These operations typically entail high tempo, high resource consumption, and high casualty rates. Large-scale combat introduces levels of complexity, lethality, ambiguity, and speed to military activities not common in other operations.

Large-scale combat operations occur in circumstances usually associated with state-on-state conflict, and they encompass divisions and corps employing joint and Army capabilities from multiple domains in a combined arms manner. Irregular warfare activities often complement large-scale combat operations, with conventional, irregular, and special operations forces conducting operations close to each other. This proximity requires cooperation between friendly forces of all types to ensure success. In other cases, irregular warfare occurs largely in a secondary joint operations area (JOA) or another theater of operations. When this occurs, the combatant commander (CCDR) ensures sufficient coordination of operations to support unity of purpose at the national level.

(Operations) IV(c). Armed Conflict 1-87

Multidomain Operations

Successful large-scale combat operations defeat enemy armed forces while establishing control over land and populations to achieve operational and strategic objectives. They may capitalize on superior military capability to quickly overwhelm a weaker enemy and consolidate gains as part of a rapid campaign. Large scale combat operations against more capable enemy forces are likely to be of longer duration, lasting months or longer.

Army forces may execute large-scale combat operations in a supporting, enabling, or advisory role, instead of constituting the bulk of ground maneuver forces. One example was OPERATION INHERENT RESOLVE, beginning in 2014, during which a U.S.-led combined joint task force supported Iraqi Security Forces and Syrian Democratic Forces in defeating the Islamic State of Iraq and Syria. In these cases, U.S. forces applied large-scale combat operations tactics in support of a partner force.

The characteristics of large-scale combat operations vary based on many factors, including the enemy. When fighting against a less capable enemy, the U.S. joint force may have significant advantages in most domains. The principal concerns during such operations include how to win rapidly at minimal cost, consolidate gains, and transition responsibility for an area to legitimate authorities. When fighting against a peer enemy, able to contest the joint force in all domains, the operational environment becomes much more difficult. Integrated air defense and long-range fires systems; cyberspace and electronic warfare capabilities; chemical, biological, radiological, and nuclear capabilities (CBRN); and global reconnaissance and surveillance networks can create parity or significant enemy advantages in one or more domains, particularly early during a conflict and when operating close to its own borders. To succeed, the U.S. joint force must create its own relative advantages, preserve combat power, and rapidly exploit what opportunities it creates. Commanders must assume risk to create opportunity and sequence their operations because they cannot defeat enemy forces in a single decisive battle.

A. Enemy Approaches to Armed Conflict

Although peer enemies mainly seek to obtain their strategic objectives during competition, they will engage in armed conflict if they view that the rewards are worth the risk. Once engaged in armed conflict, peer enemies employ combinations of threat methods to render U.S. military power irrelevant whenever possible and inflict unacceptable losses on the United States, its allies, and its partners. Russia and China employ their instruments of national power and military capabilities in distinct ways.

See pp. 1-24 to 1-27 for discussion of threat methods: information warfare, systems warfare, preclusion, isolation, and sanctuary.

Competition and potential conflict between nation-states remains a critical national security threat. **Beijing, Moscow, Tehran, and Pyongyang** have demonstrated the capability and intent to advance their interests at the expense of the United States and its allies. **China** increasingly is a near-peer competitor, challenging the United States in multiple arenas—especially economically, militarily, and technologically—and is pushing to change global norms and potentially threatening its neighbors. **Russia** is pushing back against Washington where it can—locally and globally—employing techniques up to and including the use of force. In Ukraine, we can see the results of Russia's increased willingness to use military threats and force to impose its will on neighbors. **Iran** will remain a regional menace with broader malign influence activities, and **North Korea** will expand its WMD capabilities while being a disruptive player on the regional and world stages.

See pp. 1-90 to 1-91 for an overview of the Chinese threat and pp. 1-92 to 1-93 for an overview of the Russian threat.

1-88 (Operations) IV(c). Armed Conflict

B. Relative Advantages During Armed Conflict

Ref: FM 3-0, Operations (Oct. '22), pp. 6-4 to 6-6.

Army formations most effectively achieve overmatch through the integration and synchronization of joint and multinational capabilities employed from positions in multiple domains that create cascading dilemmas and defeat the enemy's operational approach.

Physical Advantages

Friendly forces require physical advantages to defeat enemy forces and occupy land areas, exert control over lines of communications, and protect the physical infrastructure used to attain information and human advantages. Throughout armed conflict, leaders seek physical advantages that include—

- Position.
- Range.
- Speed of movement.
- Technologically superior capabilities.
- Terrain and weather.

Information Advantages

Information advantages invariably overlap with and emanate from physical and human advantages. To gain an information advantage, units first require a physical or human advantage. Army forces create and exploit information advantages by acting through the physical and human dimensions of an operational environment. Leaders combine information advantages with other advantages to understand the situation, decide, and act faster than enemy forces. Examples of information advantages during armed conflict include—

- The ability to access enemy C2 systems to disrupt, degrade, or exploit enemy information.
- Opportunities created by deception operations to achieve surprise and thwart enemy targeting.
- The ability to mask electromagnetic signatures.
- The ability to integrate and synchronize friendly forces in denied or degraded environments through use of redundant communications.
- The ability to rapidly share information with domestic and international audiences to counter enemy malign narratives.
- The ability to inform a wide range of audiences to maintain legitimacy and promote the friendly narrative.
- The ability to rapidly share and analyze information among commanders and staffs to facilitate decisions and orders.

Human Advantages

Because war is a clash between opposing human wills, the human dimension is central to war. Army formations are principally designed to achieve objectives through the threat or employment of lethal force, which has a psychological effect. Understanding an enemy force's tolerance for casualties and the political and social will to endure them is important to understanding the level of effort required to prevail against enemy forces in large-scale combat operations. Leaders do everything possible in the physical and information dimensions to reduce the enemy's will to fight. During armed conflict, human advantages include—

- Political and national will that supports strategic objectives.
- Experienced, well-trained formations.
- Leadership well versed in the mission command approach to C2.
- Adherence to the law of war.
- Unit cohesion and Soldiers with the mental and physical stamina for combat.
- Interoperability and mutual trust between allies and host-nation partners.

(Operations) IV(c). Armed Conflict 1-89

China

Ref: FM 3-0, Operations (Oct. '22), pp. 6-3 to 6-4.

China considers three aspects in the country's view of conflict: comprehensive national power, deception, and the Three Warfares. Comprehensive national power is made up of hard power and soft power. Hard power includes military capability and capacity, defense industry capability, intelligence capability, and related diplomatic actions such as threats and coercion. Soft power includes such things as economic power, diplomatic efforts, foreign development, global image, and international prestige. China views comprehensive national power as a vital measure of its global status. Ultimately, all forms of conflict—military, diplomatic, or other—must enhance China's comprehensive national power.

Deception plays a critical role in every part of the Chinese approach to conflict. People's Liberation Army planners employ stratagems to achieve their deception goals. Stratagems describe the enemy's mindset, focusing on how to achieve the desired perceptions by the opponent, and then they prescribe ways to exploit those perceptions.

China's strategic approach to conflict employs Three Warfares designed to support and reinforce the People's Liberation Army's traditional military operations. Though these approaches are called warfares, they are universally nonlethal and do not involve direct combat operations. If a battle must be fought, the Three Warfares are designed to unbalance, deceive, and coerce opponents to influence their perceptions in ways that create advantage. The Three Warfares are—

- Public opinion warfare.
- Psychological warfare.
- Legal warfare.

Public opinion warfare is China's high-level information campaign designed to set the terms of political discussion. China views this effort as capable of seizing the initiative in a conflict before any shots are fired by shaping public discourse, influencing political positions, and building international acceptance of Chinese interests.

China's psychological warfare is broadly similar to U.S. military information support operations in that it is intended to influence the behavior of a given audience. Psychological warfare is the deliberate manipulation of psychological reactions in target audiences, designed to create and reinforce attitudes and behaviors favorable to China's objectives and guide adversary behavior towards China's preferred outcomes.

Legal warfare for China is the setting of legal conditions for victory—both domestically and internationally. Legal warfare seeks to unbalance potential opponents by exploiting international or domestic law to hinder their military operations, to create legal justification for People's Liberation Army operations worldwide, and to support Chinese interests through a valid legal framework. It guides how the People's Liberation Army trains to treat prisoners of war, detainees, and civilians, and it guides how the People's Liberation Army abides by international legal conventions, codes, and laws.

During armed conflict, China employs systems warfare in combination with the other threat methods, such as preclusion, isolation, and sanctuary. China employs these threat methods throughout all domains and at all levels of warfare. Systems warfare involves—

- Bypassing enemy systems' areas of strength, thus gaining a combat advantage by approaching them asymmetrically.
- Developing systems that excel at exploiting perceived weaknesses in enemy systems
- Undermining international alliances through diplomatic efforts.
- Conducting cyberspace attacks to disable air or seaports.
- Using special operations forces to undermine civilian morale through covert operations.

1-90 (Operations) IV(c). Armed Conflict

Note. China uses the term "special operations forces" to identify their special forces units per ATP 7-100.3. Russian doctrine uses the term "special purpose forces" for their special forces units. For brevity, this manual uses "special operations forces."

Although many actors on the world stage have embraced the concepts of systems warfare, including Russia, China has woven the idea into every aspect of their warfighting capabilities and methods. The systems warfare concept consists of two basic ideas: creating purpose-built operational systems that combine key capabilities under a single command, and the use of these operational systems to asymmetrically target and exploit vulnerable components of an opponent's system. The People's Liberation Army believes that by effectively destroying, isolating, neutralizing, or offsetting key capabilities, it can degrade the enemy's will and ability to resist enough to achieve victory.

At the tactical level, systems warfare centers largely on targeting high-value battlefield systems such as radars, command and communications nodes, field artillery and air defense systems, and critical logistics support means. China relies on heavy employment of long-range fires at maximum standoff distance to target friendly joint enablers and command and control (C2) nodes. Examples of tactical system warfare include using heavy rocket artillery to defeat or destroy enemy radars and artillery systems, electronic warfare to suppress or neutralize enemy command and communications networks, and deception operations to target enemy leaders' situational understanding.

• The Chinese Communist Party (CCP) will continue efforts to achieve President's Xi Jinping's vision of making China the preeminent power in East Asia and a major power on the world stage. The CCP will work to press Taiwan on unification, undercut U.S. influence, drive wedges between Washington and its partners, and foster some norms that favor its authoritarian system. China's leaders probably will, however, seek opportunities to reduce tensions with Washington when it suits their interests.

• Beijing will press Taiwan to move toward unification and will react to what it views as increased U.S.–Taiwan engagement. We expect that friction will grow as China continues to increase military activity around the island, and Taiwan's leaders resist Beijing's pressure for progress toward unification. China's control over Taiwan probably would disrupt global supply chains for semiconductor chips because Taiwan dominates production.

• In the South China Sea, Beijing will continue to use growing numbers of air, naval, and maritime law enforcement platforms to intimidate rival claimants and signal that China has effective control over contested areas. China is similarly pressuring Japan over contested areas in the East China Sea.

- Excerpts, Director of National Intelligence, Annual Threat Assessment (Feb '22).

For more than two thousand years, China has been surrounded by enemies, adversaries, and other competitors. With a force that totals approximately two million personnel in the regular forces, the PLA views protecting Chinese sovereignty and security as a sacred duty. OPFOR1 topics and chapters include the strategic environment (defense & military strategy, strategic & operational environments, territorial disputes), force structure (PLA: Army, Navy, Marine, Air Force, Rocket Force, Strategic Support Force), system warfare, information operations, reconnaissance and security, offensive and defensive actions, antiterrorism and stability actions, and capabilities (maneuver, fire support, air defense, aviation, engineer and chemical defense, network and communications, and special operations forces).

(Operations) IV(c). Armed Conflict 1-91

Russia

Ref: FM 3-0, Operations (Oct. '22), pp. 6-2 to 6-3.

The Russian view of war is that it is often undeclared, fought for relatively limited policy objectives, and occurs across all domains. Russian leaders assess that modern conflicts are characterized by a destructive and rapid initial period of war that is more decisive than in the past. Additionally, Russia considers that non-nuclear strategic precision-guided weapons can achieve strategic effects on par with nuclear weapons. Doctrinally, Russia plans to employ nuclear weapons in response to non-nuclear attacks when those attacks threaten Russian sovereignty.

During armed conflict, Russia seeks to exert simultaneous pressure in all domains. Russian strategies intend to increase the costs of confrontation and make the objectives of the United States and its allies politically and economically unsupportable. Russia's objective is to weaken U.S. national will to continue a conflict by inflicting highly visible and embarrassing losses on U.S. forces.

Russian forces intend to win conflicts with massed and precision fires. Russian forces will attempt to set the operational conditions so that deployment of U.S. forces is ultimately counter to U.S. interests. If the U.S. does deploy forces, Russian goals are centered on creating constraints that prevent success of the United States' campaign. Russian methodologies focus on four key areas:

- **Disrupt or prevent understanding of the operational environment.** Russian information warfare activities manipulate the acquisition, transmission, and presentation of information in a way that suits Russia's preferred outcomes.

- **Target stability.** Russia may foster instability in key areas and among key groups so that regional security conditions do not support U.S. operational requirements.

- **Disaggregate partnerships.** Russia acts upon U.S. allies and partners to reduce the ability of the United States to operate in its preferred combined, joint, and interagency manner.

- **Prevent access.** Russia employs pre-conflict activities to deny access to U.S. forces, using nonlethal means initially and transitioning to lethal means if necessary. It seeks to undermine relationships, raise political stakes, manipulate public opinion, and attack resolve to constrain or deny basing rights, overflight corridors, logistics support, and concerted allied action.

As it applies instruments of national power, Russia integrates military forces and other means at selected times and locations to achieve desired objectives as part of its overall campaign. It uses offensive and defensive tactics and techniques that include acts of crime and terrorism. These actions can also be employed to manipulate population perceptions and dissuade support to U.S. military forces or other institutions. When necessary, Russia uses acts of physical violence, psychological operations, and different means of manipulating information to gain influence and develop voluntary or coerced cooperation in a target population. Concurrently, it uses indirect means to progressively degrade U.S. combat power and infrastructure resources and to otherwise psychologically influence the political, social, economic, military, and information variables of the operational environment.

Russian tactical-level units operate as combined arms forces to exploit the effects of both precision strikes and massed fires. Against lesser opponents, Russian forces employ deep maneuver when possible to defeat an enemy's will to resist early in a conflict. In other cases, they mass capabilities in pursuit of more limited objectives while fixing their adversary along a broad front. Regardless of the situation, a basic principle of Russian military actions is to use the effects of strike actions to create the conditions for military success.

1-92 (Operations) IV(c). Armed Conflict

Russia prefers to employ all available national elements of power prior to using maneuver forces, and after force-on-force operations begin, it will continue to employ these integrated national capabilities to support tactical maneuver. Russian forces also employ denial and deception (maskirovka) to mask the true intent of their operation. To execute tactics, Russian units apply intelligence methods and decision making, that are scientifically based, to—

- Understand the conditions of an operational environment that will impact operations.
- Determine the tactical functions required and calculate the required allocation of combat power needed to accomplish a mission in a specific time and location.
- Understand the psychological and cognitive issues among competing friendly forces, aggressor forces, the local population, and other actors in an operational environment.

• We expect that Moscow will remain an influential power and a formidable challenge to the United States amidst the changing geopolitical landscape during the next decade. It will continue to pursue its interests in competitive and sometimes confrontational and provocative ways, including pressing to dominate Ukraine and other countries in its "near-abroad," while exploring possibilities to achieve a more stable relationship with Washington.

• In the Middle East and North Africa, Moscow is using its involvement in Syria, Libya, and Sudan to increase its clout, undercut U.S. leadership, present itself as an indispensable mediator, and gain military access rights and economic opportunities.

• In the Western Hemisphere, Russia has expanded its engagement with Venezuela, supported Cuba, and used arms sales and energy agreements to try to expand access to markets and natural resources in Latin America, in part to offset some of the effects of sanctions.

• In the former Soviet republics, Moscow is well positioned to increase its role in the Caucasus and, if it deems necessary, intervene in Belarus and Central Asia to halt instability after widespread anti-government protests, as it did in Belarus after the fraudulent 2020 election and early this year in Kazakhstan.

• We expect Russia to continue to use energy as a foreign policy tool to coerce cooperation and force states to the negotiating table, as it recently did in 2021, when Russia stopped coal and electricity exports to Ukraine. Russia also uses its capabilities in COVID-19 vaccine development and civilian nuclear reactor construction as a soft-power tool in its foreign policy.

- Excerpts, Director of National Intelligence, Annual Threat Assessment (Feb '22).

It has been nearly thirty years since a holistic explanation of the Soviet-based Opposing Force (OPFOR) was examined in the U.S. Army Field Manual 100-2 series. Recognizing this, OPFOR SMARTbook 3: Red Team Army (Second Edition) re-examines and outlines the doctrinal operational construct and historical foundations of Soviet-era military forces from the FM 100-2 series, which is now out-of-print and largely unavailable. OPFOR3 topics and chapters include RTA overview, offensive and defensive operations, specialized warfare, tactical enabling tasks, small unit drill, urban & regional environments, rear area operations and logistics. Future editions will be revised and updated to focus centrally on modern Russian forces, operations, tactics and lessons learned in the Ukraine.

(Operations) IV(c). Armed Conflict 1-93

C. Operating as Part of the Joint Force

The Army always fights as part of a joint force, and usually as part of a multinational coalition during large-scale combat operations. Because combatant commanders (CCDRs) often assign the senior Army commander as the joint force land component commander (JFLCC), it is imperative that Army leaders from the JFLCC to brigade level understand the integration of operations on land with those in the other domains for the joint force. The Army supports the joint force by providing the capabilities and capacity to apply sustained combined arms landpower through movement, close combat, and fires at whatever scale is necessary to defeat enemies on land. It does this by employing capabilities from the land, maritime, air, space, and cyberspace domains in support of ground operations on land and employing ground-based capabilities to enable operations in the other domains.

Army forces' contributions to the joint force include—

- Establish C2 on land.
- Counter air and missile threats that deny air and maritime freedom of action with land-based systems.
- Defend and control key terrain.
- Defeat components of enemy antiaccess (A2) and area denial (AD).
- Conduct large-scale combat operations.
- Sustain large-scale combat operations.
- Consolidate gains.

Refer to JFODS6: The Joint Forces Operations & Doctrine SMARTbook, 6th Ed. (Guide to Joint, Multinational & Interorganizational Operations). Completely updated for 2023, chapters include joint doctrine fundamentals (JP 1), joint operations (JP 3-0, 2022), joint planning (JP 5-0, 2020), joint logistics (JP 4-0), joint task forces (JP 3-33), joint force operations (JPs 3-30, 3-31, 3-32 & 3-05), multinational operations (JP 3-16), and interorganizational cooperation (JP 3-08).

Establish Command and Control on Land

Establishing C2 on land requires a land component command that assigns land areas and properly defines command and support relationships between subordinate forces. A commander assigns land areas and command and support relationships based on the mission and the commander's concept of operations. A commander also considers the level of joint support available when assigning land areas and task-organizing forces. Army forces establish C2 of land areas at the direction of the joint force commander (JFC), normally the JFLCC. A JFLCC is the commander within a unified command, subordinate unified command, or joint task force (JTF) responsible to the establishing commander for recommending the proper employment of assigned, attached, or other available land forces; planning and coordinating land operations; and accomplishing assigned missions. C2 across the width and depth of the land component commander's assigned areas requires access to strategic communications. Therefore, commanders plan for the movement and placement of strategic communications systems in C2 nodes.

See JP 3-31 and FM 3-94 for more information on the JFLCC.

In addition to their assigned missions, Army forces generally consider four key issues:

- Assignment of subordinate land areas
- Concept of mutual support between subordinates
- Integration of echelons in terms of time, space, and purpose, and
- Proper task organization of the land force.

See following pages (pp. 1-96 to 1-97) for an overview and further discussion.

Doctrinal Template of Depths & Frontages

FM 3-0, Operations (Oct '22), fig. 6-1, p. 6-8.

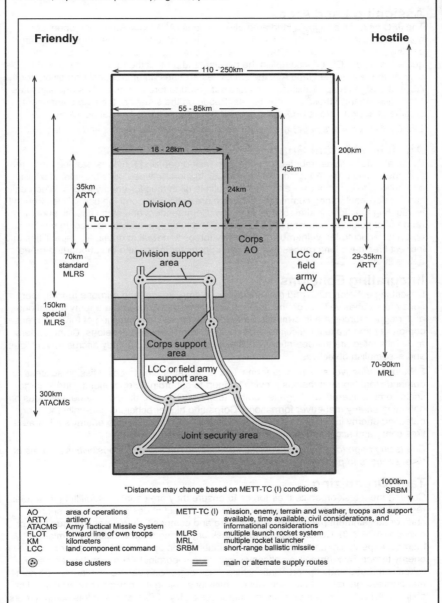

(Operations) IV(c). Armed Conflict 1-95

Establishing Command and Control

FM 3-0, Operations (Oct '22), pp. 6-6 to 6-12.

Assigning Land Areas

The JFC typically assigns a land area of operations (AO) to the land component command. A land AO does not typically encompass the entire land operational area of the JFC's joint operations area (JOA), but the size, shape, and positioning should be adequate for the JFLCC to accomplish the mission and protect the forces under the JFLCC's control. In the assigned land AO, the JFLCC establishes an operational framework for the AO that assigns responsibilities to subordinate ground force commanders and allocates capabilities to fulfill those responsibilities. Based on the situation, the land component command assigns areas (AO, zone, or sector) to subordinate tactical echelons.

See fig. 6-1 (previous page) a baseline doctrinal template for depths and frontages.

Applying Mutual Support

Commanders consider mutual support when task-organizing forces, assigning land areas, and positioning units. A high degree of mutual support between units provides flexibility and options to commanders and creates multiple dilemmas for enemy forces. When units can support each other, commanders have more options to combine capabilities from all warfighting functions and across all domains. Commanders and staffs assess supporting range of capabilities between units to understand their options with respect to mutual support and reducing the vulnerability of friendly forces to defeat in detail. Air capabilities and indirect fires have long ranges that provide options for increasing the supporting range between units.

Integrating Echelons

Describing echelon roles and responsibilities in time, space, and purpose makes operations more cohesive through the depth of an operational area where enemy formations and irregular forces are intermingled with friendly forces. Intermingling of forces increases complexity and make it difficult to retain unity of effort between echelons. Commanders and staffs integrate the operations of all echelons to ensure that they accrue advantages and accomplish objectives.

Echelons maneuver subordinate formations to defeat enemy forces. Mission success demands that higher echelons provide support to subordinate operations. During large-scale combat operations, brigade combat teams (BCTs) and divisions generally focus on defeating enemy maneuver formations. Corps and higher echelons generally focus on defeating enemy integrated air defense systems and portions of the enemy's integrated fires command according to the JFC plan and priorities.

See facing page for a notional depiction of echelon roles and responsibilities in terms of time, space, and purpose.

Task-Organizing

Commanders task organize their forces to ensure they are capable of fulfilling their roles, responsibilities, and purpose. Task-organizing is the act of designing a force, support staff, or sustainment package of specific size and composition to meet a unique task or mission (ADP 3-0). Considerations when task-organizing a force include the mission, training, experience, unit capabilities, sustainability, the operational environment, and the enemy threat. Task-organizing allocates assets to subordinate commanders and establishes command and support relationships. Task-organizing can be continuous, as commanders reorganize units for subsequent missions during the course of operations. The ability of Army forces to task-organize increases agility of formations. It lets commanders configure their units to best use available resources. It also allows Army forces to match unit capabilities to tasks. The ability of sustainment forces to tailor and task-organize ensures commanders have freedom of action to change with mission requirements.

1-96 (Operations) IV(c). Armed Conflict

Notional Echelon Roles & Responsibilities

Multidomain Operations

Land component command focus is 72 hours to 9 days and 50 to 100 km. An LCC maneuvers corps and expands freedom of action through all domains. LCCs integrate joint ISR, fires, protection, sustainment, and maneuver. LCCs coordinate with the air component command and the JFC to establish and adjust the FSCL, allocate joint capabilities from all domains to subordinate corps, employ fires and space and cyberspace effects against enemy IADS and IFC capabilities, create and maintain information advantages, and drive tactical success towards conflict termination.

Corps focus is 48 hours to 5 days and 30 to 70 km. Corps maneuvers divisions and set conditions for convergence by defeating components of the enemy IADS and IFC. Corps integrate joint capabilities from all domains at the right echelon, defeat enemy mid- and long-range fires capabilities, maintain tempo through rear area operations and sustainment, move division rear boundaries forward when necessary to allow divisions to focus on close and deep operations, and expand division efforts to consolidate gains.

Division focus is 24 to 48 hours and 20-40 km. It maneuvers brigades and sets conditions for subordinate BCTs. It sustains the tempo through deep, close, and rear operations. Divisions integrate information collection, aviation, artillery, joint EW, CAS, interdiction, and sustainment with ground maneuver, including operations to defeat enemy maneuver forces, short- and mid-range capabilities, and follow-on echelon forces.

Brigade focus is 12 to 24 hours and 5 to 25 km. Its role is to maneuver battalions and enable successful close combat. BCTs integrate information collection, fires, organic and allocated EW, and other available capabilities with ground maneuver to destroy enemy forces and seize and control key terrain.

50 - 100 km

30 - 70 km

20 - 40 km

5 - 25 km

12 - 24 hours

FLOT

BCT focus — Land

Division focus — Air — Land

24 - 48 hours

Corps focus — Space Cyber — Maritime

48 hours - 5 days

Air Land

XXX

LCC focus — Space Cyber — Maritime

72 hours - 9 days

Air Land

XXXX

= FOCUS is defeating enemy maneuver

= FOCUS is defeating enemy anti-access and area denial

Ref: FM 3-0, fig. 6-2. Notional roles and responsibilities in terms of time, space, and purpose.

(Operations) IV(c). Armed Conflict 1-97

Counter Air and Missile Threats

Enemy air and missile capabilities are a significant threat to Army forces. They increase risk to formations building combat power in assembly areas, to forces transiting lines of communications, and forces conducting rear operations. While able to attack friendly forces during offensive operations, they are particularly dangerous to command posts and any units detected while in static positions. Defeating enemy air and missile threats is necessary to create opportunities for offensive maneuver. However, the employment of maneuver forces that cause forward-positioned enemy air and missile capabilities to displace can complement counter-air operations.

Counter-air is a theater mission that integrates offensive and defensive operations to establish and maintain a desired degree of control of the air by neutralizing or destroying enemy aircraft and missiles, both on the ground and in the air. These operations may include the use of Army manned or unmanned aircraft and long-range fires, maneuver forces, special operations, space operations, cyberspace operations, and electromagnetic warfare capabilities.

Defensive counter-air operations are all defensive measures within the theater designed to neutralize or destroy enemy forces attempting to penetrate or attack through friendly airspace. Defensive counter-air encompasses active and passive defensive actions taken to destroy, nullify, or reduce the effectiveness of hostile air and missile threats against friendly forces and assets. The goal of defensive counter-air operations, in concert with offensive counter-air operations, is to provide an area from which forces can operate while protected from air and missile threats. Defensive counter-air operations must be integrated and synchronized with offensive counter-air operations and all other joint force operations. The area air defense commander, when established by the JFC, is responsible for defensive counter-air planning and operations.

The JFC designates an area air defense command with the authority to plan, coordinate, and integrate overall joint force defensive counter-air operations through the joint air operations center. Together they establish an integrated air defense system. With the support of the component commanders, the area air defense command develops, integrates, and distributes a JFC-approved joint area air defense plan.

Friendly forces conduct air and missile engagements in accordance with guidelines and rules established by the area air defense commander, who is normally the joint force air component commander. A joint force air component commander is from the service with the most air assets and the capability to plan, task, and control joint air operations in an AO, typically the Air Force or Navy.

The Army air and missile defense command (AAMDC) is the Army's lead organization for Army air and missile defense (AMD) forces in a theater. The AAMDC commander, as the theater army AMD coordinator, assists the ARFOR commander and the area air defense commander in the planning and coordination of the critical asset list and the creation of the defended asset list.

Army air defense artillery (ADA) forces conduct AMD operations for the joint force. Army forces may attack to seize key terrain for the emplacement and employment of theater air and missile defense systems by the JFC to defeat portions of the enemy air and missile capabilities. Forward deployed or early entry ADA forces defend critical assets against air attack while the JFC builds combat power.

The critical asset list identifies the most critical assets requiring protection, and it serves as the foundation for a defended asset list, which allocates available ADA forces. Integration with joint or multinational AMD components in the JOA can mitigate some shortages of AMD systems. Short-range air defense (SHORAD) provides air defense against low-altitude air threats for key joint and Army assets, primarily amphibious landing sites, ports, airfields, command posts, and crossing sites. Integrated C2 across all ADA echelons enables the most efficient allocation of limited SHORAD and early warning assets. High-to-medium altitude AMD forces defend joint and Army forces against ballistic missile threats.

1-98 (Operations) IV(c). Armed Conflict

Defend and Control Key Terrain

FM 3-0, Operations (Oct '22), pp. 6-13 to 6-15.

The joint force has enduring requirements that include defending allies with forward-stationed forces, controlling strategic lines of communications to enable the deployment of combat power, and controlling the key terrain required during joint forcible entry operations. The contributions of Army forces both protect and enable the other members of the joint force, which in turn enable greater freedom of action and opportunities for operations on land.

Defense by Forward-Stationed Forces

Armed conflict often occurs after a long period of competition and is likely to begin with some form of enemy aggression against a U.S. ally or partner, an attack on forward-stationed U.S. forces, or both. Forward-stationed forces whose combat mission involves defending as part of an allied effort may be required to conduct mobile or area defenses or retrogrades or reposition to tactical assembly areas where they can prepare for their role in future operations. Advisor teams from the theater-aligned security force assistance brigade (SFAB) may embed alongside threatened partners, providing real-time tactical intelligence and access to U.S. capabilities. Commanders and staffs ensure defending forces have the appropriate priority for Army and joint fires. Operation plans (OPLANs) and supporting subordinate plans identify main defensive positions, security areas, force requirements, the locations of tactical assembly areas, and other planned locations. Building on preparations made during competition, units occupy these positions and, when able, they make improvements to overhead cover, concealment, communications, and fields of fire. Friendly forces hinder enemy detection efforts by combining real preparations with preparations of fake positions. Defensive preparations during competition, particularly the establishment of survivability positions, preserve combat power when enemy preparatory fires commence with little or no warning.

Control Strategic Lines of Communications and Key Terrain

Controlling strategic lines of communications and key terrain is essential to the joint force's ability to project sustain combat power. Russia and China can contest joint operations across the globe, from the strategic support area in the continental U.S., along air, land, and maritime shipping lanes, to intermediate staging bases, and forward to tactical assembly areas. Because strategic lines of communications usually include key land areas, Army forces play a critical role in defending them.

Enemies integrate conventional and irregular warfare capabilities to disrupt lines of communications from the strategic support area to the close area. Enemy air, space, cyberspace, and missile capabilities are able to range targets anywhere in the world. Enemies can employ espionage or surrogates to attack infrastructure and populations in the continental United States. Enemy surface and subsurface maritime capabilities, in conjunction with unconventional approaches, will disrupt maritime lines of communications. Enemy medium- and long-range fires will contest the ability of Army forces to move combat power into forward tactical assembly areas.

Army forces control key staging areas such as airfields, railheads, and ports. They employ survivability methods and techniques appropriate to the situation to harden their positions. Depending on the operational and mission variables, these areas may be vulnerable to enemy cyberspace attacks and other methods of information warfare, espionage, terrorist attacks, special operations forces, ballistic missile attacks, and weapons of mass destruction. Although the other Services are responsible to secure the air and maritime lines of communications, Army forces may secure some land areas, especially in vicinity of airfields and ports

(Operations) IV(c). Armed Conflict 1-99

Multidomain Operations

Defeat Components of Enemy Antiaccess and Area Denial (A2&AD)

FM 3-0, Operations (Oct '22), pp. 6-15 to 6-17.

Enemy A2 and AD approaches deny friendly force protection and freedom of action. Enemies pursue A2 and AD approaches with lethal means that significantly increase the risk to forward-stationed forces and the ability to deploy and stage additional forces into tactical assembly areas. Understanding the structure and function of the enemy's integrated fires command helps friendly forces disintegrate the cohesion of enemy A2 and AD approaches and create exploitable opportunities for the joint force to conduct offensive operations.

Enemy Integrated Fires Command

An integrated fires command is a dedicated combination of C2 structures and organic and attached joint fire support units. The integrated fires command exercises centralized C2 of all allocated, dedicated fire support assets retained by its level of command. This can include aviation, artillery, naval gunfire, and surface-to-surface missile units from different commands and services. It also exercises C2 over all reconnaissance, intelligence, surveillance, and target acquisition assets dedicated to its support. An integrated fires command is tasked to engage designated operational and strategic targets. Integrated fires commands are typically associated with campaign-level headquarters. However, there are circumstances where an integrated fires command may be formed at the theater level. For example, the theater could have two separate campaigns, requiring a centralization of critical fire support assets at theater level to achieve the strategic or theater campaign objectives. Enemy forces integrate air and missile defense capabilities with the integrated fires command in different ways depending on their capabilities and the situation.

The integrated fires command executes all fire support tasks for the supported command. The integrated fires command is designed to—

- Exploit precision and massed fires through carefully integrated ground and air fire support.
- Minimize the amount of time from target acquisition to engagement.

An integrated fires command and its component systems have key vulnerabilities that Army forces can target when supporting joint force operations. Like any military system, it requires sustainment capabilities and other support that Army forces can detect and attack. It has electromagnetic signatures that enable detection from friendly joint intelligence, surveillance, and reconnaissance (ISR) from all domains. Enemy networks depend on C2 nodes Army forces can target. Perhaps most importantly, systems within an integrated fires command are comprised of land-based capabilities, including sensors, fires capabilities, and C2 nodes, all of which Army forces can attack.

Defeating Antiaccess (A2) and Area Denial (AD) Approaches

Defeating enemy A2 and AD approaches requires continuous effort, best facilitated by forward-stationed forces positioned and protected before hostilities commence. This allows air and naval capabilities to stage closer to joint force targets, increasing the number of sorties they conduct. It allows Army forces and the rest of the forward-postured joint force to retain the terrain and facilities necessary for the introduction of additional friendly forces into theater. Preserving forward-stationed forces and retaining critical terrain inside a theater during the opening stages of a conflict provides depth and operational reach of the joint force.

1-100 (Operations) IV(c). Armed Conflict

A2 and AD are two different enemy approaches that the joint force typically expects to encounter. The joint force often considers them as part of the same challenge because retaining or regaining access to geographic areas requires a cohesive joint approach through all domains. Defeating A2 and AD requires a multidomain approach that includes Army forces retaining or seizing critical terrain to establish the depth necessary for defeating enemy forces.

Commanders defeat enemy A2 and AD approaches by employing multiple attacks through multiple domains. Complementary and reinforcing networked A2 and AD capabilities are resilient against a single line of attack. An enemy operating near its border is able to reconstitute forces and capabilities from homeland sanctuaries. Maneuvering the right capabilities within range to attack critical vulnerabilities might incur too much risk when all threat systems are operating at full capability. Therefore, leaders destroy or isolate the most exposed parts of the enemy's systems over time, degrading them enough to support maneuver and create other opportunities to exploit. Destruction, isolation, and dislocation of various parts of an enemy integrated fires complex or air defense system can all contribute to its disintegration.

The main physical components of enemy A2 and AD systems are sensors, firing platforms, networks, C2, sustainment, and the forces securing them. Army forces attack these components as part of a joint operation that integrates all available capabilities. The JFLCC requests joint effects to support Army forces. The JFLCC may request space capabilities to detect enemy systems, offensive space operations for specific effects, and offensive cyberspace operations or electromagnetic capabilities to attack an enemy's networks

The JFLCC is responsible for integrating joint capabilities and synchronizing their employment and effects to achieve convergence in order to enable subordinate maneuver. Achieving convergence is a key part of the approach to defeating enemy A2 and AD.

Army forces request space and cyberspace effects to disrupt A2 and AD C2 networks and create other effects. Army formations synchronize cyberspace and electromagnetic warfare effects against the enemy's network, disrupting human and automated communications between sensors, firing units, and command posts. When Army planners identify a requirement, they request space and cyberspace effects according to unit procedures. They must understand the planning and preparation timelines required for the effects they are requesting. For example, initiating cyberspace effects takes time, and it should be part of initial OPLAN development and revisions. Many cyberspace effects can take months to generate, even though they can be delivered rapidly once developed. This is a challenge for Army echelons whose planning horizons during combat are measured in terms of hours and days. Therefore, it is important that Army leaders anticipate desired cyberspace effects well in advance of when they will need to integrate them.

Army forces employ combinations of defeat mechanisms to attack components of an enemy's A2 and AD system within the overall intent of degrading and ultimately defeating its ability to function cohesively. Subordinate echelons align their operations and objectives with conditions set by the land component command and act rapidly to exploit them. This combination of attacks and objectives ultimately defeats the enemy's preferred operational approach and renders the enemy force vulnerable to follow-on operations by the JFC. Commanders use deliberate and dynamic targeting to create opportunities to attack the enemy and create redundancies for friendly forces.

Commanders must account for the possibility that enemy forces are able to regenerate some or all of their capabilities, in some cases by repositioning forces from elsewhere. Commanders and staffs continually assess enemy A2 and AD systems and maintain enough combat power to defeat enemy regeneration efforts to avoid surprise and preserve friendly freedom of action. Defeating enemy A2 and AD approaches typically enables joint forcible entry operations and the movement of friendly forces from aerial ports of debarkation and seaports of debarkation to their tactical assembly areas.

(Operations) IV(c). Armed Conflict 1-101

Joint Forcible Entry Operations

Forcible entry is the seizing and holding of a military lodgment in the face of armed opposition or forcing access into a denied area to allow movement and maneuver to accomplish the mission (JP 3-18). A forcible entry operation may be the JFC's opening move to seize the initiative. Forcible entry operations may be used to conduct operational movement and maneuver to attain positional advantage or as part of a deception.

Commanders design their forcible entry operations to seize and hold a lodgment against armed opposition. A lodgment is a designated area in a hostile or potentially hostile operational area that, when seized and held, makes the continuous landing of troops and materiel possible and provides maneuver space for subsequent operations (JP 3-18). This requires friendly forces to be combat loaded and prepared for immediate combat operations prior to their arrival on the ground in the lodgment area. A force defends the perimeter of a lodgment until it has sufficient forces to break out and conduct offensive operations.

Conduct Large-Scale Combat Operations

During large-scale combat operations, Army forces conduct offensive, defensive, and stability operations to defeat enemy forces. Defeat of enemy forces in close combat is normally required to achieve campaign objectives and national strategic goals after the commencement of hostilities. Divisions and corps are the formations central to the conduct of large-scale combat operations, as they are organized, trained, and equipped for the deep, rear, and support operations that enable subordinate success during close combat. The ability to prevail in ground combat is a decisive factor in breaking an enemy's will to continue a conflict.

Conflict resolution requires the Army to conduct sustained operations with unified action partners as long as necessary to achieve national objectives.

See pp. 1-110 to 1-117 for discussion of defensive operations and pp. 1-118 to 1-126 for offensive operations during large-scale combat.

Sustain Large-Scale Combat Operations *(See pp. 7-15 to 7-26.)*

Large-scale combat operations require greater sustainment than other types of operations. Their high tempo and lethality significantly increase maintenance requirements and expenditure of supplies, ammunition, and equipment. Large-scale combat incurs the risk of mass casualties, which increase requirements for health service support, mortuary affairs, and large-scale personnel and equipment replacements. Large-scale combat operations demand a sustainment system to move and distribute a tremendous volume of supplies, personnel, and equipment.

Army sustainment is a key enabler of the joint force on land. Army forces provide sustainment to other elements in the joint force according to the direction of the JFC. The JFC has the overall responsibility for sustainment throughout a theater, but the JFC headquarters executes many of its sustainment responsibilities through the TSC. When directed, Army sustainment capabilities provide the bulk of Army support to other services through executive agency, common-user logistics, lead Service, and other common sustainment resources.

Capabilities from other domains enable sustainment of Army forces. Air sustainment capabilities provide responsive sustainment for high priority requirements. Maritime-enabled sustainment supports large-scale requirements. Space- and cyberspace-enabled networks facilitate rapid communication of sustainment requirements and precise distribution.

See facing page for further discussion.

1-102 (Operations) IV(c). Armed Conflict

Sustainment Operations (See chap. 7.)

FM 3-0, Operations (Oct '22), pp. 6-18 to 6-19.

Successful sustainment operations strike a balance between protecting sustainment capabilities and providing responsive support close to the forward line of troops. A well-planned and executed logistics operation permits flexibility, endurance, and application of combat power. Plans must anticipate and mitigate the risk posed by enemy forces detecting and attacking friendly sustainment capabilities. Sustainment formations pursue operations security, survivability, and protection with the same level of commitment as all other forces. While most rear and support operations are economy of force endeavors when allocating combat power in divisions and corps, the continuity and survivability of those operations are vital to deep and close operations.

Dispersion of assets and redundancy help protect sustainment formations. Dispersing sustainment formations makes it less likely that enemy long-range fires can destroy large quantities of material. Dispersion also creates flexibility, as several nodes can execute the sustainment concept without a single point of failure. However, dispersed sustainment operations complicate C2 and can be less efficient than a massed and centralized approach. Commanders balance the risk between dispersion and efficiency to minimize exposure to enemy fires while maintaining the ability to enable the supported formation's tempo, endurance, and operational reach.

Commanders must plan for the possibility of heavy losses to personnel, supplies, and equipment. Even with continuous and effective sustainment support, units may rapidly become combat ineffective due to enemy action. Commanders at all levels must be prepared to conduct reconstitution efforts to return ineffective units to a level of effectiveness that allows the reconstituted unit to perform its future mission. Reconstitution is an operation that commanders plan and implement to restore units to a desired level of combat effectiveness commensurate with mission requirements and available resources (ATP 3-94.4).

Reconstitution is a significant combined arms operation, and the actions involved in reconstitution transcend normal day-to-day force sustainment actions. Reconstitution is a command decision, typically made two echelons above the unit being reconstituted, and it is conducted across all warfighting functions, using existing systems and units to accomplish. No resources exist solely to perform reconstitution.

Reconstitution requires the full support of commanders and staffs at all echelons to achieve the necessary results commensurate with future mission success. Reconstitution efforts following a CBRN attack present even greater challenges. If units cannot decontaminate equipment, commanders must assess the risk to the mission and personnel in continuing operations with or without contaminated equipment.

Maneuver commanders, at the combatant command, ASCC, field armies, corps, or division echelons, direct reconstitution as an operation. Command guidance should address the timing, location, degree to which the unit is reconstituted, and the training required before the reconstituted unit resumes operations. As a general consideration, the more degraded or combat ineffective a unit becomes, the greater the sustainment effort and amount of individual and collective training the unit will require to return it to combat effectiveness.

Refer to SMFLS5: The Sustainment & Multifunctional Logistics SMARTbook (Guide to Operational & Tactical Level Sustainment). SMFLS5 topics include the sustainment warfighting function; sustainment operations), sustainment execution (logistics, financial management, personnel services, & health services support); sustainment planning; brigade support; division, corps & field army sustainment; theater support; joint logistics; and deployment & redeployment.

Multidomain Operations

Consolidate Gains During Armed Conflict

Army forces consolidate gains for the joint force by making temporary advantages more enduring. Consolidating gains is not a phase, but rather an imperative that achieves the ultimate purpose of campaigns and operations. Army forces consolidate gains continuously during the conduct of operations, with varying emphasis by each echelon over time. Consolidating gains initially focuses on the exploitation of tactical success to ensure enemy forces cannot reconstitute any form of resistance in areas where they were initially defeated. A small unit consolidating on an objective and preparing for enemy counterattacks can be the first part of a larger effort to consolidate gains.

Unity of purpose for consolidating gains starts at the theater strategic level, where leaders plan and coordinate the resources necessary to achieve the JFC's desired end state. They provide subordinate echelons a shared visualization of the security conditions necessary for the desired political or strategic end state. Achieving the desired end state generally requires a whole-of-government effort with unified action partners within and outside of the theater of operations. At the operational and tactical levels, land component commands and corps exploit division tactical success by maintaining contact with enemy remnants, bypassed forces, and the capabilities that enemy forces could militarize to protract the conflict. Friendly forces employ lethal and nonlethal capabilities to defeat remaining enemy forces in detail. Commanders direct information activities to reduce the will of those forces to resist and the local population to support them.

Conducting detainee operations plays a significant role in consolidating gains. Failure to secure defeated enemy forces quickly gives them the opportunity to break contact, recover the will to fight, and then reorganize resistance with whatever means remaining available to them. Capture of defeated enemy units and individuals separated or disorganized by friendly action is therefore critical. Large numbers of detainees may place a tremendous burden on operational forces as tactical forces divert combat power to process and secure them. Additionally, friendly forces must account for the detention of irregular forces and criminal actors that take advantage of a lack of civil security and civil control. Leaders assess the threat posed by irregular forces and criminal actors to determine whether or not these groups warrant detention by friendly forces.

See FM 3-63 for more information on detainee operations.

Army forces must deliberately plan and prepare for consolidating gains because it is resource intensive and requires significant coordination with unified action partners. Planning for consolidating gains includes an assessment of operational risk, available combat power, changes to task organization, and additional assets required to achieve the desired end state. Assets might include—

- Additional forces to provide area security tasks.
- Host-nation, partner, and allied security forces.
- ISR assets.
- Engineers.
- Military police.
- Explosive ordnance disposal units.
- Medical units.
- Logistics units.
- Civil affairs units.
- Psychological operations units.
- CBRN units.

See facing page for further discussion.

1-104 (Operations) IV(c). Armed Conflict

Operations to Consolidate Gains

FM 3-0, Operations (Oct '22), pp. 6-20 to 6-21.

Operations to consolidate gains can take many forms. These operations can include—

- **Offensive operations**. Forces conduct offensive operations to complete the defeat of fixed or bypassed enemy forces.
- **Area security.** Forces conduct security tasks to defeat enemy remnants, proxy or insurgent forces, and terrorists; control populations and key terrain; and secure routes, critical infrastructure, populations, and activities within an assigned area.
- **Stability operations**. Forces execute minimum-essential stability operations tasks and ensure the provision of essential governmental services (in support of host nation or in place of the host nation), emergency infrastructure reconstruction, and humanitarian relief. These may include military governance.
- **Influence local and regional audiences**. Commanders communicate credible messaging to specific audiences to prevent interference and generate support for operations or the host nation.
- **Defensive operations**. Establish security from external threats. Commanders ensure sufficient combat power is employed to prevent physical disruption from threats across the various domains seeking to reverse or subvert the military gains made by friendly forces.
- **Detainee operations**. Commanders and staffs must consider detainee operations prior to a conflict, during, and after large-scale combat operations have ceased. Detainee operations have long-lasting operational and strategic impacts.

Determining when and how to consolidate gains at the operational level and applying the necessary resources at the tactical level requires clear understanding about where to accept risk during an operation. Failure to consolidate gains generally leads to failure in achieving the desired end state, since it represents a failure to follow through on initial tactical successes and cedes the initiative to determined enemies seeking to prolong a conflict. Security is necessary for transition of responsibility to a legitimate governing authority and the successful completion of combat operations. Army forces integrate the capabilities of all unified action partners and synchronize their employment as they consolidate gains.

Although operations to consolidate gains may at times be economy of force efforts, they are critical to the long-term success of the joint force's operation. Enemy forces will continue to challenge Army gains even after their forces are defeated to gain time for a favorable political settlement, set conditions for a protracted resistance, and alter the nature of the conflict to gain relative advantages. Enemy forces will execute information warfare, exploit cultural seams, challenge security, encourage competition for resources, promote conflicting narratives, support religious divides, and create alternatives to legitimate authority. As long as enemy forces have a will to resist, they will continue to attempt to undermine friendly gains.

Consolidating gains requires a significant commitment of combat power in addition to the combat power being employed to maintain tempo and pressure on enemy main body forces. When available, Army forces rely on host-nation forces better suited for operations that involve close interaction with local populations. When a host nation or an ally is unable to provide support, Army forces are responsible for consolidating gains. During large-scale combat operations, corps should plan for a division and a division should plan for a BCT to conduct operations with the purpose of consolidating gains.

(Operations) IV(c). Armed Conflict 1-105

D. Applying Defeat Mechanisms

Ref: FM 3-0, Operations (Oct '22), pp. 6-21 to 6-26.

Defeat mechanisms are broad means by which commanders visualize and describe how they plan to defeat enemy forces. Defeat mechanisms have interactive and dynamic relationships constrained only by the resources available and imagination. Defeat mechanisms are most useful for commanders at the division-level and above to develop operational and strategic approaches to defeat enemy forces. For commanders at the brigade combat team-level and below, the defeat mechanisms may offer limited utility. Defeat mechanisms are not tactical tasks, and commanders at the lower tactical echelons do not develop or employ defeat or stability mechanisms.

Once developed, defeat mechanisms help commanders and staffs determine tactical options for defeating enemy forces in detail. Commanders translate defeat mechanisms into tactics and describe them in the concept of operations. Commanders use tactics to apply friendly capabilities against enemy forces in the most advantageous ways from as many domains as possible. Understanding how tactics support each of the defeat mechanisms improves tactical judgment.

Destroy

Destruction and the threat of destruction lies at the core of all the defeat mechanisms and makes them compelling in a specific context. It is the defeat mechanism with the most enduring effect. A commander can achieve the destruction of smaller units with massed fires. However, destruction is the most resource-intensive outcome to achieve on a large scale. Commanders use destruction as a defeat mechanism when—

- Enemy forces are not vulnerable to other means.
- The tactical situation requires the use of overwhelming combat power.
- The risk of loss is acceptable.
- It is necessary to set conditions for other defeat mechanisms.

The physical effects of destruction have significant implications in the information and human dimensions. The destruction of enemy capabilities sends the message that enemy forces may be overmatched, and defeat is imminent. Casualties and the loss of life have negative psychological impacts that can either embolden or degrade enemy morale and will to fight. Typically, significant death and destruction degrade enemy morale and will. The joint force can sometimes achieve similar results with modest and precise applications of combat power. However, modest applications of combat power may prolong the joint force's ability to achieve a decisive outcome. Excessive destruction of infrastructure can create a humanitarian crisis and create civilian casualties or suffering that undermines domestic and international support for military operations, so commanders must exercise judgment appropriate to each operational context. For moral, legal, and pragmatic reasons, commanders should take precautions to avoid death and destruction unnecessary for operational success, and they must always comply with the law of armed conflict.

At the operational level, physical destruction is rarely feasible or acceptable as the overarching defeat mechanism. Operational-level commanders choose elements of enemy forces that must be destroyed to enable the other defeat mechanisms. They synchronize Army, joint, and unified action partner capabilities to destroy critical components of enemy warfighting systems. At the tactical level, the lower the echelon the more central destruction is to its operation.

Dislocate

Dislocate as a defeat mechanism renders an enemy's position ineffective and, ideally, irrelevant. Dislocation can enable surprise. It forces enemy forces to react to the unexpected, and it imposes new dilemmas on enemy decision makers. If the dislocation

1-106 (Operations) IV(c). Armed Conflict

that occurs is great enough, enemy forces may reconsider their risk assessment and conclude that they must surrender or reposition because their position no longer offers a reasonable expectation of success. More commonly, dislocation will force an enemy to make major changes to its dispositions and give up considerable ground.

The challenge of dislocation is that enemy leaders likely understand the value of the position desired by the friendly force, and they have made efforts to protect it. Therefore, to maneuver forces into positions that are so advantageous they render the enemy dispositions ineffective, commanders must often make use of deception and assume considerable risk.

At the operational level, commanders dislocate the enemy by posturing friendly forces in multiple assembly areas that do not make any single course of action obvious, and then they threaten maneuver along multiple axes of advance that exceed the enemy force's ability to mass effects. At the tactical level, vertical and horizontal envelopments and turning movements are common forms of maneuver that can precipitate dislocation. While these tactics can create rapid success, they may come with significant risk to rear areas and flanks, operational reach, and maintaining the momentum required to achieve objectives in accordance with a specific timeline.

Isolate

Isolating an enemy force separates it from its physical, information, or human sources of support. It involves denying the enemy force access to resupply of personnel and equipment, access to intelligence, and shared understanding with adjacent units and higher echelon headquarters. Isolation denies enemy ground forces access to capabilities from other domains, forcing them to operate only in a limited area of the land domain with the resources they have on hand.

At the operational level, it is difficult to achieve complete isolation of an enemy force not already physically separated from the enemy main body. However, operational-level commands can employ capabilities to temporarily isolate units or critical capabilities from the rest of an enemy formation in one or more domains. At the tactical level, physically blocking lines of communications, controlling key terrain, fixing supporting units, and encircling an enemy force are tactics that support the achievement of isolation.

Disintegrate

To disintegrate is to attack the cohesion of the whole and involves preventing components of an enemy formation or capability from fulfilling their role as part of the overall effort. Disintegration causes the formation or capability to function less effectively, creating vulnerabilities that the friendly force can exploit. Disintegration is most effective when created by a combination of the other three defeat mechanisms, and it includes both physical and cognitive effects on enemy forces. Disintegration is typically temporary and causes enemy forces to adapt. Creating more lasting effects requires forces that are ready to exploit the opportunity provided by disintegration. Commanders ensure they have sufficient combat power for exploitation, and they synchronize their exploitation efforts with the temporal effects of disintegration.

Disintegration sets conditions for achieving operational level objectives. Destruction, isolation, and dislocation all focus on relatively limited parts of a larger enemy force in specific geographic areas, whereas the effects of disintegration can have repercussions throughout the depth and breadth of the enemy's echelons. Effective disintegration can cause collapse of coherent organized resistance for operationally significant periods.

Operational-level echelons disintegrate large enemy formations and their capabilities by attacking their individual components. Attacking operational-level C2 infrastructure impacts all enemy functions, and it is the most direct way to cause disintegration. Senior tactical echelons disintegrate enemy forces by attacking vulnerabilities that make them less able to employ a combined-arms approach to operations. Disrupting enemy communications with electromagnetic attacks or physical attacks against C2 nodes are means of doing this.

(Operations) IV(c). Armed Conflict 1-107

E. Enabling Operations

Ref: FM 3-0, Operations (Oct '22), pp. 6-26 to 6-31.

Enabling operations set friendly conditions required for most operations. Commanders direct enabling operations to support the conduct of offensive, defensive, and stability operations and defense support to civil authorities tasks. The execution of enabling operations alone does not directly accomplish the commander's end state, but enabling operations must occur to complete the mission. Examples of enabling operations are—

Reconnaissance *(See pp. 4-17 to 4-20.)*

Reconnaissance is a mission undertaken to obtain, by visual observation or other detection methods, information about the activities and resources of an enemy or adversary, or to secure data concerning the meteorological, hydrographic, or geographic characteristics of a particular area (JP 2-0). Reconnaissance occurs continuously in all domains. Reconnaissance identifies terrain characteristics, obstacles to mobility, the disposition of enemy forces, and the relevant characteristics of the civilian population. It facilitates mobility and prevents surprise. Reconnaissance prior to unit movements and occupation of assembly areas is critical to protecting friendly forces and preserving combat power. Units perform reconnaissance to make contact with enemy forces on favorable terms. Leaders at every echelon emphasize the importance of reporting and rapidly updating digital and analog systems. There are five forms of reconnaissance operations. They are route reconnaissance, zone reconnaissance, area reconnaissance, reconnaissance in force, and special reconnaissance.

Security Operations *(See pp. 4-20 to 4-24.)*

Units may perform security tasks to the front, flanks, or rear of their main body, and they must be aware of enemy threats in all domains relevant to their assigned area. The main difference between the performance of security and reconnaissance tasks is that security tasks orient on the force, area, or facility being secured, while reconnaissance tasks orient on enemy forces and terrain. Security tasks are supporting efforts. The ultimate goal of security operations is to protect main body forces from surprise and deny enemy freedom of action to collect on friendly forces. The protected force may not always be a military force; it can also be the civilian population, civil institutions, and civilian infrastructure in the unit's assigned area. The four types of security operations are area security, cover, guard, and screen.

Troop Movement

Troop movement is the movement of Soldiers and units from one place to another by any available means (ADP 3-90). Units perform troop movements using different methods, such as dismounted foot marches, mounted marches using tactical vehicles, or air, rail, and water means in various combinations. The method employed depends on the situation, the size and composition of the moving unit, the distance the unit must cover, the urgency of execution, and the condition of the troops. It also depends on the availability, suitability, and capacity of the different means of transportation. Troop movements over extended distances have extensive sustainment considerations. Troop movements are made by different methods; such as dismounted and mounted marches using organic combat and tactical vehicles; motor transport; and air, rail, and water means in various combinations.

Relief in Place

A relief in place is an operation in which, by direction of higher authority, all or part of a unit is replaced in an area by the incoming unit and the responsibilities of the replaced elements for the mission and the assigned zone of operations are transferred to the incoming unit (JP 3-07.3). Units have three techniques for conducting a relief: sequentially, simultaneously, or staggered. A sequential relief occurs when each element in the relieved unit is relieved in succession, from right to left, left to right, front to rear, or rear to front. A simultaneous relief occurs when all elements are relieved at the same time. A staggered relief occurs when each element is relieved in a sequence determined by the tactical situation, not its geographical orientation. Simultaneous relief takes the least time to execute but is more easily detected by enemy forces. Sequential or staggered reliefs can occur over a significant amount of time. These three relief techniques can occur regardless of the operational theme in which the unit is participating.

Passage of Lines

A passage of lines is an operation in which a force moves forward or rearward through another force's combat positions with the intention of moving into or out of contact with the enemy (JP 3-18). There are two types: a forward passage of lines and a rearward passage of lines. A forward passage of lines occurs when a unit passes through another unit's positions while moving toward the enemy (ADP 3-90). A rearward passage of lines occurs when a unit passes through another unit's positions while moving away from the enemy (ADP 3-90). Units perform a passage of lines to continue their attacks or perform counterattack, retrograde, and security tasks that involve advancing or withdrawing through other units' positions. A passage of lines potentially involves close combat. It involves transferring the responsibility for an AO between two units. That transfer of authority usually occurs when roughly two thirds of the passing force have moved through one or more passage points. The headquarters directing the passage of lines is responsible for determining when the passage of lines starts and ends. If not directed by higher authority, the stationary unit commander and the passing unit commanders determine—by mutual agreement—the time to pass responsibility for an area. They disseminate this information to the lowest levels of both organizations.

Mobility Operations *(See pp. 4-25 to 4-30.)*

Freedom to move and maneuver within an operational area is essential to applying combat power. Most operational environments and enemy forces present numerous challenges to movement and maneuver. Leaders overcome these challenges through the integration of combined arms mobility. Mobility tasks are those combined arms activities that mitigate the effects of obstacles to enable freedom of movement and maneuver (ATP 3-90.4). There are six primary mobility tasks: conduct breaching, conduct clearing (of areas and routes), conduct gap crossing, construct and maintain combat roads and trails, construct and maintain forward airfields and landing zones, and conduct traffic management and enforcement.

Countermobility Operations *(See pp. 4-30 to 4-34.)*

Countermobility is a set of combined arms activities that use or enhance the effects of natural and man-made obstacles to prevent the enemy freedom of movement and maneuver (ATP 3-90.8). The primary purposes of countermobility are to shape enemy movement and maneuver and to prevent the enemy from gaining a position of advantage. Countermobility is conducted to support forces operating along the range of military operations. Countermobility directly supports offensive and defensive operations. Countermobility activities include siting obstacles; constructing, emplacing, or detonating obstacles; marking, reporting, and recording obstacles; and maintaining obstacle integration.

Refer to SUTS3: The Small Unit Tactics SMARTbook, 3rd Ed., completely updated with the latest publications for 2019. Chapters and topics include tactical fundamentals, the offense; the defense; train, advise, and assist (stability, peace & counterinsurgency ops); tactical enabling tasks (security, reconnaissance, relief in place, passage of lines, encirclement, and troop movement); special purpose attacks (ambush, raid, etc.); urban and regional environments (urban, fortified areas, desert, cold, mountain, & jungle operations); patrols & patrolling.

II. Defensive Operations

Defensive operations defeat an enemy attack, buy time, economize forces, hold key terrain, or develop conditions favorable for offensive operations. Although offensive operations are usually required to achieve decisive results, it is often necessary, even advisable, to defend. Defensive operations alone do not normally achieve a decision unless they are sufficient to achieve the overall political goal, such as protecting an international border.

A. Purpose and Conditions for the Defense

One purpose of defending is to create conditions for the offense that allows Army forces to regain the initiative. Other reasons for conducting the defense include—

- Retaining decisive terrain or denying a vital area to an enemy force.
- Attritting or fixing an enemy force as a prelude to the offense.
- Countering enemy action.
- Accepting risk in one area to create offensive opportunities elsewhere.

There are many potential conditions for defensive operations. They include—

- Enemy aggression initiating armed conflict requires forward-stationed friendly forces to defend to buy time and conserve combat power until reinforced.
- Offensive operations culminate and the commander needs to build combat power while countering enemy offensive operations.
- A unit is assigned an economy of force defensive role as a supporting effort.
- The higher echelon headquarters directs a mission to defend an area, population, key infrastructure, or other key terrain in support of the overall course of action.
- U.S. forces accomplish all objectives and transition to a defense to deter future enemy aggression.

The key to a successful corps or division defense is a concept of operations that allows defensive forces to break the enemy's momentum and seize the initiative. Surprise is as important in defense as in offense, and the defensive concept should avoid obvious dispositions and techniques. When executing a defense, commanders orchestrate combat power from all available domains to synchronize effects at a decisive place and time that results in the enemy's defeat. Commanders decide where to concentrate combat power and where to accept risk. Success may require that a defending unit exploit opportunities to seize the initiative, such as a spoiling attack or counterattack.

Time is often the most important resource for defending forces. The enemy chooses the time and location for its attack, so the amount of time friendly units have to prepare a defense is often unknown and usually inadequate. Defending corps and divisions must have a sense of urgency to complete their planning, coordinating, rehearsing, and conducting information collection. Their subordinate units need time to develop engagement areas by preparing battle positions, pre-positioning sustainment assets, and emplacing obstacles. Taken together this means that strict adherence to priorities of work and priorities of effort is critical to time management.

Defending commanders seek to create more time to prepare an effective defense. A corps or division commander may task-organize and resource a security force for employment in the security area to guard or cover main battle area forces as a means to create additional preparation time and prevent surprise. Commanders may also launch spoiling attacks, raids, or feints to disorganize enemy preparations and gain more time to prepare. A defender continually attacks enemy forces in depth with joint and Army fires and aviation to attrit the enemy force and disrupt its scheme of maneuver. Friendly conventional and special operations forces can slow an enemy

1-110 (Operations) IV(c). Armed Conflict

B. Types of Defensive Operations

Ref: ADP 3-90, Offense and Defense (Jul '19), pp. 4-3 to 4-4 and FM 3-0, Operations (Oct '22), pp. 6-34 to 6-35.

Friendly forces use three types of defensive operations to deny enemy forces advantages:

- Area defense focuses on terrain
- Mobile defense focuses on the movement of enemy forces
- Retrograde focuses on the movement of friendly forces

Although on the defense, the commander remains alert for opportunities to attack the enemy whenever resources permit. Within a defensive posture, the defending commander may conduct a spoiling attack or a counterattack, if permitted to do so by the mission variables of mission, enemy, terrain and weather, troops and support available, time available, and civil considerations (METT-TC).

A. Area Defense

The area defense is a type of defensive operation that concentrates on denying enemy forces access to designated terrain for a specific time rather than destroying the enemy outright. The focus of an area defense is on retaining terrain where the bulk of a defending force positions itself in mutually supporting, prepared positions. Units maintain their positions and control the terrain between the position of enemy forces and the terrain they desire. The decisive operation focuses fires into engagement areas, possibly supplemented by a counterattack.

B. Mobile Defense

The mobile defense is a type of defensive operation that concentrates on the destruction or defeat of the enemy through a decisive attack by a striking force. The mobile defense focuses on defeating or destroying enemy forces by allowing them to advance to a point where they are exposed to a decisive counterattack by a striking force.

C. Retrograde

The retrograde is a type of defensive operation that involves organized movement away from the enemy. An enemy force may compel these operations, or a commander may perform them voluntarily. The higher echelon commander of a force executing a retrograde must approve the retrograde before its initiation. A retrograde is not conducted in isolation. It is always part of a larger scheme of maneuver designed to regain the initiative and defeat the enemy.

Refer to FM 3-0, pp. 6-26 to 6-46, for detailed discussion of defensive tasks at the corps and division levels. Corps and division commanders combine area and mobile defense tasks based upon availability of assets, terrain, their higher echelon commander's concept of operations, and enemy capabilities. Corps and divisions may also conduct retrograde operations. Both the area and mobile defenses contain static and dynamic elements. The dynamic elements involve the maneuver of combat forces and the movement of their supporting combat multipliers.

Refer to SUTS3: The Small Unit Tactics SMARTbook, 3rd Ed., completely updated with the latest publications for 2019. Chapters and topics include tactical fundamentals, the offense; the defense; train, advise, and assist (stability, peace & counterinsurgency ops); tactical enabling tasks (security, reconnaissance, relief in place, passage of lines, encirclement, and troop movement); special purpose attacks (ambush, raid, etc.); urban and regional environments (urban, fortified areas, desert, cold, mountain, & jungle operations); patrols & patrolling.

attack by complicating the enemy force's movements and supply. Uncertainty as to when or where enemy forces will attack requires a commander to maintain a larger reserve.

A successful defense requires the integration and synchronization of all available assets. The defending commander assigns missions, allocates forces (including the reserve), and apportions functional and multifunctional support and sustainment resources within the construct of main and supporting efforts. The commander determines where to concentrate defensive efforts and where to accept risks based on the results of intelligence preparation of the battlefield. This determination includes accounting for the enemy's reconnaissance and surveillance efforts and the vulnerability of friendly troop concentrations to massed enemy fires. Commanders strive to counter enemy attacks by accounting for its capabilities in each domain that can influence friendly forces.

C. Characteristics of the Defense

Successful defenses share some important characteristics. They include—

- **Disruption**—deceiving or destroying enemy reconnaissance forces, breaking up combat formations, separating echelons, and impeding an enemy force's ability to synchronize its combined arms.

- **Flexibility**—developing plans that anticipate a range of enemy actions and allocate resources accordingly.

- **Maneuver**—achieving and exploiting a position of physical advantage over an enemy force.

- **Mass and concentration**—creating overwhelming combat power at specific locations to support the main effort.

- **Depth**—engaging multiple enemy echelons, enemy long-range fires, sustainment, and C2.

- **Preparation**—preparing the defense before attacking enemy forces arrive.

- **Security**—conducting security, protection, information activities, operations security, and cyberspace and electromagnetic warfare tasks.

D. Enemy Offense

Enemy tactics, capabilities, and probable courses of action all inform defensive planning. Defending commanders must see the terrain and their own forces from the enemy perspective in all relevant domains to anticipate both threat courses of action and friendly weaknesses that an enemy force could exploit. Defending commanders identify probable enemy objectives and possible enemy avenues of approach to achieve them. Understanding enemy capabilities in all domains is critical to devising the most effective friendly defensive schemes. Identifying enemy limitations helps determine opportunities to exploit friendly advantages.

At the initiation of offensive operations, enemy forces seek to disrupt and destroy friendly C2, apply overwhelming firepower, and rapidly penetrate friendly rear areas. They use information warfare to attack information systems, deceive, and protect their own information systems. Enemy forces use electronic attack and long-range strikes against C2 systems, information collection systems, air defense systems, and aircraft. The enemy's goal is to weaken or destroy the joint force's decision-making, air and missile defense, and air combat capabilities, which creates opportunities for enemy forces to achieve early decisive success.

A key to an enemy force's ability to achieve early dominance is to strike first. General considerations that influence the enemy force's decision to initiate offensive operations include when the enemy determines the joint force is unprepared for an attack, when the enemy identifies a weakness in the joint force that it can exploit, or when the enemy is fully prepared and can commit to operations at any time. The enemy's decision is tied to factors that provide them with the greatest opportunity to win the first battle, which they believe greatly improves their chances for success in later operations.

1-112 (Operations) IV(c). Armed Conflict

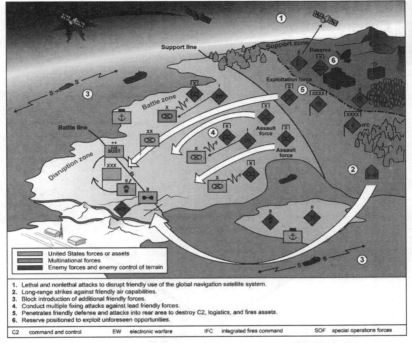

Ref: FM 3-0 (Oct. '22), fig. 6-4. Notional enemy offensive operation.

An enemy force seeks to establish early advantages in the air, space, and cyberspace domains to set conditions for dominance in the maritime and land domains. An enemy force aims to prevent the joint force from introducing additional forces into the conflict region and to disrupt the friendly logistics systems enabling forward-positioned forces. Enemy forces concentrate their long-range attacks on C2 nodes, logistics bases, and assembly areas to disrupt the joint force's defensive operations and to degrade the joint force's ability to prepare for offensive operations.

When enemy leaders commit forces into ground combat during offensive operations, they typically attempt to conceal the location of their main effort with multiple fixing attacks on the ground, allowing them to isolate friendly forward units. Enemy forces use fires and electronic attack to disrupt critical friendly command posts, radars, and fire direction centers.

Generally, enemy forces seek to reinforce success, massing capabilities at a vulnerable point to achieve large force-ratio advantages to enable a rapid penetration of friendly defenses. Enemy leaders use mobile forces to exploit a penetration rapidly to the maximum possible depth to make the overall friendly defensive position untenable. Enemy forces seek advantages in both volume and range of fires to simultaneously mass fires at the point of penetration to enable rapid closure and breakthrough, fix other friendly elements along the forward line of own troops, and target key friendly C2 and logistics nodes along the depth of the defense. Enemy forces prefer to use fires to fix, move around friendly battle positions when possible, and move through destroyed units when necessary. Enemy forces seek to maneuver tactically to a depth that achieves operational objectives in support of their overall strategic purpose. Enemy forces employ reconnaissance, electronic warfare, information warfare, and other capabilities at their disposal to both enable and exploit initial tactical gains. These are likely to include chemical weapons.

(Operations) IV(c). Armed Conflict 1-113

E. Defensive Operational Framework Considerations

In the defense, commanders typically retain the deep and rear areas, but they divide the close area into two distinct portions: the security area and main battle area. Commanders use this approach to synchronize operations, including those in air, space and cyberspace, to defeat an enemy force throughout its depth. Figure 6-5 on page 6-36 depicts friendly defensive operations across a notional operational framework.

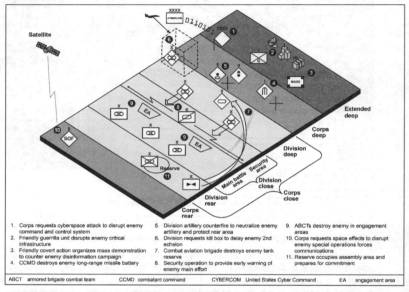

Ref: FM 3-0 (Oct. '22), fig. 6-5. Notional operational framework during defensive operations.

Extended Deep Operations

Operational- and strategic-level deep operations are typically outside an assigned land AO for a corps or division, but parts fall within the senior Army formation's area of interest and area of influence. During defensive operations, Army forces may request JFC or other strategic effects against strategic infrastructure or sanctuary areas that are enabling enemy offensive operations.

Strategic leaders may task Army long-range fires, cyberspace, space, and other global capabilities to support attacking targets in the extended deep area to set conditions for friendly defensive operations. Long-range artillery and ground-based missile capabilities can range enemy long-range missile batteries, manufacturing and economic nodes, critical infrastructure such as airfields and ports, strategic communications nodes, and strategic sustainment and reserve locations. Special operations forces, either unilaterally or combined with indigenous forces, can attack vulnerable targets, influence populations, and motivate or support established networks, such as local militias and resistance groups, to gather valuable information, oppose enemy forces, and weaken popular support to enemy offensive operations. Friendly-backed indigenous forces, or guerilla forces, can also damage or degrade critical infrastructure to disrupt enemy sustainment activities.

Deep Operations During Defensive Operations

Deep operations are essential to the effectiveness of the defense. The commander uses them to attrit, isolate, disrupt, and disorganize attacking formations and to create windows of opportunity in which to act decisively against lead enemy echelons. Deep operations are the commander's means of ensuring success in the main battle area by limiting the enemy's options, disrupting the enemy's ability to mass fires against friendly battle positions and denying the ability of subsequent echelons to support the lead echelon. Commanders use rocket artillery, rotary-wing aviation, UASs, special operations forces, space and cyberspace capabilities, electromagnetic warfare, and influence activities to conduct deep operations. Commanders request joint fires to divert, disrupt, delay, or destroy the enemy's long-range capabilities before they are in direct contact with Army forces. The battlefield coordination detachment, as the senior Army operational commander's liaison with the air component, serves a central role in relaying requests for joint fires to the joint air operations center.

Deep operations begin to disrupt the cohesion of the enemy attack as early as possible, before the enemy closes with maneuver forces in the security area. Psychological effects degrade the will of enemy forces to fight, interfere with enemy decision making, and disrupt enemy attacks as they begin, especially when combined with attacking C2 nodes, fires, and other key capabilities. Deception to conceal the location and disposition of defending forces and limit the effectiveness of enemy fires delays the enemy's ability to mass effects. Army forces in the defense conduct deep operations to—

- Isolate enemy forces in the security area and main battle area from follow-on echelons.
- Disrupt the enemy's ability to support committed forces with fires and logistics.
- Guard against interference with the commitment of friendly reserves.
- Attrit sufficient enemy combat power to achieve favorable force ratios in the main battle area.
- Transition to the offense.

As part of conducting deep operations, corps and division commanders maintain a current intelligence picture of enemy forces throughout their area of interest. They coordinate for and integrate effects created by special operations core activities to achieve operational and strategic objectives and disrupt enemy forces, capabilities, and infrastructure that support enemy offensive operations. Commanders also ensure that deep operations remain focused on setting conditions for forces in the security and main battle areas while avoiding becoming overly fixated on close operations.

Security Area Operations

During the defense, the security force occupies an assigned area far enough forward of the forward edge of the battle area to protect main battle area units from surprise. Security forces provide early warning to give main battle area units time to reposition forces against enemy maneuver and to mitigate the effects of enemy medium-range fires.

Commanders have the option of employing a screen, guard, or cover in a security area, depending on the mission variables. When deciding which option to use, commanders consider the following:

- The depth, breadth, and terrain of the security area relative to number of forces available.
- The capabilities of the security force relative to the threat it faces, particularly its mobility.
- The amount of time the commander needs the security force to provide forces in the main battle area.

(Operations) IV(c). Armed Conflict 1-115

Commanders determine the location, orientation, and depth of the security area based on the terrain available and the purpose of the defense. Units identify enemy avenues of approach and named areas of interest. The depth of the security area determines the time available to react to approaching enemy forces. Occupying a deep security area allows the security force to destroy enemy reconnaissance assets without compromising critical observation posts or positions. It prevents enemy forces from easily penetrating the security area and prevents gaps when observation posts or units displace or are lost. Security forces cover less depth along a broad front because they have fewer resources to position in depth and potentially fewer opportunities to exploit terrain. Shallow security areas may require units to task-organize and resource their security forces to perform guard operations that provide more reaction time for the main body.

The task organization of the security force depends on its role in the overall concept for defense. Covering forces and guard forces require more reinforcement than screening forces. Fires, engineers, aviation, and other attachments increase the ability of guard or covering forces to slow and disorganize the enemy, degrade the enemy's security forces, and gain additional time for the defending commander.

The security force's rear boundary is normally the battle handover line. Handover of the battle from forces in the security area to forces in the main battle area requires close coordination to avoid confusion and fratricide. The security force must retain freedom to maneuver to conduct a rearward passage of lines. Main battle area forces establish contact points, passage lanes, and routes through the main battle area. Control of indirect fires passes to main battle area units as the security force moves through the passage lanes. Regardless of size or echelon, they must be able to alert the main battle area commander that the enemy is approaching on particular avenues of approach and keep the enemy's leading units under observation.

Main Battle Area Operations

The main battle area is where the commander intends to deploy the bulk of the unit's combat power and defeat an attacking enemy force. The commander positions forces in the main battle area to block enemy penetrations, choosing terrain that puts enemy forces at the greatest possible disadvantage. Commanders assign sectors to subordinate forces in the main battle area as a means of controlling subordinate maneuver unit operations. Defensive sectors should align with major enemy avenues of approach. The force responsible for the most dangerous sector is typically the main effort. Commanders employ their reserves in the main battle area to halt an enemy attack, to destroy penetrating enemy formations, or to counterattack to regain the initiative.

Restrictive terrain, choke points, and natural obstacles such as rivers usually favor an area defense oriented on key terrain and avenues of approach. Open, less restrictive terrain usually favors a mobile defense, which orients on the enemy. Most defenses at the division and corps levels offer opportunities for a combination of mobile and area defense.

Spoiling attacks and counterattacks can be used to disrupt enemy forces and prevent them from massing combat power or exploiting success in the main battle area. Future operations cells conduct contingency branch and sequel planning to counter potential enemy penetrations of the main battle area, typically by repositioning forces and establishing decision points for committing the reserve.

Reserve

Commanders base the size of their reserve on the level of uncertainty about the enemy force's capabilities and intentions. A commander's concept of operations describes the size, composition, location, and priorities of planning for the reserve. The more uncertainty that exists, the larger the reserve is. The purpose of the reserve during the defense is to maintain a hedge against uncertainty and counter enemy success, usually by blocking a penetration or enveloping it from a flank. When planning a defense, commanders cannot typically be strong everywhere, and therefore they accept risk about unlikely enemy courses of action. A reserve mitigates the risk the commander assumes. Commanders may use reserves to counterattack enemy vulnerabilities, such as exposed flanks or support units, or to defeat isolated parts of the enemy forward echelon. In some cases, a commander must use the reserve to reinforce battle positions in the main battle area to hold critical terrain, block penetrating enemy forces, or to react to threats against the division or corps rear area. Units position their reserve force for maximum flexibility.

Timing is critical to counterattacks. Commanders must anticipate the circumstances requiring commitment of their reserve and rehearse its commitment to the main battle area. Rehearsals help validate the reserve's response plan and increase the speed with which it can respond. Commanders make the decision to commit a reserve promptly, with sound understanding of the movement and deployment timelines from its assembly area to the main battle area in existing terrain and weather conditions. If committed too soon, reserves may not have the desired effect or be in a good position for commitment against a more dangerous situation later. Committed too late, the reserve may not be able to influence the situation enough to meet the overall intent. Movement control and air defense in the rear area is vital to getting the reserve into the battle on time and in good order.

Rear Operations During Defensive Operations

Rear operations maintain freedom of action in the security and main battle areas and prevent culmination. The rear command post enables this freedom of action by planning and directing sustainment, conducting terrain management, providing movement control, and providing area security of the rear area. Rear operations ensure prompt delivery of commodities in high demand during defensive operations, particularly ammunition. Depending on the enemy situation, commanders commit maneuver units to secure rear operations, although all forces conducting rear operations must maintain local security and conduct survivability tasks. Leaders must consider the terrain, protection, and sustainment requirements for higher headquarters or joint enablers that may be located in the rear area.

Because rear operations divert combat power from other priorities, commanders must weigh the need for this diversion against other potential consequences, and they must be prepared to assume risk based upon both mission analysis and running estimates. Assuming risk may be as simple as suspending all but critical operations in the rear area for a time, so that units operating in the rear area can concentrate on self-defense at critical junctures.

Transition to Offense

The ultimate goal of defensive operations is to defeat the enemy's attacks and transition, or threaten to transition, to the offense. Units must deliberately plan for transitions to identify and establish the necessary friendly and enemy conditions for a successful transition. As friendly forces meet their defensive objectives, forces consolidate and reorganize for offensive operations or prepare to facilitate forward passages of lines for fresh formations. Units should do everything possible to prevent my forces from reinforcing their forward echelons, consolidating, or reorganizing while friendly forces prepare for follow-on operations.

Commanders transition to the offense when they assess they have enough combat power to maintain pressure on the enemy. They do not wait for perfect conditions, and they sometimes must push tired formations to attack because the opportunity to complete the defeat of enemy formations reduces the risk of future casualties fighting the same enemy formations after they have recovered. Commanders continuously assess the effects of battle on their formations relative to their opponents, and they let that assessment guide how hard they pursue enemy forces and how high of a tempo they sustain.

(Operations) IV(c). Armed Conflict 1-117

III. Offensive Operations

The key to successful offensive operations is to achieve all desired objectives prior to culmination. This requires the force in the offense to have some combination of relative advantage in the physical, information, or human dimensions. Typically, offensive operations require advantages in multiple domains, but commanders may achieve those advantages through deception operations and surprise rather than the physical means of combat power alone.

A. Purpose and Conditions for the Offense

The purpose of the offense is to defeat or destroy enemy forces and to gain control of terrain, resources, or population centers. Offensive operations take something from an enemy force. They are characterized by aggressive initiative on the part of subordinate commanders, by rapid shifts of the main effort to create and exploit opportunity, by momentum, and by the deepest, most rapid possible destruction of enemy defensive schemes and the capabilities that enable them.

B. Characteristics of the Offense

The high risk, tempo, and physical toll of offensive operations require high levels of unit training, morale, and cohesion. Successful offenses share these characteristics:

- **Audacity**—the ability to assume risk to create opportunity with bold action.
- **Concentration**—orchestrating forces or effects to create and exploit opportunity. (Concentrating effects is called "mass.")
- **Surprise**—taking action that catches enemy forces off guard.
- **Tempo**—maintaining a pace of operations that is faster than the enemy's, but not so fast that it cannot be sustained for as long as necessary to achieve all assigned objectives.

Refer to ADP 3-90 for more information on the characteristics of the offense.

C. Enemy Defense

The purposes of enemy defensive operations are to set military conditions to resume offensive operations or defend until the enemy achieves a favorable political outcome. The enemy employs two types of defenses generally, a maneuver defense and an area defense. A maneuver defense trades terrain for the opportunity to destroy portions of an opponent's formation and render the opponent's combat system ineffective. In an area defense, the enemy denies key areas to friendly forces. In most situations against a peer or superior opponent, enemy forces are willing to surrender terrain to preserve their major combat forces, since the loss of those forces threatens the survival of the enemy's state or regime. Figure 6-6 on page 6-42 depicts a notional enemy maneuver defense.

The enemy establishes a defensive system on favorable terrain and employs capabilities throughout the depth of the battlefield. The enemy's goal is to resist the joint force's ability to attack in depth by creating layered standoff that makes integrated action between ground forces and the rest of the U.S. joint force impossible. In the disruption zone, enemy forces attack the joint force with long-range fires and limited objective attacks by ground forces to preempt or disrupt the joint force's planned attack. Enemy forces employ aviation, artillery, and ballistic missiles in long-range attacks against the joint force's C2 systems, long-range fires capabilities, attack helicopters, logistics bases, and assembly areas. The enemy uses special operations forces, guerilla forces, and proxy forces in limited objective attacks to harass and disrupt the friendly forces' preparations for offensive operations.

In the main battle zone, the enemy force designs its defensive system to defeat penetrations of its main defensive lines and envelopments by the joint force's ground, airborne, or air assault forces. Along its defensive lines, the enemy attempts to slow

1-118 (Operations) IV(c). Armed Conflict

D. Types of Offensive Operations

Ref: ADP 3-90, Offense and Defense (Jul '19), pp. 3-3 to 3-4 and FM 3-0, Operations (Oct '22), pp. 6-43 to 6-45.

The four types of offensive operations are movement to contact, attack, exploitation, and pursuit. The types of offensive operations describe friendly force arrangements by purpose.

Movement to Contact

Movement to contact is a type of offensive operation designed to develop the situation and to establish or regain contact. The goal of a movement to contact is to make initial contact with a small element while retaining enough combat power to develop the situation and mitigate the associated risk. A movement to contact creates favorable conditions for subsequent tactical actions. Commanders conduct a movement to contact when an enemy situation is vague or not specific enough to conduct an attack. A movement to contact may result in a meeting engagement. Meeting engagements are combat actions that occur when an incompletely deployed force engages an enemy at an unexpected time and place. Once an enemy force makes contact, the commander has five options: attack, defend, bypass, delay, or withdraw. Subordinate variations of a movement to contact include search and attack and cordon and search operations.

Attack

An attack is a type of offensive operation that destroys or defeats enemy forces, seizes and secures terrain, or both. Attacks incorporate coordinated movement supported by fires. They may be part of either decisive or shaping operations. A commander may describe an attack as hasty or deliberate, depending on the time available for assessing the situation, planning, and preparing. A commander may decide to conduct an attack using only fires, based on an analysis of the mission variables. An attack differs from a movement to contact because in an attack commanders know at least part of an enemy's dispositions. This knowledge enables commanders to better synchronize and employ combat power.

Variations of the attack are ambush, counterattack, demonstration, feint, raid, and spoiling attack. See following pages (pp. 1-120 to 1-121) for an overview and discussion.

Exploitation

An exploitation is a type of offensive operation that usually follows a successful attack and is designed to disorganize the enemy in depth. Exploitations seek to disintegrate enemy forces to the point where they have no alternative but to surrender or retreat. Exploitations take advantage of tactical opportunities. Division and higher echelon headquarters normally plan exploitations as branches or sequels.

Pursuit

A pursuit is a type of offensive operation designed to catch or cut off a hostile force attempting to escape, with the aim of destroying it. There are two variations of the pursuit: frontal and combination. A pursuit normally follows a successful exploitation.

Refer to SUTS3: The Small Unit Tactics SMARTbook, 3rd Ed., completely updated with the latest publications for 2019. Chapters and topics include tactical fundamentals, the offense; the defense; train, advise, and assist (stability, peace & counterinsurgency ops); tactical enabling tasks (security, reconnaissance, relief in place, passage of lines, encirclement, and troop movement); special purpose attacks (ambush, raid, etc.); urban and regional environments (urban, fortified areas, desert, cold, mountain, & jungle operations); patrols & patrolling.

(Operations) IV(c). Armed Conflict 1-119

Multidomain Operations

Subordinate Forms of the Attack
Ref: FM 3-90-1, Offense and Defense (Mar '13), pp. 3-23 to 3-31.

Subordinate forms of the attack have special purposes and include the ambush, counterattack, demonstration, feint, raid, and spoiling attack. The commander's intent and the mission variables determine which of these forms of attack commanders employ. Commanders can conduct each of these forms of attack, except for a raid, as either a hasty or a deliberate operation.

An attack is an offensive operation that destroys or defeats enemy forces, seizes and secures terrain, or both. Movement, supported by fires, characterizes the conduct of an attack. However, based on his analysis of the factors of METT-TC, the commander may decide to conduct an attack using only fires. An attack differs from a MTC because enemy main body dispositions are at least partially known, which allows the commander to achieve greater synchronization. This enables him to mass the effects of the attacking force's combat power more effectively in an attack than in a MTC.

Special Purpose Attacks

 I Ambush

 II Raid

 III Counterattack

 IV Spoiling Attack

 V Demonstration

 VI Feint

Ref: FM 3-90-1, Offense and Defense, Vol 1, p. 3-23.

Special purpose attacks are ambush, spoiling attack, counterattack, raid, feint, and demonstration. The commander's intent and the factors of METT-TC determine which of these forms of attack are employed. He can conduct each of these forms of attack, except for a raid, as either a hasty or a deliberate operation.

Refer to SUTS3: The Small Unit Tactics SMARTbook, 3rd Ed., completely updated with the latest publications for 2019. Chapters and topics include tactical fundamentals, the offense; the defense; train, advise, and assist (stability, peace & counterinsurgency ops); tactical enabling tasks (security, reconnaissance, relief in place, passage of lines, encirclement, and troop movement); special purpose attacks (ambush, raid, etc.); urban and regional environments (urban, fortified areas, desert, cold, mountain, & jungle operations); patrols & patrolling.

A. Ambush

An ambush is a form of attack by fire or other destructive means from concealed positions on a moving or temporarily halted enemy. It may include an assault to close with and destroy the engaged enemy force. In an ambush, ground objectives do not have to be seized and held.

Refer to SUTS3, pp. 6-3 to 6-16 for further discussion on the ambush.

B. Raid

A raid is a form of attack, usually small scale, involving a swift entry into hostile territory to secure information, confuse the enemy, or destroy installations. It ends with a planned withdrawal from the objective area on mission completion. A raid can also be used to support operations designed to rescue and recover individuals and equipment in danger of capture.

Refer to SUTS3, pp. 6-17 to 6-22 for further discussion on the raid.

C. Spoiling Attack

A spoiling attack is a form of attack that preempts or seriously impairs an enemy attack while the enemy is in the process of planning or preparing to attack. The objective of a spoiling attack is to disrupt the enemy's offensive capabilities and timelines while destroying his personnel and equipment, not to secure terrain and other physical objectives. A commander conducts a spoiling attack whenever possible during friendly defensive operations to strike the enemy while he is in assembly areas or attack positions preparing for his own offensive operation or is temporarily stopped. It usually employs heavy, attack helicopter, or fire support elements to attack enemy assembly positions in front of the friendly commander's main line of resistance or battle positions.

D. Counterattack

A counterattack is a form of attack by part or all of a defending force against an enemy attacking force, with the general objective of denying the enemy his goal in attacking. The commander directs a counterattack—normally conducted from a defensive posture—to defeat or destroy enemy forces, exploit an enemy weakness, such as an exposed flank, or to regain control of terrain and facilities after an enemy success. A unit conducts a counterattack to seize the initiative from the enemy through offensive action. A counterattacking force maneuvers to isolate and destroy a designated enemy force. It can attack by fire into an engagement area to defeat or destroy an enemy force, restore the original position, or block an enemy penetration. Once launched, the counterattack normally becomes a decisive operation for the commander conducting the counterattack.

E. Demonstration

A demonstration is a form of attack designed to deceive the enemy as to the location or time of the decisive operation by a display of force. Forces conducting a demonstration do not seek contact with the enemy.

F. Feint

A feint is a form of attack used to deceive the enemy as to the location or time of the actual decisive operation. Forces conducting a feint seek direct fire contact with the enemy but avoid decisive engagement . A commander uses them in conjunction with other military deception activities. They generally attempt to deceive the enemy and induce him to move reserves and shift his fire support to locations where they cannot immediately impact the friendly decisive operation or take other actions not conducive to the enemy's best interests during the defense.

The principal difference between these forms of attack is that in a feint the commander assigns the force an objective limited in size, scope, or some other measure. Forces conducting a feint make direct fire contact with the enemy but avoid decisive engagement. Forces conducting a demonstration do not seek contact with the enemy. The planning, preparing, and executing considerations for demonstrations and feints are the same as for the other forms of attack.

(Operations) IV(c). Armed Conflict 1-121

and disrupt friendly forces with a combination of obstacles, prepared positions, electronic warfare, and favorable terrain. The enemy's basic goal is to fix friendly forces with maneuver units and destroy them with massed fires in a layered defensive approach. The layered approach starts with identifying friendly units with long-range air, space, and cyberspace capabilities and then targeting them with fires before they are in range to maneuver during close combat. Ideally, they would attrit friendly forces to the point that they lack the combat power to exploit any initial successes they achieve.

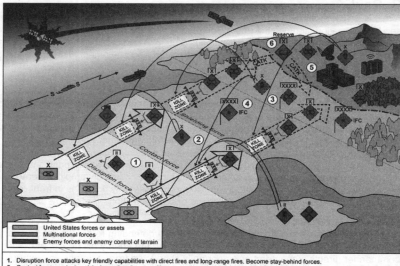

Ref: FM 3-0 (Oct. '22), fig. 6-6. Notional enemy maneuver defense.

Enemy forces continuously improve positions in ways that make attacks against them more costly and allow enemy forces to commit the minimum amount of ground combat power forward. Enemy forces use deception, dispersion, and repositioning to avoid easy acquisition by friendly information collection systems. Enemy forces typically employ significant reserves to counterattack penetrations and attempted envelopments. Integrated air defense systems arrayed in depth provide protection and freedom of movement for enemy maneuver units and fires systems.

Forward-positioned enemy forces focus on providing observed fires for long-range surface-to-surface systems and fixing friendly forces long enough to be engaged effectively by those systems. Enemy forces are likely to conduct a maneuver defense, whenever they are able, by using a series of subsequent battle positions to achieve depth. An enemy commander seeks to use fires and obstacles to prevent decisive engagement of the defending ground forces as they reposition, while causing friendly forces to move as slowly as possible under continuous fire. An enemy force can be expected to employ significant electronic warfare, reconnaissance, surveillance (including UASs), and cyberspace capabilities as part of this defensive effort. Peer enemies can employ chemical weapons, and some can employ tactical nuclear weapons, to prevent the culmination of their defending forces.

E. Offensive Operational Framework Considerations

Within the context of the higher echelon commander's operational framework and the general phasing scheme, commanders must design attacks that defeat enemy forces across all echelons, while enabling subordinate disciplined initiative. When designing attacks, commanders divide the task of defeating an enemy force and maintain an integrated approach through deep, close, and rear operations. Commanders account for air, space, cyberspace, and when relevant, maritime capabilities across the operational framework. Figure 6-7 illustrates friendly offensive operations across a notional operational framework.

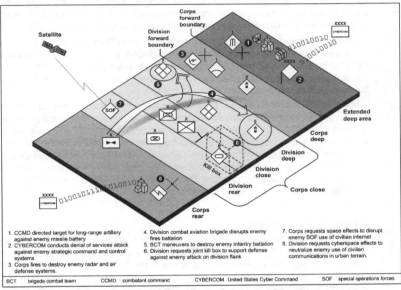

Ref: FM 3-0 (Oct. '22), fig. 6-7. Notional operational framework during offensive operations.

Extended Deep Area During Offensive Operations

The extended deep area is an important part of any formation's area of interest because it contains enemy capabilities that can inflict damage on friendly forces and affect friendly forces' operational reach and endurance. Enemy strategic C2 nodes and long-range fires capabilities, including ground-based missiles and aviation assets, are generally located in the extended deep area out of the corps or division's AO. The situation in the extended deep area influences future operations, and it may become part of an assigned area when offensive operations progress, forward boundaries advance, and units move around the battlefield. Commanders integrate the effects and activities of special operations forces and partner irregular forces in extended deep areas. Army forces request joint effects in the deep area, and they may be tasked by the JFC to employ long-range fires against targets in the extended deep area. Disrupting enemy decision making or destroying enemy long-range fires capabilities limits the enemy's ability to mount a coordinated defense and can destroy enemy forces reconstituting or preparing for counterattacks.

Deep Operations During Offensive Operations

Deep operations focus on parts of an assigned area that are not in direct fire contact with the main body of the formation but may be in the future. At division and corps, which have assigned and attached long-range fires capabilities, the deep area

(Operations) IV(c). Armed Conflict 1-123

extends beyond the forward line of troops to a distance that corresponds with the ability of Army and joint capabilities to reach. Divisions and corps integrate special operations forces, depending on C2 relationships, with deep operations to degrade the enemy's will to fight, destroy high-payoff targets, and disrupt enemy defensive infrastructure and sustainment. Special operations forces operating in deep areas require control measures to synchronize actions and mitigate risk.

Commanders synchronize joint and organic lethal and nonlethal effects in the deep area to disrupt enemy echelons, neutralize reserves, and destroy key capabilities including counterbattery radars and fires systems that can impact close operations. For example, a commander may combine massed fires against the enemy reserve with military information support operations, attacking that unit's resupply operations to psychologically isolate it from the main body and erode its will to fight.

Corps and divisions may use their indirect fires capabilities to support close operations and to conduct deep operations. Commanders employ reconnaissance and security forces, both air and ground, to make initial contact, detect targets, and facilitate fires in support of deep operations. They may shift priority of fires from deep to close operations in support of the main effort.

Commanders may also employ reconnaissance and security forces to conduct deep operations. Commanders may task-organize dedicated security forces to make contact with the smallest possible ground formations and to develop the situation before the main body can be decisively engaged. Normally an attacking unit does not need extensive forward security forces; most attacks are launched from positions already in contact with the enemy, which reduces the usefulness of a separate forward security force. An exception occurs when the attacking unit is transitioning from the defense to an attack, and it had previously established a forward security area as part of the defense.

Close Operations

Close operations occur where forces at divisions and lower echelons maneuver and where forces at the BCT and lower echelons conduct most direct fire engagements. BCTs and lower echelon formations destroy or render enemy forces combat ineffective through movement and fires during close operations. Close combat at the BCT-level and below relies heavily on the warfighting skill and determination of leaders and small-unit teams. Battle drills and creating effects from massed indirect fire play a critical role in success during close combat, since immediate action or reaction can determine success and failure.

When commanders achieve their initial objectives in the close area, they transition to the next phase of the operation, or they execute a branch or sequel should the situation require a change to the original plan. A branch may require friendly forces to defeat a counterattacking enemy force on the division's flank with joint capabilities or reserves. When commanders defeat most enemy defensive forces, they may transition some of their focus to consolidating gains. For example, when lead friendly echelons fix and bypass enemy forces to maintain momentum, follow-on friendly forces must defeat the bypassed enemy forces to prevent them from disrupting friendly lines of communications, negatively influencing local populations, or breaking out. Commanders initiate operations to consolidate gains as early as possible, while still in close proximity to enemy units, to mitigate the risks those enemy units breaking contact could generate later.

Close operations require graphic control measures to synchronize the application of combat power, to ensure integration of subordinate units, to maintain maximum pressure on the enemy, and to mitigate the risk of fratricide within the friendly force. There is no arbitrary benefit to minimizing or maximizing the number of graphic control measures for a particular operation. The optimum number of graphics depends on the mission variables. Having many graphics provides options and agility to the plan, and this can be especially helpful in air-ground coordination or facilitating rapid changes of direction. Leaders balance the complexity of control measures. They should be simple enough for bold execution and detailed enough to enable agility and adaptation based on the situation. The best approach for employing control measures is to ensure they maximize subordinate unit freedom of action and prevent subordinate units from having more tasks than they can reasonably accomplish.

Commanders and staffs maintain situational awareness about adjacent friendly and enemy units in their area of interest. Although commanders and staffs have a plan for where a zone of attack will be and can use it to create initial boundaries and other graphic control measures, enemy forces operate independently of them. This means that while friendly forces have some constraints or restrictions in terms of employing capabilities outside of their assigned area, enemy forces do not. This requires friendly forces to understand what enemy capabilities outside their assigned area can influence friendly operations and closely coordinate responses with friendly units to the flanks and rear.

Rear Operations During the Offense

Rear operations encompass a wide range of activities directly controlled through the rear command post of a corps or division. Rear operations include sustainment operations, support area security, risk mitigation of areas not assigned to subordinate units, terrain management, movement control, coordinating and synchronizing protection capabilities, consolidating gains, and conducting stability operations as required. Corps and division reserves typically occupy positions in the rear area before their commitment, and they must receive priority of movement through the rear area when they are repositioned or committed. Rear operations contribute to the operational reach, tempo, freedom of action, endurance, and momentum of the whole formation.

Rear operations must adapt to the forward progress of the attacking divisions or BCTs. This may require moving support units, the rear command post, and sustainment activities forward as an attack progresses. The rear command post typically assumes responsibility for much of the new land area gained by the attacking forces. The division rear command post must be prepared to assume responsibility for the land areas left by forward-moving forces. In particular, it must be ready to control and repair routes, manage terrain for follow-on support units, and assure that the division reserve can move without obstruction. These activities may also require the rear command post to direct the reduction of explosive ordnance threatening critical infrastructure or key resources.

Tactical units move through rear areas on their way to being integrated into corps or division operations. These units report through the rear command post and occupy assembly areas designated by the rear command post until they are able to integrate with the main command post, should they be employed as part of close or deep operations. Long-range fires, ADA, CBRN decontamination, aviation, and reserve units occupy assembly areas and other positions in the rear area. Their requests for fires and other enablers, their use of ground and airspace, and their other operations are coordinated through the rear command post, unless otherwise directed by the main command post.

Commanders typically economize combat power reserved to protect their rear operations. Bypassed enemy forces, enemy special operations forces, and irregular forces pose a significant threat to rear operations. Typically, divisions assign a BCT to provide security and consolidate gains in rear areas. Maneuver enhancement brigades, with augmentation, employ a tactical combat force to defeat Level III threats throughout the support areas, including major sustainment nodes and main supply routes.

Rear operations play a key role during transitions. For example, a transition between defense and offense increases the amount of fuel, maintenance, and movement support required for operations. A transition to stability operations requires greater humanitarian aid supplies, force protection and construction material, and contract support.

(Operations) IV(c). Armed Conflict 1-125

As stability becomes a greater focus of operations, commanders must assess the resources available against the mission to determine how best to conduct the minimum-essential stability operations tasks and what risk they can accept. The land component commander can establish a civil affairs task force to operate in the rear area and form a transitional military authority or support civil administration that facilitates the ultimate consolidation of gains in support of the desired strategic end state. The purpose of the civil affairs task force is to concentrate stability operations in an AO under a commander's main effort. Additionally, commanders may require Soldiers from all branches to conduct civil security related tasks such as transitional public security tasks until local security forces are able to conduct these tasks. The speed with which transition from U.S. forces to local authorities occurs is dependent on the ability to plan and control elements aimed at developing and legitimizing governance at the tactical and operational levels.

F. Transition to Defense and Stability

When offensive operations culminate before enemy forces are defeated, friendly forces rapidly transition to the defense. Commanders may deliberately transition to the defense when enemy forces are incapable of fully exploiting an opportunity, or when they believe they can build combat power to resume the offense before enemy forces can react effectively. Depending on where culmination occurs, friendly forces may have to reposition forces on defensible terrain and develop a form of defense and scheme of maneuver based on an assessment of the mission variables.

Successful offensive operations end because Army forces have achieved their assigned objectives. A successful offense can also require a transition to a defensive posture dominated by stability operations and a strategic environment moving toward post-conflict political goals. These operations have the goal of transitioning responsibility for security and governance to legitimate authorities other than U.S. forces.

As a transition to stability operations occurs, leaders focus on stability tasks and information activities to inform and influence populations and conduct security force assistance. Effective collaboration with diplomatic and humanitarian organizations enhances the ability to achieve stability mechanisms. Army forces play a key role in enabling the joint force to establish and conduct military governance until a civilian authority or government is given control of their assigned areas.

G. Transition to Post-Conflict Competition

Army forces conclude armed conflict by establishing conditions that are favorable to the United States on the ground. Army forces support these conditions throughout armed conflict by consolidating gains and prosecuting operations with desired end state in mind. As hostilities end, stability tasks dominate operations with the purpose of transitioning responsibilities to legitimate authorities in a secure environment. Army forces provide the joint force with the option of establishing a military transitional government before transitioning full governing responsibility to host-nation or other provisional governments.

Standards for transitioning governance responsibility depend on the credibility, capability, and capacity of the governing organization to maintain the favorable conditions established during armed conflict. Strategic leaders determine the broad conditions for transition at the outset of operations and refine them based on how the situation changes. Army forces play a key role in understanding the host-nation culture, understanding critical infrastructure, assisting strategic leaders in the development of realistic transition goals and timings, and determining the duration and scale of U.S. commitments required to maintain stability.

1-126 (Operations) IV(c). Armed Conflict

Chap 1

V. Army Operations in Maritime Environments

Multidomain Operations

Ref: FM 3-0, Operations (Oct. '22), chap. 7.

I. Overview of the Maritime Environment

Previous conflicts have proven the critical role of land forces in maritime theaters. In almost all cases it is land that makes a maritime area important to a combatant commander. Land masses near or surrounded by water create maritime choke points, enable force projection in and out of maritime areas, and contain the majority of the world's population. The ultimate objective of conflict is typically not control over vast expanses of open water, but rather the land and people who control it. Planning and training for the unique considerations of operations in these environments is critical, as is an integrated planning approach with the rest of the joint force. A maritime operational environment adds coordination and synchronization requirements for Army echelons that typically operate and train in land operational environments. Army and joint force planning must reflect an understanding of the dynamic nature of the threats and constraints to land forces in maritime regions. Army movement and maneuver between land masses is almost entirely dependent on joint capabilities. Control of critical land masses is essential to the sustainment and protection of joint operations in a maritime operational environment. This interdependence has been historically crucial to success during armed conflicts against peer threats in maritime theaters, and it continues to be crucial in the foreseeable future.

A. Considerations Unique to the Maritime Environment

Threats to Army forces in a maritime theater include those in any other type of theater. Additionally, commanders and staffs must take into account unique joint and enemy courses of action in planning and executing operations in a maritime environment.

Joint Force Considerations in a Maritime Environment

The heavily interdependent nature of joint operations means that the defeat of one part of the joint force puts the other parts at significant risk. The failure of Army forces to retain key terrain or protect air and naval bases while preserving their own combat power could result in the loss of air and maritime superiority, which in turn could lead to the ultimate defeat in detail of unsupported Army forces in an entire area of responsibility (AOR).

Army forces require a joint common operational picture (COP) of friendly forces and their operations, including those of allies and partners, in all domains. An inaccurate joint or partner COP could cause flawed assumptions and situational understanding that decreases effective decision making.

As with other environments, planning land operations in maritime environments should address relevant factors affecting friendly and enemy operations. The products and tools typically gathered to plan and portray the unique characteristics of a maritime environment include riverine and coastal navigation charts or tidal reports and observations from local fishing communities. Additional characteristics to consider include—

- Coastal terrain and soil compositions.
- Commerce and trade along navigable waterways.
- Maritime-specific infrastructure.

(Operations) V. Army Operations in Maritime Environments 1-127

- Navigable bodies of water.
- Population densities and variations along shores and near navigable waterways.
- Tidal flow, surf conditions, and current directions.
- Natural obstacles adjacent to terrain (including sandbars, shoals, mud flats, and dunes).

Enemy Courses of Action Unique to the Maritime Environment

Enemy forces have the ability to take multiple actions unique to the maritime environment to hinder joint operations. They use capabilities from all domains to interdict lines of communications between the strategic support area and forward-positioned Army forces, putting forces at risk of isolation and beyond the supporting distance from other joint force elements. Enemy forces target shipping that carries the bulk of Army heavy equipment into theater and the pre-positioned equipment already in theater as part of their preclusion and isolation approaches. While Army forces can take measures to protect equipment already in theater, they are dependent upon the other Services for the protection of people, equipment, and supplies in transit.

Threats may employ asymmetric tactics to reinforce their area denial (AD) approaches throughout a region, capitalizing on the vulnerability of isolated friendly locations lacking adequate protection. Locations that cannot be reinforced or supported by other elements of the joint force are particularly vulnerable. Attacks by enemy special operations and naval forces are difficult for Army forces to detect and counter without proper positioning and preparation. Anticipating possible enemy courses of action in one region that would support a broad theater-wide outcome should drive friendly priorities of planning and preparation.

Enemy forces may use proxy forces to destabilize regional partners, further delaying Army forces from safely accessing basing and lines of communications. Proxy forces can facilitate enemy reconnaissance, surveillance, and disruption of joint operations. They can also affect access to sea-lanes and airports for transportation, hindering the resupply of island bases, and limiting the ability to reinforce forward-positioned friendly forces through attacks on bases and base clusters.

Enemy surface-to-surface and surface-to-air fires systems are critical to the layered defense and early warning systems necessary to attack forward-positioned friendly forces with little warning. These systems simultaneously prevent reinforcement or support of those friendly forces. Enemies position robust and integrated air and missile defenses, early warning surveillance radars, and electronic warfare capabilities that range elements of the U.S. joint force put them at risk during competition, crisis, and conflict. Enemy medium range ballistic missiles, cruise missiles, anti-ship missiles, and air and naval forces also put friendly forces at risk. This combination of systems warfare approaches threatens land-based forces even when they do not directly target them, since land-based forces require the support of air and maritime forces for sustainment, early warning, and protection.

Army-Specific Considerations

Commanders and staffs account for multiple considerations when planning and executing operations in a maritime environment. Counter-reconnaissance by Army forces on an island requires a 360-degree approach to avoid surprise. Responding to threats requires highly mobile capabilities that mass effects against enemy forces quickly. Although threats can approach from any direction to achieve surprise, there are constraints in terms of where enemy forces are able to land during an amphibious or air assault and range constraints that might cause aircraft to favor certain approaches. Intelligence preparation of the battlefield requires an understanding of air and maritime avenues of approach and their relationship to a particular land mass or grouping of land masses.

1-128 (Operations) V. Army Operations in Maritime Environments

B. Physical Characteristics of the Maritime Environment

Ref: FM 3-0, Operations (Oct. '22), pp. 7-1 to 7-3.

Habitable land masses vary in size and geology. They include land masses as large as Australia to small islets that make up larger atoll systems, such as the Marianas islands in Micronesia or Alaska's Aleutian Islands chain. Regardless of their size, not all islands are suitable for extended occupation without externally provided water and other supplies. While most islands are natural, formed by tectonic or volcanic action, manmade islands created for military and other purposes are also found within certain regions, such as the South China Sea. Currently, 22 nations in the world are recognized as archipelagic states, and these unique maritime nations can include several cultures, religions, languages, and geopolitical histories within their own territorial boundaries.

In a predominantly maritime environment, any land that can be occupied to attain a physical position of relative advantage by friendly or enemy forces can become key terrain. The largely maritime domains challenge planners to account for operational reach and the impact of space and time on reinforcement of existing forward-stationed forces and allies. The distance between bases throughout the maritime regions hinders mutually supporting operations and sustainment from supporting echelons in the theater. Remoteness and distance increases vulnerability to amphibious raids by enemy special operations forces and attacks from long-range aircraft or missiles, and it increases the risk of physical isolation by air and naval forces.

Littoral Regions

Maritime environments include littoral regions, divided into two segments: seaward and landward. Seaward segments include the area from open ocean to the shore, which must be controlled to support operations ashore. Landward segments are those areas inland from the shore that can be supported and defended directly from the sea. Maritime littoral regions are divided into five categories:

- **Enclosed and semi-enclosed seas**—bodies of water surrounded by a landmass and connected to either an ocean or another enclosed sea by a connecting body of water, such as a strait.
- **Islands**—single land masses surrounded by a body of water.
- **Archipelagoes**—groups of islands.
- **Open seas**—unenclosed bodies of water, typically outside of territorial boundaries.
- **Marginal seas**—portions of open seas or oceans that bound land masses such as peninsulas, islands, and archipelagos.

Operations in these environments require information concerning tides in the local area, average wave heights, and daily wind forecasts, as tides and winds may unexpectedly impact wave heights. Tidal flow and currents impact operations for small or shallow draft vessels. Tidal changes can also affect vehicle mobility when crossing unimproved beachheads or operations without causeways or engineer support. Extreme weather plus the corrosive effects of salt water can rapidly degrade the maintenance readiness of equipment already at the extent of long logistics lines of communications.

Arctic Region

The Arctic region is significantly influenced by maritime considerations. The Arctic encompasses part of the areas of responsibility of three different geographic combatant commands, eight countries, and all time zones. Extreme temperatures, long periods of darkness and extended daylight, high latitudes, seasonally changing terrain, and rapidly changing weather patterns define Arctic conditions, and they all have impacts on the operational and mission variables.

(Operations) V. Army Operations in Maritime Environments 1-129

Army forces positioned at existing bases within the range of adversary long-range fires establish primary and alternate survivability positions for themselves while providing air and missile defense (AMD) and local security to airbases and ports. They may also be required to defend against amphibious assaults, and airborne or air assault operations by enemy forces seeking to control a particular land area or destroy critical infrastructure. Army forces operating on islands with austere infrastructure and resources require significant sustainment during prolonged operations.

Land areas able to accommodate significant military forces in strategically or operationally important areas are scarce in maritime environments, and what is available is already a known point for enemy planners. This makes surveillance of friendly activities simpler, affecting the ability to achieve surprise or avoid being surprised. It also makes the process of enemy observation, information collection, and subsequent targeting faster and more effective. The implications, with regard to friendly forces concealing and protecting themselves, are significant.

C. Planning Considerations for a Maritime Environment

Planning for operations on land in maritime environments requires a high level of complex detail to coordinate the movement and landing of troops, equipment, and supplies by air and surface means. The success of Army operations in the maritime environment are uniquely dependent on unity of effort and integrated, collaborative planning with joint headquarters and subordinate echelons.

The joint force commander (JFC) for major maritime operations is typically a senior naval officer. The JFC ensures unity of effort across the joint force and ensures the task and purpose of Army operations nest with the overall joint operational concept. Despite the physical distances and relative isolation typically associated with land operations in a largely maritime environment, Army commanders must maintain situational understanding through joint collaboration to ensure their operations or activities do not have a negative effect on other operations or units. This includes maintaining situational awareness during movement, which requires direct and continuous integration of Army personnel into the command and control (C2) nodes of Marine Corps, Navy, and Air Force units. This may also include integration with multinational amphibious or maritime units.

Planning for operations in maritime environments requires collaboration between Army forces and other relevant components of the joint force command and integration across warfighting and joint functions at each echelon. Ideally, planning should be conducted by commanders and their staffs in the same location, which in some cases may be aboard a ship or in the air. When this is not practical, the exchange of liaison officers facilitates planning functions. Planning efforts—particularly in crisis situations—are conducted in parallel and collaboratively across the involved echelons and Services.

II. Operational Framework

Applying the operational framework in maritime environments requires commanders and staffs to consider the impacts of maritime surface areas and integration with maritime forces. It also requires a different appreciation about what constitutes deep, close, and rear operations in relation to each other. The physical separation of forces by bodies of water affects considerations of mutual support since many operations are likely to involve noncontiguous AOs. Army echelons may have responsibilities for information collection of maritime surface areas and for providing fires into maritime and littoral regions in support of other Services. Army forces may be required to defend against enemy amphibious assaults, requiring tactical level coordination with Navy and Marine Corps forces.

See facing page for further discussion.

1-130 (Operations) V. Army Operations in Maritime Environments

Applying the Operational Framework

Ref: FM 3-0, Operations (Oct. '22), pp. 7-7 to 7-9.

Assigning a Joint Operations Area (JOA)

The Unified Command Plan designates the AOR. Within that AOR, a larger maritime environment might have several joint operations areas to facilitate C2 and resource prioritization. For example, United States Indo-Pacific Command may designate a JOA for operations in the South China Sea and a JOA for operations in Korea. Both JOAs require unique C2, movement and maneuver, sustainment, intelligence, fires, and protection planning and resources. This also includes the designation of an appropriate land component headquarters and staff to facilitate joint integration specific to that JOA and the particular Army operations within it. A JOA is established for operations within an AOR that is specialized or limited in its scope or duration, and a JFC directs military operations as a joint task force (JTF).

Designating a Joint Security Area (JSA)

Within the JOA, the JFC designates numerous joint security areas (JSAs). In a maritime environment, JSAs can be separated by considerable distances, and they probably will not be with areas that are actively engaged in combat. A theater army headquarters or theater sustainment command (TSC) may be required to conduct theater sustainment operations from a single designated JSA or from multiple locations, depending on mission requirements. Army forces may also be designated to secure the JSA and critical intermediate staging bases.

Assigning an Area of Operations (AO)

An AO is the operational area defined by a commander for land and maritime forces. The JFC assigns land areas of operations. The designation of subordinate AOs in a maritime environment enables freedom of action, maintains tempo, and maximizes available combat power. Larger island land masses may allow Army forces to operate with a contiguous AO, with the unit boundaries directly adjacent to each other. Smaller archipelagic island chains may require a noncontiguous AO and may even leave some islands within a designated AO completely unoccupied by friendly forces depending on the operational requirements and threat.

Deep Operations in Maritime Environments

Deep operations in a maritime environment may focus on defeating enemy antiaccess (A2) and AD capabilities to set conditions for joint offensive operations using information collection, special operations forces, and fires. They may also focus on reconnaissance and security activities in support of joint defensive operations using the same capabilities.

Close Operations in Maritime Environments

Close operations in a maritime environment may appear much the same as a traditional land-based approach discussed previously. Seizing key terrain requires Army forces to conduct offensive operations, which can include airborne, air assault, and amphibious assaults. Close operations also include defense of islands and island-based nodes that facilitate joint operations, such as airfields and ports.

Rear Operations in Maritime Environments

Rear operations in a maritime environment include those necessary to set and sustain the theater and facilitate combat operations for the joint force. These operations include conducting protected reception, staging, onward movement, and integration (RSOI) and conducting theater sustainment. Sustainment and protection operations are associated with support areas, enabling the building and preservation of combat power.

(Operations) V. Army Operations in Maritime Environments 1-131

III. Operational Approach

In a predominantly maritime JOA, naval and air components are typically the key components of the JFC's operational approach. Army forces develop a nested operational approach that reflects and supports the JFC plan. This section details what Army forces consider to successfully nest their operational approach with the joint force.

A. Establish Command and Control

Given the size of an AOR for most maritime environments and the distance between land areas, there may be multiple active JOAs, each with separate headquarters. Maritime environments impose significant challenges for theater army signal and sustainment architecture. Subordinate ARFORs rely on maritime and space capabilities to overcome these challenges.

B. Defend and Control Key Terrain

Friendly forces are stationed or positioned in dispersed forward positions in maritime theaters, and often have been there for decades to facilitate joint operations throughout a maritime theater. Their locations are typically key or even decisive terrain, which requires that Army forces must also be able to defend and control that terrain. Their ability to do so is a function of survivability that depends on active and passive defense measures, reinforcement with critical weapons systems, hardening and camouflaging of command posts, securing critical C2 networks, sustainment, and the employment of protection-oriented land forces. Providing active and passive AMD, to include early warning, reduces the effectiveness of enemy long-range fires or attack. Forward-positioned forces must be prepared to fight outnumbered and from exposed terrain, specifically islands, for as long as is required. This increases the need for security throughout the AOR and forward-positioned sustainment capabilities. Army forces enhance the protection of their own less mobile assets by continually improving the survivability of their positions.

Retaining critical island terrain through an effective defense, one that includes counterreconnaissance and security operations, is vital for the success of the JFC's objectives to deny enemy forces a relative advantage. Retaining key land masses enables joint freedom of action for operations in the air and maritime domains, since land-based capabilities can maintain a persistent physical presence that lessens requirements on air and naval forces to secure avenues of approach. Defeating enemy C2 systems is key to defeating layered standoff, A2 weapons, early warning, and enemy reconnaissance and surveillance capabilities. The ability to maintain key terrain for access and security enables the JFC to employ key long-range fires and protection capabilities. This enables regional access during armed conflict critical for maritime freedom of navigation and air superiority. Denying enemy access to limited terrain, sea channels, airspace, and cyberspace by occupying key terrain is crucial to creating an advantage. As there is no easy exfiltration from an exposed island base or maneuver to positions outside of enemy fires ranges, the physical reinforcement of bases may be critical to the success of Army forces in enabling joint force success.

1-132 (Operations) V. Army Operations in Maritime Environments

Protection Support to the Defense

Ref: FM 3-0, Operations (Oct. '22), pp. 7-10 to 7-12.

The ability to protect and augment the security of the other Services is critical in maritime theaters because of the relative isolation of the forces they position forward and the exposure of those forces to threat capabilities that could be employed with few indications or warnings. The following tasks directly support theater setting and are critical to the success of joint maritime operations:

- AMD, to include counter-unmanned aircraft systems (UASs).
- Littoral defense.
- Area security (base and base cluster defense).
- Chemical, biological, radiological, and nuclear (CBRN) defense.
- Explosive ordnance disposal support.
- General, combat, and geospatial engineering support.

Refer to FM 3-01 or ATP 3-01.15 for detailed information about the planning and integration of AMD.

Security Operations in Maritime Environments

The JFC designates the area commander for base and lines of communication security, and most of the security tasks are typically the responsibility of land force commanders. Collaborative planning and integration between ground forces providing base security and naval and air forces providing sea lines of communications security is crucial to maintaining freedom of navigation for combat operations. Army forces provide security support to all bases within their designated JOA. Brigade combat teams (BCTs), military police brigades, and maneuver enhancement brigades are suitable for this function. This responsibility can include bases commanded by organizations that are not part of the area commander's forces, such as multinational allies or other joint services. The JFC may institute standard force protection policies for all commands and bases within an AOR to ensure unity of effort.

Successful security of key terrain and infrastructure depends on an integrated and aggressive plan consisting of dedicated security forces and responsive sustainment and protection forces (including medical, ADA, and engineer). The theater army synchronizes the base security plans, integrates them into the overall JFC's intent, and allocates additional forces for securing sustainment nodes and command posts, key terrain, or critical infrastructure necessary for combat operations. Remote island bases have the same security planning considerations, but they vary in their tactical application based on the environment. These considerations include, but are not limited to—

- Defense against sea-based attacks, such as enemy special operations forces or naval fires.
- The integration and application of coastal-based indirect fire systems and fire support planning.
- Integrated planning for aviation support, including movement and protection of forward arming and refueling points.
- Coastal and harbor security support and integration of host-nation or local security forces.
- Integration of coastal barrier systems, sea-based obstacles, and sea-based or harbor mines.

Refer to JP 3-10 for more information on joint security coordination command posts. Refer to ADP 3-37 for more information regarding the fundamentals of protection and protection planning.

(Operations) V. Army Operations in Maritime Environments 1-133

C. Defeat Components of Enemy Antiaccess and Area Denial and Enable Joint Offensive Operations

Army forces support joint defeat of enemy integrated air defense, fires and strike complexes, surveillance, and reconnaissance, and integrated C2 networks to enable success during joint operations. AMD and fire support are two of the Army's critical contributions to these efforts, enabling ground, naval, and air forces to maintain access to the various regions of an operational environment.

Provide Theater Air and Missile Defense

Friendly ADA forces may protect maritime access points, such as shipping channels or ports, to enable the flow of combat power and theater sustainment. Considerations for employing ADA capabilities during AMD operations in a maritime environment are—

- Defeat enemy air and missile threats encountered in strategic and tactical operations, including medium-range, short-range, and close-range ballistic missiles; cruise missiles; UASs; rockets, artillery, and mortars; tactical air-to-surface missiles; and fixed-wing and rotary-wing aircraft.

- Integrate and maintain tactical data linkages to other Service and multinational forces conducting AMD operations in the JOA, weapons systems, sensors, effectors, and C2 nodes at each echelon.

- Provide early warning for air and missile attacks and disseminate attack warnings.

- Provide extended-range surveillance of the airspace and detect, acquire, track, classify, discriminate, and identify aerial objects from near-ground level to high altitudes in difficult terrain and in adverse weather conditions.

- Contribute to airspace management and control functions by identifying, coordinating, integrating, and deconflicting Army assets in the JOA airspace.

The number and dispersion of critical assets across numerous islands may exceed the ability of available ADA forces to defend against the air and missile threats. The employment of passive AMD measures by tactical formations is critical to their survivability, particularly on remote islands that may not have access to sufficient active ADA capabilities or reinforcement from a neighboring island chain. Access to early warning networks with over-the-horizon and redundant communications increases reaction time and further mitigates risks for remote locations primarily using passive defense measures.

Refer to FM 3-01 for more detailed information on planning considerations associated with AMD operations.

Forcible Entry Operations

In a maritime environment, Army forces are likely to conduct two complex forms of forcible entry operations: airborne or air assault and amphibious landing. Forcible entry operations seize and hold lodgments against armed opposition to set conditions for follow on operations. To set favorable conditions for success during forcible entry, commanders and staffs must—

- Visualize the entry location and understand the impact of other domains on forcible entry operations.

- Control air and maritime areas to protect the force and preserve lines of communications leading up to and during the entry.

- Disrupt enemy influence during entry.

- Isolate the lodgment from reinforcement by enemy forces.

- Maintain access to the lodgment throughout the duration of operations to build and sustain combat power.

1-134 (Operations) V. Army Operations in Maritime Environments

- Manage the lodgment to integrate other supporting operations.
- Seize and maintain the initiative to achieve surprise throughout the entry operations.

The JFC decides whether to conduct the forcible entry as a concurrent or integrated operation. Concurrent forcible entry operations occur when a combination of amphibious assault, airborne, air, or ground assault forcible entry operations are conducted simultaneously, but as distinct operations with separate AOs and objectives. Integrated forcible entry operations occur when amphibious assault, airborne assault, air assault, and ground forcible entries are conducted simultaneously within the same AO against mutually supporting objectives. Large island land masses would likely involve integrated forcible entry involving U.S. Marines and potentially allied or partner forces. Smaller archipelagic island chains could require concurrent forcible entries that enable security operations or subsequent main and supporting efforts against larger objectives.

Forcible entry operations by ground, sea, or air all use the same phasing model to facilitate coordination and synchronization. These phases are preparation and deployment, assault, stabilization of the lodgment, introduction of follow-on forces, and termination or transition operations. Planning for forcible entry also includes planning for—

- Movement planning over extended lines of communications over water.
- Information collection against an enemy with layered and integrated early warning.
- Management of transitions when lines of communications and networks are extended over water.
- Insertion of special operations forces.
- Ensuring air superiority over the joint landing area.
- Coordination for initial and reinforcing entry forces for the initial assault and main assault.
- Establishment and operation of any potential intermediate staging bases.
- Lodgment security, organization, and expansion from shore to an island interior.

Refer to JP 3-18 for more detailed discussion and information about Army forces roles, responsibilities, and planning considerations for joint forcible entry operations.

Amphibious Operations

Planning for an amphibious operation is continuous, and it requires collaborative, parallel, and detailed planning by all participating forces. The organization of any amphibious operation should be sufficiently flexible to meet the planned objectives in each phase of the operation and account for unforeseen developments. Sound planning provides for unity of effort through unity of command, centralized planning and direction, and decentralized execution. JFCs may decide to establish a functional component command to integrate planning and reduce JFC span of control. This improves the efficiency of information flow, weapons systems management, component interaction, unity of effort, or control over the scheme of maneuver. Regardless of approach, the JFC designates command relationships for the commanders of the amphibious task force and landing forces. The designation of the supported and supporting role of the amphibious force commanders is important, as it establishes main and supporting efforts and prioritizes resources.

Opposed landings in a maritime environment are one of the most difficult and dangerous military operations, so achieving the element of surprise should be pursued by all available means.

See following pages (pp. 1-136 to 1-137) for an overview and further discussion.

(Operations) V. Army Operations in Maritime Environments 1-135

Multidomain Operations

Amphibious Operations

Ref: FM 3-0, Operations (Oct. '22), pp. 7-14 to 7-15.

Planning for an amphibious operation is continuous, and it requires collaborative, parallel, and detailed planning by all participating forces. The organization of any amphibious operation should be sufficiently flexible to meet the planned objectives in each phase of the operation and account for unforeseen developments. Sound planning provides for unity of effort through unity of command, centralized planning and direction, and decentralized execution. JFCs may decide to establish a functional component command to integrate planning and reduce JFC span of control. This improves the efficiency of information flow, weapons systems management, component interaction, unity of effort, or control over the scheme of maneuver. Regardless of approach, the JFC designates command relationships for the commanders of the amphibious task force and landing forces. The designation of the supported and supporting role of the amphibious force commanders is important, as it establishes main and supporting efforts and prioritizes resources.

Opposed landings in a maritime environment are one of the most difficult and dangerous military operations, so achieving the element of surprise should be pursued by all available means. Overt activities that threaten the element of surprise should be kept to a minimum and conducted as close as practical to the arrival of joint fire support assets in the JOA. Deception operations may facilitate surprise if they portray a course of action the enemy expects or a timeline for execution different than planned. Deception operations are a necessity during assault breaching of a defended beachhead. Large lane sizes demand large numbers of weapons and multiple aircraft passes to clear zones prior to the assault. This may draw immediate attention to the landing force unless alternate lanes are brought under fire as well

The landing force typically consists of maneuver, protection, and tactical echelon sustainment forces. The JFC designates the landing forces commander. If Army forces are part of the landing force, they must be task-organized with appropriate combat and sustainment capabilities to support the landing force. Army forces also provide intra-theater ship-to-shore transport, including landing craft, cargo handling, logistics, traffic control, and engineering capabilities.

Note. If the JFC organizes planning along functional lines, functional component commanders normally exercise OPCON over their parent Services' forces and tactical control (TACON) over other Services' forces attached or made available for tasking.

Setting conditions prior to the execution of any amphibious operation is critical. Enabling operations that set conditions include supporting and pre-landing operations. These require detailed integration at all echelons and highly synchronized employment of Army and joint capabilities that can reinforce success and ensure landing force survivability under conditions where reinforcement or support may not be readily available. Planning for forcible entry also includes planning for—

- Force concealment as part of a larger joint military deception operation.
- Maritime clearance of mines in the vicinity of transportation lanes, landing beaches, and shore-based sustainment nodes.
- Maritime and hydrographic reconnaissance for landing beaches and sea approach lanes.
- Accurate prediction of weather and tidal conditions.
- Seabasing of supporting aviation resources, such as aeromedical evacuation or attack aviation.
- Assault breaching and beach clearance of anti-landing obstacles, which may include mines.

1-136 (Operations) V. Army Operations in Maritime Environments

- Status of civil and local national inhabited areas, to both conceal landing forces and prevent civilian casualties.

Pre-landing operations take place between the commencement of the action phase (arrival of the amphibious force into the operational area) and the ship-to-shore movement. There is rarely a clear transition between support and pre-landing operations, and this must be planned for and clearly communicated across echelons prior to execution. Some planning considerations during this phase include—

- Obstacle clearance (including perimeter and main barrier minefields and the engineer and beach barriers that canalize landing forces) and marking usable sea and shore channels for follow-on forces.
- Integration of naval fire support.
- Integration of air support, including electromagnetic warfare and airspace between the landing force and amphibious task force.
- Clearance of fires on landing areas.
- Ammunition and fuel expenditure prior to landing.
- Loss and recovery of equipment prior to landing.
- Loss of personnel and recovery of casualties prior to landing.
- Resupply and rearming schedule for amphibious and landing forces.
- Landing force requirements to support other forces prior to and after the landing assault.
- Organization and location of reserve in a similar manner as the assault force.

During an amphibious operation, a subsidiary landing outside the designated landing area is normally conducted by elements of the amphibious task force to support the main landing. Subsidiary landings should be planned and executed by commanders with the same precision as the main landing. Amphibious re-embarkation for follow-on operations may require additional support from specialized units or other Service forces and additional logistics support to replace lost or damaged equipment and depleted supplies. For tactical echelons, combat rubber raiding craft can provide flexibility to amphibious operations, including movement to reconnaissance objectives and subsidiary landings sites, movement of forces inland from a coast along waterways, or recovery of casualties to support vessels. Additionally, Army planning must account for casualties and the continued medical treatment of patients, potential for CBRN decontamination, requirement to transport remains or provide mortuary services, transportation and transfer of enemy detainees, and logistics to support basic sustenance ashore.

Refer to JP 3-02 for more information about the fundamentals of planning and executing joint amphibious operations. Refer to MCTP 13-10M for more detailed information and considerations on amphibious embarkation. Refer to ATP 3-17.2 for tactical considerations on movement from ship to shore and the C2 of shore and beach parties.

Refer to The MAGTF Operations & Planning SMARTbook (Guide to Planning & Conducting Marine Air-Ground Task Force Operations). Topics and chapters include Marine Corps roles & forces, the Marine Air-Ground Task Force (MAGTF), expeditionary operations, Marine Corps operations (ROMO, offense, defense, tactical operations, reconnaissance & security, tactical tasks, etc), planning considerations, the Marine Corps Planning Process (MCPP & R2P2), integrating processes (IPB, CM, D3A, ORM, IM), and the six warfighting functions.

(Operations) V. Army Operations in Maritime Environments 1-137

D. Sustain Large-Scale Combat Operations in Maritime Environments *(See pp. 7-15 to 7-26.)*

Setting and sustaining a maritime theater includes RSOI of personnel and equipment and the protection of forward-positioned forces critical to the security of key strategic assets, such as theater AMD, airfields, ports, and sea lanes. Army watercraft are essential to sustainment in maritime environments as they are designed to perform missions specifically related to intratheater movement of combat power and sustainment. While capable of deploying over strategic distances, Army watercraft are not strategic lift platforms, but they are a critical link between strategic lift and land-oriented tactical movements. Army and Navy engineering assets are critical to the establishment and maintenance of port facilities.

Establish Protected Reception, Staging, Onward Movement, and Integration

During armed conflict in a maritime operational environment, theater opening and RSOI vary to meet mission requirements of the JFC. The theater engineer command is responsible for developing and maintaining the necessary infrastructure (including ports and roadways) that support RSOI and follow-on operations. Multiple islands requiring multiple ports will increase the demand for specialized engineer and sustainment formations to construct, develop, maintain, and operate them.

Rather than gaining efficiencies from a consolidated location, the theater army may be required to support multiple JSAs and RSOI sites. Using multiple RSOI locations increases the overall signal, sustainment, protection, and maneuver requirements for the theater. In a maritime environment, reception, staging, and integration activities may need to occur prior to onward movement into a JOA. RSOI may occur from sea bases. Seabasing is the deployment, assembly, command, projection, sustainment, reconstitution, and reemployment of joint power from the sea without reliance on land bases within the operational area (JP 3-02).

Refer to JP 3-02 for more information on considerations for planning and executing staging operations from sea bases.

Conduct Theater Sustainment Operations

Theater sustainment in a maritime environment is a highly collaborative process. When directed to provide management of common sustainment functions that include other Services, the TSC leads the joint sustainment planning board. Army forces provide theater and port opening functions for joint forces to maintain strategic and operational reach. Theater sustainment plans must account for—

- Maritime movement of Army pre-positioned stocks (APS).
- Joint logistics over-the-shore (JLOTS).
- Intertheater transportation by joint assets.
- Intratheater transportation for personnel during and after RSOI.
- Classes of supply and access to field services.
- Field and sustainment maintenance in remote locations.
- Distribution.
- Operational contract support.
- General engineering (ports and airfields).
- Sustaining and mitigating impact of limited island infrastructure.
- Mortuary affairs.

Refer to ATP 3-35.1 for more information on planning for APS stocks in combat operations. Refer to JP 4-01.6 for more detailed information about JLOTS.

1-138 (Operations) V. Army Operations in Maritime Environments

Chap 1

VI. Contested Deployment

Ref: FM 3-0, Operations (Oct. '22), app. C.

I. Force Projection and Threat Capabilities

Army forces cannot expect to deploy without being challenged by the threat. For decades, U.S. military forces conducted uncontested and generally predictable deployments from home stations to operational theaters because threat actors lacked the capability to significantly affect deploying units at home station or while in transit to a theater of operations. This is no longer the case. Peer threats possess the capability and capacity to observe, disrupt, delay, and attack U.S. forces at any stage of force projection, including while still positioned at home stations in the United States and overseas. Commanders and staffs must therefore plan and execute deployments with the assumption that friendly forces are always under observation and in contact.

A peer threat's ability to impact U.S. military operations prior to arrival in an operational area extends beyond directly targeting unit personnel and equipment. The Army relies on various interdependent infrastructures, the majority of which it does not own or operate, making its domestic operations heavily reliant on external resources. This includes the use of civilian transportation infrastructure to move from installations to ports of embarkation, and it also includes home station military dependencies on civilian infrastructure for power, communications, fuel, water, and other life support.

During armed conflict, Army forces should expect deployments to be contested by enemy actions in all domains. Army forces will require greater emphasis on protection functions to conserve combat power and should expect to provide forces to support homeland defense and DSCA operations. Defending U.S. territory against attacks by state and non-state actors through an active, layered defense while simultaneously seeking to project forces in a conflict with a peer enemy requires coordination across organizations, agencies, and jurisdictions at the local, state, and federal levels.

Threat actions to contest a deployment are most visible during crisis and armed conflict, but they can also occur during competition. Army forces deploy globally as part of operations during competition to meet national objectives, assure allies and partners, and deter adversary malign actions. Adversary abilities to disrupt these deployments create risks that leaders must assess and mitigate during movement planning and execution. While a conventional attack on U.S. forces conducting operations during competition is unlikely, the greater the perceived danger to their vital national interests, the greater the chance a peer threat will contest U.S. military force projection. Leaders account for this intensified risk during planning and conduct training to improve their units' resilience and ability to mitigate risk, coordinate with appropriate partner organizations, and respond effectively.

Refer to SMFLS5:The Sustainment & Multifunctional Logistics SMARTbook (Guide to Operational & Tactical Level Sustainment), chapter nine for 44 pages on deployment operations from ATP 3-35 and JP 3-35. Topics include Predeployment, Movement; (RSO&I) Reception, Staging, Onward movement, Integration; and redeployment. See also p. 4-10 and 7-32.

II. Movement Phase

Ref: FM 3-0, Operations (Oct. '22), p. C-4 to C-6.

Fort to Port

As part of the strategic support area, home station installations, Reserve Centers, National Guard Armories, and other designated points of origin are where force projection begins. They present targets that enemy forces may attack to delay, disrupt, and degrade force flow into theater. Additional vulnerabilities are present along all routes of movement, and at all potential sea and aerial ports of embarkation. Army forces at all echelons must comprehensively assess emerging threat capabilities that will impede deployment in a contested environment. To the greatest extent possible, formations should account for being under constant observation through strict operations security, including the safeguarding of information on specific deployment timelines and locations and maintaining dispersion of critical assets. The effects of attacks on critical military, national, or private infrastructure could halt or delay unit deployment operations before units have departed the United States.

Contested deployments are a national issue, and they require coordination with a large number of civilian unified action partners to overcome the challenges peer threats can create. However, moving Army forces from military installations to ports of embarkation is also a local and regional challenge. When routine deployment is not possible, installations and units should have a plan to mitigate deployment disruptions.

Deployment disruption mitigation planning requires collaboration between the deploying unit, the installation, appropriate federal, state, and local agencies (both government and law enforcement), and U.S. Army Reserve and National Guard elements. Installations are responsible for building these relationships and understanding how threats will likely affect their local areas. Installations do this by modifying their threat working groups to incorporate relevant military, government, and other local and regional stakeholders. The working group shares information about threats that promote civil unrest, cyber threats that impact critical transportation infrastructure, and other threat activities that impact deployment operations. Key planning and training considerations are—

- The local, state, and federal authorities able to mitigate deployment disruptions.
- Coordination and relationship building with local, state, and federal civilian law enforcement agencies to ensure effective movement control from fort to port.
- Understanding about critical infrastructure vulnerable to sabotage and unsuited for the movement of heavy equipment along surface lines of communication, both road and rail.
- Planning to use alternate railheads and marshalling yards and multiple lines of communication to reach ports of embarkation.
- Developing alternate surface transportation options to deliver unit equipment to a sea port of embarkation when rail service is degraded or disrupted.
- Establishment of fuel, maintenance, and rest locations along lines of communications.
- Implementation of a communication plan that informs the public while maintaining operations security.
- Establishing specific cyber defenses for systems and associated data used to support movement.

Port to Port

Ports of embarkation in the strategic support area, whether in the United States or overseas, are likely targets of cyberspace attack, space capability degradation or denial, and other impacts designed to reduce capabilities or capacity as U.S. forces conduct deployment operations into other theaters. During armed conflict, enemy forces have the range and capacity to target ports with long-range fires, special operations forces, and other capabilities.

Port authorities open, close, and manage port operations based on their pre-existing requirements, not based on what is necessarily most advantageous to a given unit's deployment requirements. This might include requirements for other Army forces, the joint force, or other government or commercial entities designated as priorities for a period of time. Early, frequent, and detailed coordination with port authorities helps mitigate potential disruptions to deployment operations. Installations and units incorporate port officials into deployment readiness exercises and other training events to improve mutual understanding and effectiveness.

Adversary actions may directly or indirectly cause port officials to close their port or reduce operations. If this occurs, and if authority is granted to use another port, commanders consider the impacts to both their unit and the broader military effort, and to civilian requirements, before recommending or deciding to move assets to another port of embarkation. Port operations conducted by other government agencies or civilian officials may take priority over those that Army forces need to conduct based on current local, state, or national requirements. Additionally, other ports may become congested or disrupted by the time Army forces arrive. Adhering to the original plan while adapting to challenges at the port is generally less disruptive to the overall deployment effort than moving to another port.

Depending on the level of conflict and assessed threat to their interests, peer enemies may attempt to disrupt or destroy unit equipment while in transit. If all of a unit's equipment is placed on one transport this could lead to the catastrophic loss of land component capability for the JFC. Spreading unit equipment across multiple transport ships increases the likelihood that some will arrive and be available for employment. While this is not the most efficient or expeditious method to transport Army equipment, it helps ensure that the JFC receives some employable level of Army tactical and operational capability.

During armed conflict, Army forces conduct additional protection activities based on enemy capabilities. Effective targeting by enemy lethal and nonlethal effects will drive protection and mitigation actions for unit personnel and equipment. Army forces maybe called on to provide additional protection capabilities to support port authorities. Unit commanders and staffs should balance protection requirements, both at the port and in-transit, against requirements to get as many critical capabilities to the required operational theater as quickly as possible and requirements to have combat-ready units arrive at ports of debarkation for employment by JFCs.

Whether conducting operations at a port during competition, crisis, or armed conflict, Army forces coordinate with the relevant authorities to mitigate potential complications at the port. This includes—

- Coordinating for products that provide a general layout of the port and flow of port operations.
- Obtaining an understanding of transport ship loading.
- Understanding port authority structure and decision making.
- Understanding reliance of the port on local infrastructure to conduct operations and identifying potential redundancies (for example, if power is lost can port gantry cranes load containers).
- Planning to train Soldiers on port equipment, such as material handling equipment.

(Operations) VI. Contested Deployments 1-141

While a threat's use of lethal capabilities to target the U.S. homeland or deploying forces is unlikely before a crisis or the commencement of armed conflict, it may choose to use other methods to surveil and disrupt Army forces during competition. Russia, China, and other threats possess wide-ranging capabilities to conduct cyberspace attacks, disrupt space capabilities, and conduct information warfare to influence the perceptions and behavior of target audiences. Attribution for these malign activities is challenging. Adversaries take steps to deliberately obscure the source of these activities, and they take full advantage of the ambiguity provided by operating below the threshold of armed response. Cyberspace attacks can be used to compromise government, private sector, and military capabilities, potentially targeting military dependencies on civilian infrastructure. Disruption of space capabilities can be used to hamper communication and navigation capabilities of both military forces and the civilian infrastructure they rely upon. Information warfare, including dissemination of or support for disinformation and misinformation, can be used to attempt to fracture bonds among elements of society. This may include seeking to create or exacerbate divisions between the military, government, private sector, and the public, both nationally and in local municipalities.

Peer threats use a variety of means to understand and predict U.S. and allied force projection, including open source and cyberspace collection. Army leaders should understand threat collection capabilities to reduce the chance of effective detection. Different ways threat forces are able to collect on a division's deployment are described in this notional scenario.

Threat forces will conduct information warfare operations to slow or otherwise degrade force projection. These campaigns can vary in scope and size, and may target local communities, Service members, Department of Defense (DOD) Civilians, contractors, and Soldiers' family members. This includes, but is not limited to—

- Targeted threats through social media, email, or other means designed to frighten and distract deploying Soldiers and their families.
- Cyberspace attacks against Soldier and family member banks and credit agencies, cutting off or disrupting access to personal funds.
- Cyberspace attacks against civilian infrastructure (including transportation, supply, fuel, and navigation) used to support military operations.
- Targeted strikes against defense communications infrastructure to disrupt communications between units, installations, and other unified action partners that assist in deployment.
- Disinformation dissemination and misinformation support designed to—
  - Undermine the legitimacy of, or otherwise reduce support for, U.S. Government action.
  - Incite civil unrest in local communities and along rail and road lines of communications that deploying forces need or plan to use to reach ports of embarkation.
  - Reduce trust in future official communications, from government, law enforcement, or military officials, by releasing disinformation that appears genuine but contains incorrect or confusing information.

Threat information warfare operations can be conducted at a very low cost compared to conventional warfare methods, and they have global reach. At low levels, they can be used to hamper military deployments. With sufficient scale or precision, they have the potential to completely halt effective unit deployment operations. Targeted disinformation and threats delivered via social media to the family members of every Soldier in a unit could potentially be devastating without prior planning, preparation, and trust building.

Commanders and their staffs must understand the potential effects of adversarial disinformation operations on units and leaders. Targeted adversary or enemy activi-

1-142 (Operations) VI. Contested Deployments

III. RSOI During Contested Deployments

Ref: FM 3-0, Operations (Oct. '22), pp. C-6 to C-7.

Historically, even during armed conflict, Army forces enjoyed high degrees of sanctuary in rear areas to receive and organize forces before moving them forward. However, the long-range strike capabilities of peer adversaries mean that sanctuary to conduct unimpeded RSOI operations in rear areas can no longer be assumed. It is likely that strikes by peer threats will degrade or destroy port and other transportation infrastructure vital to U.S force projection. This could cause Army forces to arrive in a disaggregated manner and disrupt RSOI operations. While integrating U.S. forces in theater could be challenging, Army planners must also consider host-nation requirements for logistics infrastructure. The host nation's response to an attack on its infrastructure, including its military mobilization, can affect freedom of movement for U.S. forces. All of these challenges may require JFCs to alter their operational plans or stay in a defensive posture for an extended period until sufficient combat power is built to enable offensive operations.

The theater army has primary responsibility for conducting RSOI for the entire joint land force. Army equipment may arrive in a piecemeal fashion across numerous ports. Commanders must establish secure communications across the distributed footprint, which allows staff coordination for unit personnel to meet their equipment and facilitate ship offloading. Units provide port support teams with the right personnel and capabilities to expedite port operations, such as licensed vehicle operators and communications.

If ports are unavailable, are severely degraded, or do not have the draft required for deep draft strategic sealift vessels, the JFC may consider joint logistics over-the-shore (JLOTS) operations. JLOTS provides the JFC with a limited capability to discharge strategic sealift ships. JLOTS requires a significant amount of lead time to build the needed conditions for successful operations, and it entails significant risk that strategic leaders will need to balance against its potential benefits.

After reception is complete, staging occurs. Staging during force projection contested by a peer threat requires understanding of threat standoff and friendly protection capabilities. Units must disperse and seek concealment to improve survivability. Operation plans (OPLANs) should take into account the need for expanded assembly areas to increase distances between vehicles and between unit assembly areas. As deploying units assemble, efforts focus on preparing for future operations and integrating into the joint force.

Onward movement during contested deployments requires units to execute movement and sustainment along multiple, dispersed routes. Road, rail, and other lines of communication must be assessed and classified for use by arriving forces. Units can avoid aggregation, enhance survivability, and achieve balance by directing personnel, equipment, materiel, and information flow at a rate that can be accommodated at every point along the entire network, from origin to destination.

To facilitate rapid onward movement and overcome the likely degradation of Global Positioning System (known as GPS) and other enabling transportation technology, units conduct convoy briefs, have paper maps, and conduct detailed route planning. This helps prepare Soldiers for the unique considerations of the host-nation transportation infrastructure, and it mitigates vulnerability to threat attacks. Theater-level personnel may augment or conduct separate briefings to share information about current conditions and threat tactics.

RSOI concludes with integration. Effective and efficient integration operations can reduce force vulnerability by ensuring units quickly assess vulnerabilities and counter potential threats to forces, infrastructure, and information systems as they transfer capabilities to an operational commander's force. When forces are fully integrated, operational control (OPCON) is transferred to the gaining unit.

See p. 4-10 for additional information on RSOI operations.

ties in the information dimension could rapidly degrade the performance of Soldiers, impacting their readiness. It could also degrade civilian performance and affect the critical infrastructure they manage. Leaders combat this through public communications both prior to and during deployment operations, coordination with relevant public affairs personnel, and Service member and family preparation. This preparation can include incorporating response strategies for disinformation dissemination into exercises and other training.

As U.S. forces transition from competition to crisis or armed conflict, threat actors will increase the intensity and lethality of their tactics. This could include infrastructure sabotage by pre-positioned agents, cyber or information attacks broader in scope (such as targeting an oil pipeline supplying a large region rather than only a specific port), or long-range precision strikes using a variety of munitions. Concurrently, they posture for, and may eventually escalate through, nonlethal and lethal actions of increasing intensity to improve stand-off and prevent power projection from the U.S. homeland and other basing and staging areas. Threat actors may also strike transport vessels along sea lines of communication while these vessels are en route to a seaport of debarkation.

Peer threats may choose to support proxy forces or influence unwitting groups, including irregular forces, saboteurs, sympathetic civil organizations, and criminals. These groups may be used to prevent timely deployment operations by denying access to roads or facilities with crowds, protests, or looting. Use of these forces may also allow for direct action against U.S. targets while masking culpability. Threat actors may design these activities to affect the economy and global trade in addition to the political-military balance in the United States or overseas. Additionally, other state and non-state actors may exploit the situation with attacks in pursuit of their own objectives. These attacks may be conducted within the United States or allied nations, in the theater into which Army forces are preparing to deploy, or in other, unrelated regions.

Leaders anticipate adversary activities in all domains while preparing for or conducting deployment operations. Disruptions may not be preventable. They can, however, be mitigated through training, preparation, and coordination with unified action partners. Effective mitigation in planning, preparation, and execution ensures the Army provides the required forces to combatant commanders (CCDRs) and other joint force commanders (JFCs).

Refer to AR 525-93 for Army deployment policies and responsibilities.

Homeland Defense and Defense Support Of Civil Authorities (DSCA)

While Army forces are preparing to deploy during crisis response or armed conflict against a peer threat, other units may be tasked to support homeland defense or DSCA. The circumstances that lead to national authorities directing the deployment of Army forces may also necessitate operations to simultaneously defend the U.S. homeland or support civil authorities. Additionally, the act of deploying forces may incite enemy attacks on the homeland, causing some forces to be mobilized for homeland defense or DSCA while others deploy forward to the operational theater.

Homeland defense and DSCA operations are conducted in a complex operational environment that contains layers of different jurisdictions (federal, state, territorial, tribal, and local), many agencies and organizations, the private sector, and several allies and multinational partners. Interorganizational coordination and synchronization with governmental and nongovernmental entities may assume a level of importance not matched in most overseas theaters of operations.

Chap 2 — I. Combat Power (Generating & Applying)

Ref: FM 3-0, Operations (Oct. '22), chap. 2.

Combat Power

Combat power is the total means of destructive and disruptive force that a military unit/formation can apply against an enemy at a given time (JP 3-0). It is the ability to fight. The complementary and reinforcing effects that result from synchronized operations yield a powerful blow that overwhelms enemy forces and creates friendly momentum.

Army forces integrate capabilities and synchronize **warfighting functions** to generate **combat power** and apply it against enemy forces. Successful application of combat power requires leaders to understand the enemy and understand friendly capabilities. A broad understanding of the strategic environment *(pp. 1-23 to 1-36)* and threat methods *(pp. 1-24 to 1-27)* provides a basis for understanding specific enemy situations. Leaders must understand how Army forces enable joint operations through multiple domains and the basic roles of Army echelons. They must also understand how the joint force enables the Army to integrate capabilities through all domains to generate more effective landpower.

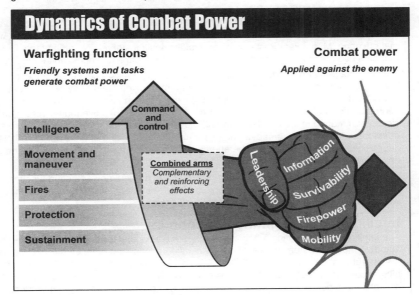

I. Warfighting Functions

A warfighting function is a group of tasks and systems united by a common purpose that commanders use to accomplish missions and training objectives (ADP 3-0). The warfighting functions are command and control (C2), movement and maneuver, intelligence, fires, sustainment, and protection. The purpose of warfighting functions is to provide an intellectual organization for common critical capabilities available to commanders and staffs at all echelons and levels of war. *See following pages (pp. 2-2 to 2-3).*

Warfighting Functions (Overview)

Ref: FM 3-0, Operations (Oct. '22), pp. 2-1 to 2-3.

Command and Control Warfighting Function *(See chap. 3.)*

The command and control warfighting function is the related tasks and a system that enable commanders to synchronize and converge all elements of combat power (ADP 3-0). The primary purpose of the C2 warfighting function is to assist commanders in integrating the other warfighting functions (movement and maneuver, intelligence, fires, sustainment, and protection) effectively at each echelon, and to apply combat power to achieve objectives and accomplish missions.

The C2 system includes people, processes, networks, and command posts. All elements of the system are critical in supporting effective decision making and the tempo required to defeat enemy forces. C2 supports the creation and exploitation of information advantages through the activities of developing situational understanding, decision making, and operating networks.

C2 synchronizes the systems and capabilities that comprise the other warfighting functions. Strategy, operational art, planning, operational approaches, operational frameworks, risk assessment, and decision making are all part of C2. C2 reflects leader action and how Army forces achieve unity of effort and unity of purpose during operations.

Movement and Maneuver Warfighting Function *(See chap. 4.)*

The movement and maneuver warfighting function is the related tasks and systems that move and employ forces to achieve a position of relative advantage over the enemy and other threats (ADP 3-0). Direct fire and close combat are inherent in maneuver. The movement and maneuver warfighting function includes tasks associated with force projection. Movement is necessary to position and disperse the force as a whole or in part when maneuvering. Maneuver directly gains or exploits positions of relative advantage. Commanders use maneuver for massing effects to achieve surprise, shock, and momentum.

Effective maneuver requires some combination of reconnaissance, surveillance, and security operations to provide early warning and protect the main body of the formation. Every Soldier on the battlefield is a potential sensor that makes key contributions to information collection and the development of intelligence. Effective maneuver requires close coordination of fires and movement. Movement and maneuver contribute to the development of information advantages through the positioning of units able to employ capabilities in close proximity to the enemy, as well as by physically establishing the facts on the ground that an enemy or adversary cannot refute.

Maneuver requires sustainment. The movement and maneuver warfighting function does not include routine transportation of personnel and materiel that support operations, which falls under the sustainment warfighting function.

Intelligence Warfighting Function *(See chap. 5.)*

The intelligence warfighting function is the related tasks and systems that facilitate understanding the enemy, terrain, weather, civil considerations, and other significant aspects of the operational environment (ADP 3-0). Intelligence involves analyzing information from all sources, which includes the other warfighting functions, and conducting operations to collect information. The integration of intelligence into operations facilitates understanding of an operational environment and assists in determining when and where to employ capabilities against adversaries and enemies. Intelligence likewise facilitates responses by Army forces to other situations, such as public health crises and events precipitating noncombatant evacuation. The intelligence warfighting function provides support to force generation, situational understanding, targeting and information operations, and information collection. The intelligence warfighting function fuses the information collected through reconnaissance, surveillance, security operations, and intelligence operations.

Commanders drive intelligence and intelligence drives operations. Army forces execute intelligence, surveillance, and reconnaissance (ISR) through the operations and intelligence processes, with an emphasis on intelligence analysis and information collection.

Timely, accurate, relevant, and predictive intelligence enables decision making, tempo, and agility during operations. Due to the fog and friction of warfare, commanders must fight for intelligence and share it with adjacent units and across echelons.

Fires Warfighting Function *(See chap. 6.)*

The fires warfighting function is the related tasks and systems that create and converge effects in all domains against the adversary or enemy to enable operations across the range of military operations (ADP 3-0). These tasks and systems create lethal and non-lethal effects delivered from both Army and joint forces and other unified action partners. The fires warfighting function does not entirely encompass, nor is it wholly encompassed by, any particular branch or function. Many of the capabilities that contribute to fires also contribute to other warfighting functions, often simultaneously. For example, an aviation unit may simultaneously execute missions that contribute to the movement and maneuver, fires, intelligence, sustainment, protection, and C2 warfighting functions. Space and cyberspace capabilities can provide commanders with options to defeat, destroy, disrupt, deny, or manipulate enemy networks, information, and decision making.

Sustainment Warfighting Function *(See chap. 7.)*

The sustainment warfighting function is the related tasks and system that provide support and services to ensure freedom of action, extended operational reach, and prolong endurance (ADP 3-0). Sustainment employs capabilities from all domains and enables operations through each domain. Sustainment determines the limits of depth and endurance during operations. Sustainment demands joint and strategic integration, and it should be meticulously coordinated across echelons to ensure continuity of operations and that resources reach the point of employment.

Sustainment employs an integrated network of information systems linking sustainment to operations. As a result, commanders at all levels see an operational environment, anticipate requirements in time and space, understand what is needed, track and deliver what is requested, and make timely decisions to ensure responsive sustainment. Because the situation is always changing, sustainment requires leaders capable of improvisation. Because sustainment operations are often vulnerable to enemy attacks, sustainment survivability depends on active and passive measures and maneuver forces for protection.

Protection Warfighting Function *(See chap. 8.)*

The protection warfighting function is the related tasks, systems, and methods that prevent or mitigate detection, threat effects, and hazards to preserve combat power and enable freedom of action. Protection encompasses everything that makes Army forces hard to detect and destroy. Protection requires commanders and staffs to understand threats and hazards throughout the operational environment, prioritize their requirements, and commit capabilities and resources according to their priorities. Commanders balance their protection efforts with the need for tempo and resourcing the main effort. They may assume risk in operations or areas that may be vulnerable, but that are considered low enemy priorities for targeting or attack. Commanders account for threats from space, cyberspace, and outside their assigned area of operations (AO) as they develop protection measures. Protection results from many factors, including operations security, dispersion, deception, survivability measures, and the way forces conduct operations. Planning, preparing, executing, and assessing protection is a continuous and enduring activity. Defending networks, data, and systems; implementing operations security; and conducting security operations contribute to information advantages by protecting friendly information. Prioritization of protection capabilities is situationally dependent and resource informed.

(Combat Power) I. Overview 2-3

II. Dynamics of Combat Power

Combat power is the total means of destructive and disruptive force that a military unit/formation can apply against an enemy at a given time (JP 3-0). It is the ability to fight. The complementary and reinforcing effects that result from synchronized operations yield a powerful blow that overwhelms enemy forces and creates friendly momentum. **Army forces deliver that blow through a combination of five dynamics.** The dynamics of combat power are—

Dynamics of Combat Power

- **A** Leadership
- **B** Firepower
- **C** Information
- **D** Mobility
- **E** Survivability

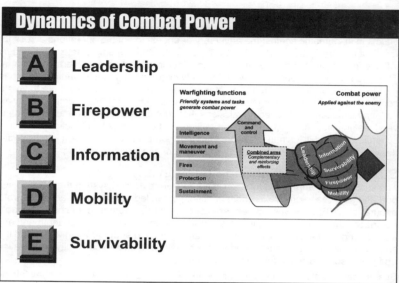

All warfighting functions contribute to generating and applying combat power. Well sustained units able to move and maneuver bring combat power to bear against the opponent. Joint and Army indirect fires complement and reinforce organic firepower in maneuver units. Survivability is a function of protection tasks, the protection inherent to Army platforms, and schemes of maneuver that focus friendly strengths against enemy weaknesses. Intelligence determines how and where to best apply combat power against enemy weaknesses. C2 enables leadership, the most important qualitative aspect of combat power.

A. Leadership *(See pp. 2-7 to 2-12.)*

Leadership is the most essential dynamic of combat power. Leadership is the activity of influencing people by providing purpose, direction, and motivation to accomplish the mission and improve the organization (ADP 6-22). It is the multiplying and unifying dynamic of combat power, and it represents the qualitative difference between units. Leadership drives C2 but is also dependent upon it. The collaboration and shared understanding inherent in the operations process prepare leaders for operations, expand shared understanding, hone leader judgment, and improve the flexibility that leaders apply to the other dynamics of combat power against enemy forces.

Commanders communicate their will to their formations through leadership. Sound leadership manifests as an unrelenting will to accomplish the mission, the ability to understand and adapt to changing conditions, and the motivation to persevere through hardship. Leadership inspires individuals to push past their perceived breaking point, and to fight for their unit and fellow Soldiers under the most difficult circumstances. It provides the intangible qualitative difference in how much combat power a formation can generate against enemy forces.

B. Firepower (See chap. 6)

Firepower is the primary source of lethality, and it is essential to defeating an enemy force's ability and will to fight. Leaders generate firepower through direct and indirect fires, using mass, precision, or, typically, a combination of the two. Intelligence enables the identification and selection of targets and objectives for the application of lethal force. Movement and maneuver enable the positioning of fires capabilities where they can be most lethal.

Firepower facilitates maneuver by suppressing enemy fires and disrupting or preventing the movement of enemy forces. Firepower exploits maneuver by neutralizing enemy forces when they react, destroying equipment and people, and degrading the will of enemy forces to fight.

Leaders increase firepower by using capabilities from all domains in combinations that overwhelm an enemy targets, while they rely on conventional unguided munitions against enemy units and area targets. Large-scale combat requires large reserves of both precision and unguided munitions and the sustainment capacity to move them to forward locations. Air, maritime, space, and cyberspace-based fires enhance the firepower of ground forces. Similarly, ground-based firepower complements firepower from other domains. A multidomain approach to firepower requires understanding the techniques for controlling and integrating joint fires. This includes requesting and integrating space and cyberspace capabilities, electromagnetic attack capabilities, and air capabilities.

C. Information (See pp. 2-13 to 2-18.)

Information contributes to the disruption and destruction of enemy forces. It is central to the application and amplification of combat power. It enables decision making and influences enemy perceptions, decision making, and behavior. Information, like leadership, provides a qualitative advantage to friendly combat power when it can be acted upon more quickly and effectively than the enemy.

Army forces collect data and information for analysis and process it to understand situations, make decisions, and direct actions that apply combat power against enemy forces. Army forces must fight for information about enemy forces while protecting their own information. Friendly counterintelligence, counterreconnaissance, and security operations prevent enemy access to friendly information. Offensively, commanders fight for information about enemy forces and terrain through continuous reconnaissance and surveillance and offensive tasks such as movement to contact or reconnaissance in force.

Army forces also use information to enhance the effects of destructive or disruptive physical force to create psychological effects that disrupt morale, cause human error, and increase uncertainty. Using information to manipulate shock and confusion amplifies the psychological effects of lethality and other dynamics of combat power.

Employing information to confuse, manipulate, or deceive can induce threats to act in ways that make them more vulnerable to destruction by Army forces. Employing information creatively can enable Army forces to achieve surprise, cause enemy forces to misallocate or expend combat power, or mislead them as to the strength, readiness, locations, and intended missions of friendly forces.

D. Mobility (See pp. 4-25 to 4-30.)

Mobility is a quality or capability of military forces which permits them to move from place to place while retaining the ability to fulfill their primary mission (JP 3-36). Mobility encompasses the capability of a formation to move and apply capabilities in specific terrain under specific conditions relative to enemy forces. Exploiting mobility requires intelligence of an enemy force's disposition, composition, strength, and course of action. This understanding allows leaders to assess their mobility in rela-

(Combat Power) I. Overview 2-5

tion to adversary or enemy forces. Maneuver and fires increase relative mobility by fixing enemy units, reducing obstacles, and providing obscuration.

The environment impacts mobility and the level of combat power a unit can produce. For example, an armored brigade combat team's (BCT's) mobility is limited in dense jungle or urban terrain, but it increases in steppes, in deserts, and on modern roads. Weather affects mobility when it degrades route conditions, or when it increases risks to fixed-and rotary-wing aviation operations. Space-based environmental monitoring provides real-time understanding of the impacts of weather on terrain and mobility. Enemy forces also influence conditions that affect mobility. For example, enemy standoff approaches can isolate land forces operating on islands in maritime environments by destroying maritime transportation capabilities and denying friendly air support.

Mobility is a function of how quickly units can move in specific terrain under specific conditions. At the tactical level, Army forces exploit mobility to conduct information collection, posture forces in advantageous locations, position fires to range enemy forces, and move classes of supply around an AO. During offensive operations, mobility enables forces to concentrate and then disperse rapidly, achieve surprise, attack enemy forces in unexpected locations, exploit opportunity, and evade enemy fires. During defensive operations, mobility enables counterattacks and the ability to rapidly shift resources between fixed positions. The ability to conduct gap crossings and passage of lines are other operations that can facilitate mobility.

E. Survivability *(See pp. 8-6 to 8-7.)*

Survivability is a quality or capability of military forces which permits them to avoid or withstand hostile actions or environmental conditions while retaining the ability to fulfill their primary mission (ATP 3-37.34). It represents the degree to which a formation is hard to kill. Survivability is relative to a unit's capabilities and the type of enemy effects it must withstand, its ability to avoid detection, and how well it can deceive enemy forces. Survivability is also a function of how a formation conducts itself during operations. For example, an infantry BCT's survivability against indirect fire is contingent on it not being detected, being dispersed, digging in, and adding overhead cover when stationary. An armor BCT's survivability is a function of logistics, security, and avoiding situations that constrain its mobility or freedom of action.

Leaders assess survivability as the ability of a friendly force to withstand enemy effects while remaining mission capable. Armor protection, mobility, tactical skill, avoiding predictability, and situational awareness contribute to survivability. Enforcement of operations security techniques and avoiding detection while initiating direct fire contact on favorable terms also increases survivability. Situational awareness regarding the nine forms of contact and minimizing friendly signatures contributes to survivability.

To increase survivability, units employ air defense systems, reconnaissance and security operations, modify tempo, take evasive action, maneuver to gain positional advantages, decrease electromagnetic signatures, and disperse forces. Dispersed formations improve survivability by complicating targeting and making it more difficult for enemy forces to identify lucrative targets. Tactical units integrate procedures for the use of camouflage, cover, concealment, and conducting electromagnetic protection—including noise and light discipline. During large-scale combat operations, survivability measures may include radio silence, communication through couriers, or alternate forms of communication. Space-based missile warning systems provide early warning of adversary artillery and missile attacks, allowing friendly forces to seek cover. Application of chemical, biological, radiological, and nuclear (CBRN) defense measures increase survivability in CBRN environments.

See ATP 3-37.34 for more information on survivability.

Chap 2

II. Leadership (as a Dynamic of Combat Power)

Ref: FM 3-0, Operations (Oct. '22), chap. 8.

Leadership is the most essential dynamic of combat power. Leadership is the activity of influencing people by providing purpose, direction, and motivation to accomplish the mission and improve the organization (ADP 6-22).

I. The Art of Command and the Commander

The commander's chief responsibilities during operations are to develop effective tactics and to lead and direct the unit in executing them. Soldiers want and willingly follow leaders who win. The most important quality for commanders or other leaders, however, is competence. Good leaders understand tactics, the abilities of their own and enemy units, what is possible in any situation. Combined arms competence is the combat commander's greatest contribution to winning. Effective commanders lead and inspire their subordinates to fight and win based upon that competence, while ensuring unity of purpose and unity of effort. They do this through a combination of processes, the staffs and subordinate leaders they empower, and their personal example and influence. While the commander is the primary leader of a formation, all leaders play an important role in carrying out their higher echelon commander's intent as they provide command and control (C2) to their own subordinates during operations.

Commanders and their subordinate leaders determine success or failure through the decisions they make, the examples they set, the actions they inspire, and their will to win. The commander is the focal point for the employment of staff processes that inform the judgment critical for rapid, effective decision making. Commanders must possess the confidence and skill to rapidly make difficult decisions, accepting risk to create and exploit fleeting opportunities. They conceive and implement sound operational and tactical solutions with the speed and force necessary to win. They also seek staff officers and subordinates who are proficient at their specialties and use them to make the unit tactically effective.

Commanders have a moral responsibility to ensure their units are prepared for combat. They prepare them through realistic training that reinforces individual and small-unit task and battle drill proficiency, collective proficiency in the execution of mission essential tasks under the most demanding conditions, and the development of staffs able to integrate the warfighting functions and joint capabilities in combined arms approaches against enemy forces. Commanders resource training and protect subordinates' training time. They establish the command climate within their unit, direct staff and subordinates during operations, and continually assess all aspects of their performance. They build trust, develop their subordinates' competence and confidence, encourage subordinate commanders to think critically, and demand that they demonstrate initiative within the commander's intent. Lastly, they prepare subordi-

Refer to TLS6: The Leader's SMARTbook, 6th Ed. (Military Leadership & Training in a Complex World) for complete discussion of Military Leadership (ADP 6-22); Leader Development (FM 6-22); Counsel, Coach, Mentor (ATP 6-22.1); Army Team Building (ATP 6-22.6); Military Training (ADP 7-0); Train to Win in a Complex World (FM 7-0); Unit Training Plans, Meetings, Schedules, and Briefs; Conducting Training Events and Exercises; Training Assessments, & After Action Reviews (AARs).

nates to assume responsibility one and two echelons above their current assignment so that they can effectively assume control of the larger formation when required to because of casualties or lost communications.

Commanders make decisions informed by judgment grounded in their experience, expertise, intuition, and self-awareness. Judgment is the single most important leadership attribute applicable to selecting the critical time and place to act, assigning missions, prioritizing, managing risk, and allocating resources. Thorough knowledge of military science, a strong ethical sense, and an understanding of enemy and friendly capabilities across domains form the basis of the judgment required of commanders. Judgment becomes more refined as commanders become more experienced, and good judgment becomes even more essential in ambiguous or uncertain operational environments. Self-awareness and experience help commanders become more expert about assessing situations and receiving assistance from their staffs. Self-awareness is required to assess one's self and to earn the trust of those one commands. Judgment allows commanders to distinguish between risk acceptance essential to successful operations and potentially disastrous rashness. The commander's judgment and experience help shape information requirements and staff priorities during operations. Subordinate commanders and staffs collaborate to provide the most relevant information in the most effective format for the commander to make sound decisions.

II. Applying the Art of Command

Command is more art than science because it requires commanders exercise their judgment, leverage their experience, and use their intuition when leading their units. Commanders apply the art of command by providing leadership, delegating authority, allocating resources, and making decisions. Subordinates operating within the commander's intent facilitate unity of effort to create and exploit relative advantages. Higher echelon leaders provide subordinate echelons access to information and the authority to use Army and joint multidomain capabilities through delegation. Leaders delegate appropriate authority to subordinates based on an assessment of their competence, talents, and experience.

A. Commander's Intent *(See p. 3-7.)*

The commander's intent is a clear and concise expression of the purpose of an operation and the desired objectives and military end state (JP 3-0). The commander's intent succinctly describes what constitutes success for the operation. It facilitates unity of effort as it allows subordinates to understand what is expected of them, what constraints apply, and most importantly, why the mission is being conducted. Understanding why a mission is conducted is important both in maintaining unity of effort and in bolstering the morale and will of subordinates. Soldiers who understand why they are called upon for a specific mission, and how that mission aligns with the higher echelon commander's intent and concept of operations, are more committed to the success of that specific mission.

Commanders communicate their intent two echelons down to ensure subordinates understand the boundaries within which they may exercise initiative while maintaining unity of effort. They likewise understand the commander's intent two echelons above them. A clear and concise commander's intent should be easily remembered and understood even without an order. Commanders collaborate with subordinates to ensure their commander's intent is understood. Soldiers who understand the commander's intent are more able to exercise disciplined initiative in unexpected situations than those who do not.

B. Initiative

Leaders at every echelon exercise initiative as they follow orders and adhere to a course of action or scheme of maneuver. When the enemy does something

2-8 (Combat Power) II. Leadership in Operations

Commander Presence on the Battlefield

Ref: FM 3-0, Operations (Oct. '22), pp. 8-2 to 8-4.

Commanders provide leadership in combat to inspire their Soldiers, especially in challenging situations. Command presence is the influence commanders have on those around them, through their physical presence, communications, demeanor, and personal example. Commanders establish command presence through personal interaction with subordinates, either physically or virtually through C2 systems, and demonstrating their character, competence, dignity, strength of conviction, and empathy prior to and throughout operations.

Commanders go where they can best influence operations, assess units, and improve unity of effort. Where commanders place themselves on the battlefield is one of the most important decisions they can make. Commanding forward allows commanders to effectively assess and manage the effects of operations on their formations through face-to-face interactions. It allows them to gather information about actual combat conditions, but it must be balanced against the requirement to be where the best overall situational awareness can be maintained for the entire formation. As far as operational conditions allow, leadership should be exercised up front at critical times and places without interfering with subordinate leader prerogatives, becoming unreachable by other elements of the unit, or making it simpler for the enemy to target multiple echelons of leadership in one place.

At the battalion level and below, commanders lead by personal example, acquire much information themselves, and communicate face to face with those they direct. Typically, they position themselves well forward to influence the main effort during different phases of an operation. However, even at these levels, commanders cannot provide direct leadership for their whole unit given the challenges of maintaining continuous communications when units are dispersed or in contested electromagnetic environments.

At higher levels, echeloned command posts are central to effective C2. During operations, commanders must assess the situation up front as often as possible without being disruptive to the focus of subordinate commanders. They deliberately plan and organize their C2 approach to mitigate their loss of broad situational understanding during battlefield circulation by the development of subordinate commanders, and staff officers empowered to make decisions on the commander's behalf to exploit opportunities and respond to changing circumstances without needing to ask permission.

Commanders convey importance and focus the efforts of the command by how they communicate, regardless of where they are physically located. A calm and authoritative tone of voice generates a sense of presence, as does a crisp and efficient manner of providing guidance; both require practice to master. No matter their location, effective commanders encourage their troops, sense their morale, and inspire through personal example.

Imperative: Understand and Manage the Effects of Operations on Units and Soldiers

Continuous operations rapidly degrade the performance of people and the equipment they employ, particularly during combat. In battle, Soldiers and units are more likely to fail catastrophically than gradually. Commanders and staffs must be alert to small indicators of fatigue, fear, indiscipline, and reduced morale, and they must take measures to deal with these before their cumulative effects drive a unit to the threshold of collapse. Staffs and commanders at higher echelons must take into account the impact of prolonged combat on subordinate units, which causes efficiency to drop, even when physical losses are not great.

Provided here as an example. See pp. 1-44 to 1-45 for a listing of all imperatives.

(Combat Power) II. Leadership in Operations 2-9

unexpected, or a new threat or opportunity emerges that offers a greater chance of success, subordinate leaders take action to adjust to the new situation and achieve their commander's intent. Disciplined initiative requires a bias towards action rather than waiting on new orders. The mission command approach to C2 demands subordinates exercise initiative in the absence of orders, when current orders no longer apply, or when an opportunity or new threat presents itself. The cumulative effect of multiple subordinate commanders, leaders, and individual Soldiers exercising initiative produces agility in Army formations. It enables Army forces to seize the operational initiative as they pose multiple dilemmas on the enemy.

Commanders and subordinate leaders develop subordinates capable of exercising effective initiative during both training events and operations. Leaders create opportunities during training events in which subordinates must take action on their own. Commanders must foster a climate that encourages initiative during training, before the unit is committed to combat. Accepting risk and underwriting good faith mistakes establishes a unit culture that allows subordinates to learn and gain the experience they need to operate under their own responsibility. Endurance requires that the burdens of combat leadership be shared between leaders at each echelon, which is not possible without a leadership culture that demands disciplined initiative.

C. Drive the Operations Process *(See pp. 3-9 to 3-22.)*

Commanders employ the operations process to incorporate coalition and joint partners, empower subordinate initiative, and to ensure authorities and risk acceptance are delegated to the appropriate echelon required for the situation. Staffs and subordinate headquarters earn the commander's trust by providing relevant information, anticipating needs, and directing supporting actions.

The major components of the operations process are planning, preparing, executing, and continuously assessing the operation. Planning normally begins upon receipt of orders from a higher echelon headquarters and continues through the execution of the operation. The commander and staff continually assess operations and revise the plan through fragmentary orders. Commanders, assisted by their chiefs of staff or executive officers, drive the preparation for an operation by allocating time, prioritizing resources, and supervising preparation activities such as rehearsals to ensure their forces are ready to execute operations. During execution, commanders and staffs focus their efforts on translating plans into direct action to achieve objectives in accordance with higher commander's intent.

D. Discipline

Discipline underpins the effective application of initiative during operations. For many parts of an operation, the benefits of exercising disciplined initiative outweigh the cost in synchronization. At other times, when synchronization or a specific process is critical to successful execution, ill-advised initiative can be costly. Neither commanders nor subordinates are independent actors when exercising initiative. Subordinates consider at least two factors when deciding how to exercise initiative, assessing each in terms of the circumstances affecting it:

• Whether the benefits of the action outweigh the risk of desynchronizing the overall operation.

• Whether the action will further attainment of the desired end state.

Refer to ADP 6-0 for more information on disciplined initiative.

E. Accepting Risk to Create and Exploit Opportunities

Risk, uncertainty, and chance are inherent to all military operations. The pace and tempo of large-scale combat operations demands a high level of expertise for commanders and staffs. Commanders delegate risk acceptance to subordinates through their commander's intent and other guidance. Based on this guidance, refined as necessary by the commander and staff as the situation changes, leaders accept risk during operations by massing effects in one area while taking economy of force measures in other areas to create opportunities that can be exploited. Understanding how much or what kind of risk to accept requires the ability to accurately see friendly and enemy forces and an operational environment in the context of a particular situation. Leaders and staff officers who are competent in all aspects of their specialties assist the commander in weighing the risk of various decisions and courses of action, balancing short-term versus long-term risk to the mission and the force. This analysis, coupled with imagination and the courage to act boldly, provides opportunities that outweigh the risk incurred. Successful commanders are aware of the effects of cumulative risk over time, and therefore they continuously assess it throughout an operation.

Opportunities to exploit a relative advantage are fleeting during large-scale combat operations against a capable enemy that adapts quickly. Delaying action while setting optimal conditions, waiting for perfect intelligence, or achieving greater synchronization may end up posing a greater danger than swift acceptance of significant risk now. Leaders' judgment of risk must be thoroughly informed by their understanding of their operational environment, particularly how actions in other domains impact their forces and how their forces can generate effects outside the land domain. Leaders maintain a common operational picture with their counterparts in higher and lower echelons, since any risk accepted at one echelon may impose additional risk at other echelons.

Contingency planning enables commanders to swiftly decide and act when the unexpected occurs. By considering multiple enemy courses of action, available friendly capabilities, and developing appropriate branch and sequel plans, the staff and the commander develop a deeper appreciation for how an operation could unfold and build flexibility into their plans. Contingency planning also guards against confirmation bias, where situational cues are interpreted in ways that match preconceived ideas. Effective planning expands understanding of a situation and operational environment, allowing the commander and staff to be better postured to rapidly decide, adjust, or simply be proactive.

Commanders designate commander's critical information requirements (CCIRs) linked to decision points during operations and the execution of branch and sequel plans developed during contingency planning. CCIRs help subordinate leaders prioritize allocation of information collection assets and other limited resources, including time. Detailed contingency planning and the development of CCIRs ensure that commanders and staffs are proactive rather than reactive in responding to enemy activity. CCIRs potentially become irrelevant as the situation changes; therefore, commanders, staffs, and other leaders must regularly review and update CCIRs to ensure relevance throughout the conduct of operations. This is a forcing function for questioning earlier assumptions made during planning that may no longer be valid.

See ADP 2-0 and ADP 6-0 for more detailed descriptions of what constitutes CCIR.

Commanders and staffs remain alert to exceptional information during operations. Exceptional information results from an unexpected event or opportunity, or a new threat, which directly affects the success of operations. It would have been a CCIR if it had been foreseen. Identifying exceptional information requires initiative from subordinates, shared understanding of the situation, a thorough understanding of the commander's intent, and the exercise of judgement based on experience.

(Combat Power) II. Leadership in Operations 2-11

F. Command and Control During Degraded or Denied Communications

U.S. adversaries, including Russia and China, have demonstrated the ability to contest communications in the electromagnetic spectrum and degrade friendly C2. Degraded connectivity to a secure communications network poses risks to situational understanding, C2, and ultimately mission accomplishment. Army forces must be prepared to continue operations and achieve mission objectives when out of contact with higher echelons or adjacent units. They should assume intermittent rather than continuous connectivity during the course of operations. A mission command approach to C2 empowers subordinate leaders to act within the commander's intent with degraded communications, reporting their new situation when able to do so.

Exercising C2 with degraded connectivity does not begin when communications break down or when a crisis develops. It must be trained and rehearsed as part of an overall cultural norm designed to make friendly forces more difficult to detect. Commanders ensure that staffs are trained on analog and manual C2, and that units have rehearsed reliable primary, alternate, contingency, and emergency communications plans. During extended periods of communication breakdown, C2 becomes more difficult as shared understanding of a situation deteriorates over time. To maintain C2 with degraded communication, personnel should be trained and proficient in—

- Employing all available C2 systems.
- Operating dispersed command posts.
- Maintaining an other-than-digital common operational picture (COP).
- Managing information with analog processes.
- Monitoring communications channels and crosstalk across echelons.
- Maintaining manual running staff estimates.
- Conducting command post battle drills.

III. Adapting Formations for Missions and Transitions

The conduct of successful operations requires leaders and units that are able to anticipate changes and quickly adapt their formations, dispositions, or activities to meet those changes. Anticipating and adapting to changes begins with commanders, but all leaders help create agile and adaptive units, develop subordinates, inspire resilience in their people, and maintain mission focus in the face of adversity.

Changing conditions and transitions may impact the teamwork and cohesion of a formation; both require adaptation and leader attention. Some examples include—

- Task organization changes.
- New or changing guidance.
- Periods of intense privation and fatigue.
- Mission transitions.
- Mission failures or setbacks.
- Reconstitution.

Commanders are responsible for developing subordinate leaders and units that are capable of adapting to the environment and the dynamic nature of operations. Training is the primary opportunity to do so, and all leaders must work to be experts at training formations and developing subordinates during that training. Successful adaptation and leader development depends on a command climate and learning environment that encourages subordinates at all levels to become subject matter experts, think independently, and take the initiative. Using a mission command approach, leaders foster a sense of shared commitment and involvement in decision making.

2-12 (Combat Power) II. Leadership in Operations

III. Information
(as a Dynamic of Combat Power)

Chap 2

Combat Power

Ref: FM 3-13, Information Operations (Dec '16) and FM 3-0, Operations (Oct. '22), chap. 2.

I. Information

All military activities produce **information**. Informational aspects are the features and details of military activities observers interpret and use to assign meaning and gain understanding. Those aspects affect the perceptions and attitudes that drive behavior and decision making. The JFC leverages informational aspects of military activities to gain an advantage; failing to leverage those aspects may cede this advantage to others. Leveraging the informational aspects of military activities ultimately affects strategic outcomes.

Information contributes to the disruption and destruction of enemy forces. It is central to the application and amplification of combat power. It enables decision making and influences enemy perceptions, decision making, and behavior. Information, like leadership, provides a qualitative advantage to friendly combat power when it can be acted upon more quickly and effectively than the enemy.

Army forces collect data and information for analysis and process it to understand situations, make decisions, and direct actions that apply combat power against enemy forces. Army forces must fight for information about enemy forces while protecting their own information. Friendly counterintelligence, counterreconnaissance, and security operations prevent enemy access to friendly information. Offensively, commanders fight for information about enemy forces and terrain through continuous reconnaissance and surveillance and offensive tasks such as movement to contact or reconnaissance in force.

Army forces also use information to enhance the effects of destructive or disruptive physical force to create psychological effects that disrupt morale, cause human error, and increase uncertainty. Using information to manipulate shock and confusion amplifies the psychological effects of lethality and other dynamics of combat power.

Employing information to confuse, manipulate, or deceive can induce threats to act in ways that make them more vulnerable to destruction by Army forces. Employing information creatively can enable Army forces to achieve surprise, cause enemy forces to misallocate or expend combat power, or mislead them as to the strength, readiness, locations, and intended missions of friendly forces.

II. Information Operations (IO)

Information Operations (IO) is the integrated employment, during military operations, of **information-related capabilities** in concert with other lines of operation to influence, disrupt, corrupt, or usurp the decision-making of adversaries and potential adversaries while protecting our own (JP 3-13).

See p. 2-15 for an overview of the information-related capabilities (IRCs) and also on p. 6-14, as related to employment of fires across the domains.

The Purpose of Information Operations

The purpose of IO is to **create effects in and through the information environment that provide commanders decisive advantage over enemies and adversaries.** Commanders achieve this advantage in several ways: preserve and facilitate decision making and the impact of decision making, while influencing, disrupting or degrading enemy or adversary decision making; get required information faster and with greater accuracy and clarity than the enemy or adversary; or influence the attitudes and behaviors of relevant audiences in the area of operations having an impact on operations and decision making.

(Combat Power) III. Information 2-13

III. Information (as one of the Joint Functions)

Ref: JP 3-0, Joint Campaigns and Operations (Jun '22), pp. III-16 to III-27.

There are seven joint functions common to joint operations: C2, information, intelligence, fires, movement and maneuver, protection, and sustainment.

The elevation of information as a joint function impacts all operations and signals a fundamental appreciation for the military role of information at the strategic, operational, and tactical levels within today's complex operational environment.

The information function encompasses the management and application of information to support achievement of objectives; it is the deliberate integration with other joint functions to change or maintain perceptions, attitudes, and other elements that drive desired relevant actor behaviors; and to support human and automated decision making. The information function helps commanders and staffs understand and leverage the prevalent nature of information, its military uses, and its application during all military operations. This function provides JFCs the ability to preserve friendly information and leverage information and the inherent informational aspects of military activities to achieve the commander's objectives. The information joint function provides an intellectual framework to aid commanders in exerting one's influence through the timely generation, preservation, denial, or projection of information.

All military activities have an informational aspect since most military activities are observable in the IE. Informational aspects are the features and details of military activities observers interpret and use to assign meaning and gain understanding. Those aspects affect the perceptions and attitudes that drive behavior and decision making. The JFC leverages informational aspects of military activities to gain an advantage in the OE; failing to leverage those aspects in a timely manner may cede this advantage to an adversary or enemy. Leveraging the informational aspects of military activities can support achieving operational and strategic objectives. The information function also encompasses the use of friendly information to influence foreign audiences and affect the legitimacy, credibility, and influence of the USG, joint force, allies, and partners. Additionally, JFCs use friendly information to counter, discredit, and render irrelevant the disinformation, misinformation, and propaganda of other actors.

The information joint function helps commanders and their staffs understand and leverage the pervasive nature of information, its military uses, and its application across the competition continuum, to include its role in supporting human and automated decision making. Information planners should consider coordination activities not only within the information joint function but also among all other joint functions. The information joint function organizes the tasks required to manage and apply information during all activities and operations. The three tasks of the information joint function stress the requirement to incorporate information as a foundational element during the planning and conduct of all operations.

A. Joint Force Capabilities, Operations, and Activities for Leveraging Information

In addition to planning all operations to derive benefit from the inherent informational aspects of physical actions and influence relevant actors, the JFC also has additional means with which to leverage information in support of objectives. Leveraging information involves the generation and use of information through tasks to inform relevant actors; influence relevant actors; and/or attack information, information systems, and information networks. Planning for operations in the information environment (OIE) provides the means for the integrated employment of military information. The JFC uses various forces, operations, and activities to reinforce the actions of assigned or attached forces, support lines of operation (LOOs) or lines of effort (LOEs), or as the primary activity in an LOE to drive the behavior of selected target audiences or decision makers.

See p. 2-18 for examples of information use across the competition continuum.

2-14 (Combat Power) III. Information

B. Information-Related Capabilities (IRCs)

Ref: FM 3-13, Information Operations (Dec '16), p. 1-3.

> An **information-related capability (IRC)** is a tool, technique, or activity employed within a dimension of the information environment that can be used to create effects and operationally desirable conditions (JP 1-02). The formal definition of IRCs encourages commanders and staffs to employ all available resources when seeking to affect the information environment to operational advantage. For example, if artillery fires are employed to destroy communications infrastructure that enables enemy decision making, then artillery is an IRC in this instance. In daily practice, however, the term IRC tends to refer to those tools, techniques, or activities that are inherently information-based or primarily focused on affecting the information environment.

The information-related capabilities (IRCs) include—
- Military deception
- Military information support operations (MISO)
- Soldier and leader engagement (SLE), to include police engagement
- Civil affairs operations
- Combat camera
- Operations security (OPSEC)
- Public affairs
- Cyberspace electromagnetic activities (see following pages)
- Electronic warfare (see following pages)
- Cyberspace operations
- Space operations
- Special technical operations

All unit operations, activities, and actions affect the information environment. For this reason, whether or not they are routinely considered an IRC, a wide variety of unit functions and activities can be adapted for the purposes of conducting information operations or serve as enablers, to include:
- Commander's communications strategy or communication synchronization
- Presence, profile, and posture
- Foreign disclosure
- Physical security
- Physical maneuver
- Special access programs
- Civil military operations
- Intelligence
- Destruction and lethal actions

Refer to INFO1: The Information Operations & Capabilities SMARTbook (Guide to Information Operations & the IRCs). INFO1 chapters and topics include information operations (IO defined and described), information in joint operations (joint IO), information-related capabilities (PA, CA, MILDEC, MISO, OPSEC, CO, EW, Space, STO), information planning (information environment analysis, IPB, MDMP, JPP), information preparation, information execution (IO working group, IO weighted efforts and enabling activities, intel support), fires & targeting, and information assessment.

(Combat Power) III. Information 2-15

C. Cyberspace Operations (CO) and Electromagnetic Warfare (EW)

Ref: JP 3-12, Cyberspace Operations (Jun '18).

United States armed forces operate in an increasingly network-based world. The proliferation of information technologies is changing the way humans interact with each other and their environment, including interactions during military operations. This broad and rapidly changing operational environment requires that today's armed forces must operate in cyberspace and leverage an electromagnetic spectrum that is increasingly competitive, congested, and contested.

Cyberspace *(See pp. 1-19, 1-35 and 6-13.)*

Cyberspace reaches across geographic and geopolitical boundaries and is integrated with the operation of critical infrastructures, as well as the conduct of commerce, governance, and national defense activities. Access to the Internet and other areas of cyberspace provides users operational reach and the opportunity to compromise the integrity of critical infrastructures in direct and indirect ways without a physical presence. The prosperity and security of our nation are significantly enhanced by our use of cyberspace, yet these same developments have led to increased exposure of vulnerabilities and a critical dependence on cyberspace, for the US in general and the joint force in particular.

Cyberspace Operations (CO)

Cyberspace Operations (CO) are the employment of cyberspace capabilities where the primary purpose is to achieve objectives in or through cyberspace. CO comprise the military, national intelligence, and ordinary business operations of DOD in and through cyberspace. Although commanders need awareness of the potential impact of the other types of DOD CO on their operations, the military component of CO is the only one guided by joint doctrine and is the focus of JP 3-12. CCDRs and Services use CO to create effects in and through cyberspace in support of military objectives. Military operations in cyberspace are organized into missions executed through a combination of specific actions that contribute to achieving a commander's objective. Various DOD agencies and components conduct national intelligence, ordinary business, and other activities in cyberspace.

The Relationship of Cyberspace Operations to Operations in the Information Environment

Cyberspace is wholly contained within the information environment. CO and other information activities and capabilities create effects in the information environment in support of joint operations. Their relationship is both an interdependency and a hierarchy; cyberspace is a medium through which other information activities and capabilities may operate. These activities and capabilities include, but are not limited to, understanding information, leveraging information to affect friendly action, supporting human and automated decision making, and leveraging information (e.g., military information support operations [MISO] or military deception [MILDEC]) to change enemy behavior. CO can be conducted independently or synchronized, integrated, and deconflicted with other activities and operations.

While commanders may conduct CO specifically to support information-specific operations, some CO support other types of military objectives and are integrated through appropriate cells and working groups. The lack of synchronized CO with other military operations planning and execution can result in friendly force interference and may counter the simplicity, agility, and economy of force principles of joint operations.

2-16 (Combat Power) III. Information

Cyberspace Missions

All actions in cyberspace that are not cyberspace-enabled activities are taken as part of one of three cyberspace missions: OCO, DCO, or DODIN operations. These three mission types comprehensively cover the activities of the cyberspace forces. The successful execution of CO requires integration and synchronization of these missions. Military cyberspace missions and their included actions are normally authorized by a military order (e.g., execute order [EXORD], operation order [OPORD], tasking order, verbal order), referred to hereafter as mission order, and by authority derived from DOD policy memorandum, directive, or instruction. Cyberspace missions are categorized as OCO, DCO, or DODIN operations based only on the intent or objective of the issuing authority, not based on the cyberspace actions executed, the type of military authority used, the forces assigned to the mission, or the cyberspace capabilities used.

Cyberspace Missions

- **A** DODIN Operations
- **B** Offensive Cyberspace Operations (OCO)
- **C** Defensive Cyberspace Operations (DCO)

Most DOD cyberspace actions use cyberspace to enable other types of activities, which employ cyberspace capabilities to complete tasks but are not undertaken as part of one of the three CO missions: OCO, DCO, or DODIN operations.

Cyberspace Actions

- **A** Cyberspace Security
- **B** Cyberspace Defense
- **C** Cyberspace Exploitation
- **D** Cyberspace Attack

Refer to CYBER1: The Cyberspace Operations & Electronic Warfare SMARTbook (Multi-Domain Guide to Offensive/Defensive CEMA and CO). Topics and chapters include cyber intro (global threat, contemporary operating environment, information as a joint function), joint cyberspace operations (CO), cyberspace operations (OCO/DCO/DODIN), electronic warfare (EW) operations, cyber & EW (CEMA) planning, spectrum management operations (SMO/JEMSO), DoD information network (DODIN) operations, acronyms/abbreviations, and a cross-referenced glossary of cyber terms.

(Combat Power) III. Information 2-17

D. Information Use Across the Competition Continuum (Examples)

Ref: JP 3-0, Joint Campaigns and Operations (Jun '22), pp. III-24 to III-26.

Cooperative use of Information. During day-to-day activities, the joint force integrates information in SC and FHA activities by:

- Assuring and maintaining allies, widening/publicizing combined exercises and other PN cooperation activities, encouraging neutral actors that the joint force is the partner of choice or that they should remain neutral, and reminding partners of benefits to maintain their support.
- Informing enemies and adversaries of benefits to friendly multinational force membership and collective defense, informing enemies and adversaries that the joint force is committed to its allies and security agreements.

Competitive use of Information. During competition, the joint force conducts activities against state or non-state actors with incompatible interests that are below the level of armed conflict. Competition can include military operations such as CO, special operations, demonstrations of force, CTF, and ISR and often depends on the ability to leverage the power of information through OIE. Expect additional time to coordinate and obtain approval from DOD or other USG departments and agencies to use information due to increased risk. Specific information tasks may include:

- Informing allies and partners of malign influence and antagonistic behavior.
- Declassifying and sharing images that reveal or confirm enemy or adversarial behavior, recommending allies and partners communicate to relevant audiences within their areas of influence, and educating the joint force and allies about online disinformation activities to build understanding and resilience against propaganda.
- Influencing adversary's audiences to prevent escalation to armed conflict by demonstrating joint force resolve, strength, and commitment, as well as the costs and expectations of response actions.
- Targeting adversarial information, networks, and systems by temporarily denying communication or Internet access, disrupting jamming of Internet access to its internal population, and partnering with private-sector communication companies to remove inappropriate enemy and adversarial recruiting and fundraising advertisements.

Use of Information in Armed Conflict. In addition to the above tasks, the joint force can use information defensively or offensively. JFCs can employ information as independent activities, integrated with joint force physical actions, or in support of other instruments of national power. Many of these information activities require additional authorities as they present larger strategic risks or risks to the joint force, though capabilities like PA, which has the preponderance of public communication resources and rarely requires additional authorities in armed conflict.

- **Defensive purposes.** Basic defense activities include protecting data and communications, movements, and locations of critical capabilities and activities. PA can assist in countering adversary propaganda, misinformation, and disinformation. MILDEC can help mask strengths, magnify feints, and distract attention to false locations. DCO can defeat specific threats that attempt to bypass or breach cyberspace security measures. EW can protect personnel, facilities, and equipment from any effects of friendly, neutral, or enemy use of the EMS. The management of EM signatures can mask friendly movements and confuse enemy intelligence collectors. Finally, well-coordinated communication and messaging activities not only minimize OPSEC violations but also increase the consistency and alignment of joint force words, actions, and images. Conflicting messages or remaining silent allows adversaries and enemies to exploit or monopolize the media and propagate their agenda.
- **Offensive purposes.** Offensive information activities decrease enemy and adversary effectiveness, increase Ally and partner support and effectiveness, and reduce interference from neutral audiences.

2-18 (Combat Power) III. Information

Chap 3: Command & Control Warfighting Function

Ref: ADP 6-0, Mission Command (Jul '19) and ADP 3-0, Operations (Jul '19), p. 5-3.

I. Command & Control Warfighting Function

The command and control warfighting function is the related tasks and a system that enable commanders to synchronize and converge all elements of combat power (ADP 3-0). The primary purpose of the command and control warfighting function is to assist commanders in integrating the other elements of combat power to achieve objectives and accomplish missions. The command and control warfighting function consists of the command and control warfighting function tasks and the command and control system.

Command and Control Warfighting Function

The related tasks and a system that enables commanders to synchronize and converge all elements of combat power.

Tasks
- Command forces
- Control operations
- Drive the operations process
- Establish the command and control system

Command and Control System
- People
- Processes
- Networks
- Command posts

Ref: ADP 6-0 (Jul '19), Figure 1-2. Combat power model.

The command and control warfighting function tasks focus on integrating the activities of the other elements of combat power to accomplish missions. Commanders, assisted by their staffs, integrate numerous processes and activities within their headquarters and across the force through the mission command warfighting function:

- Command forces
- Control operations
- Drive the operations process *(See pp. 3-9 to 3-18.)*
- Establish the command and control system *(See p. 3-8.)*

Refer to BSS6: The Battle Staff SMARTbook, 6th Ed. for further discussion. BSS6 covers the operations process (ADP 5-0); commander's activities; Army planning methodologies; the military decisionmaking process and troop leading procedures (FM 7-0 w/Chg 2); integrating processes (IPB, information collection, targeting, risk management, and knowledge management); plans and orders; mission command, C2 warfighting function tasks, command posts, liaison (ADP 6-0); rehearsals & after action reviews; and operational terms and military symbols (ADP 1-02).

(Command & Control) Warfighting Function 3-1

Mission Command: Command and Control of Army Forces

Ref: ADP 6-0, Mission Command (May '12), preface and introduction.

ADP 6-0, Mission Command: Command and Control of Army Forces, provides a discussion of the fundamentals of mission command, command and control, and the command and control warfighting function. It describes how commanders, supported by their staffs, combine the art and science of command and control to understand situations, make decisions, direct actions, and lead forces toward mission accomplishment.

This revision to ADP 6-0 represents an evolution of mission command doctrine based upon lessons learned since 2012. The use of the term mission command to describe multiple things—the warfighting function, the system, and a philosophy—created unforeseen ambiguity. Mission command replaced command and control, but in practical application it often meant the same thing. This led to differing expectations among leadership cohorts regarding the appropriate application of mission command during operations and garrison activities. Labeling multiple things mission command unintentionally eroded the importance of mission command, which is critical to the command and control of Army forces across the range of military operations. Differentiating mission command from command and control provides clarity, allows leaders to focus on mission command in the context of the missions they execute, and aligns the Army with joint and multinational partners, all of whom use the term command and control.

Command and control—the exercise of authority and direction by a properly designated commander over assigned and attached forces—is fundamental to the art and science of warfare. No single specialized military function, either by itself or combined with others, has a purpose without it. Commanders are responsible for command and control. Through command and control, commanders provide purpose and direction to integrate all military activities towards a common goal—mission accomplishment. Military operations are inherently human endeavors, characterized by violence and continuous adaptation by all participants. Successful execution requires Army forces to make and implement effective decisions faster than enemy forces. Therefore, the Army has adopted mission command as its approach to command and control that empowers subordinate decision making and decentralized execution appropriate to the situation.

Mission command requires tactically and technically competent commanders, staffs, and subordinates operating in an environment of mutual trust and shared understanding. It requires building effective teams and a command climate in which commanders encourage subordinates to take risks and exercise disciplined initiative to seize opportunities and counter threats within the commander's intent. Through mission orders, commanders focus their subordinates on the purpose of an operation rather than on the details of how to perform assigned tasks. This allows subordinates the greatest possible freedom of action in the context of a particular situation. Finally, when delegating authority to subordinates, commanders set the necessary conditions for success by allocating resources to subordinates based on assigned tasks.

Commanders need support to exercise command and control effectively. At every echelon of command, commanders are supported by the command and control warfighting function—the related tasks and a system that enables commanders to synchronize and converge all elements of combat power. Commanders execute command and control through their staffs and subordinate leaders.

ADP 6-0 provides fundamental principles on mission command, command and control, and the command and control warfighting function. Key updates and changes to this version of ADP 6-0 include—

- Combined information from ADP 6-0 and ADRP 6-0 into a single document.
- Command and control reintroduced into Army doctrine.

3-2 (Command & Control) Warfighting Function

- An expanded discussion of command and control and its relationship to mission command.
- Revised mission command principles.
- Command and control system reintroduced, along with new tasks, and an updated system description.
- Expanded discussion of the command and control system.

Mission Command (Logic Map)

Nature of War

Military operations are inherently human endeavors representing a contest of wills, characterized by violence and continuous adaption by all participants, conducted in dynamic and uncertain operational environments to achieve a political purpose.

Operations must account for the nature of war. As such the Army's operational concept is...

Unified Land Operations

The simultaneous execution of offense, defense, stability, and defense support of civil authorities across multiple domains to shape operational environments, prevent conflict, prevail in large-scale ground combat, and consolidate gains as part of unified action.

The Army's operational concept is enabled by....

Mission Command

The Army's approach to command and control that empowers subordinate decision making and decentralized execution appropriate to the situation.

Enabled by the principles of...
Competence | Mutual trust | Shared understanding | Commander's intent
Mission orders | Disciplined initiative | Risk acceptance

Command and control is fundamental to all operations...

Command and Control

Command and control is the exercise of authority and direction by a properly designated commander over assigned and attached forces in the accomplishment of a mission.

Elements of Command	**Elements of Control**
• Authority	• Direction
• Responsibility	• Feedback
• Decision making	• Information
• Leadership	• Communication

Executed through...

Command and Control Warfighting Function

The related tasks and a system that enables commanders to synchronize and converge all elements of combat power.

Tasks

- Command forces
- Control operations
- Drive the operations process
- Establish the command and control system

Command and Control System

- People
- Processes
- Networks
- Command posts

Ref: ADP 6-0 (Jul '19), Introductory figure-1. Logic map.

(Command & Control) Warfighting Function 3-3

II. Mission Command

> *Never tell people how to do things. Tell them what to do and they will surprise you with their ingenuity.*
>
> *- General George S. Patton, Jr.*

Army operations doctrine emphasizes shattering an enemy force's ability and will to resist, and destroying the coherence of enemy operations. Army forces accomplish these things by controlling the nature, scope, and tempo of an operation and striking simultaneously throughout the area of operations to control, neutralize, and destroy enemy forces and other objectives. The Army's command and control doctrine supports its operations doctrine. It balances coordination, personal leadership, and tactical flexibility. It stresses rapid decision making and execution, including rapid response to changing situations. It emphasizes mutual trust and shared understanding among superiors and subordinates.

Mission command is the Army's approach to command and control that empowers subordinate decision making and decentralized execution appropriate to the situation. Mission command supports the Army's operational concept of unified land operations and its emphasis on seizing, retaining, and exploiting the initiative.

The mission command approach to command and control is based on the Army's view that war is inherently chaotic and uncertain. No plan can account for every possibility, and most plans must change rapidly during execution to account for changes in the situation. No single person is ever sufficiently informed to make every important decision, nor can a single person keep up with the number of decisions that need to be made during combat. Subordinate leaders often have a better understanding of what is happening during a battle, and are more likely to respond effectively to threats and fleeting opportunities if allowed to make decisions and act based on changing situations and unforeseen events not addressed in the initial plan in order to achieve their commander's intent. Enemy forces may behave differently than expected, a route may become impassable, or units could consume supplies at unexpected rates. Friction and unforeseeable combinations of variables impose uncertainty in all operations and require an approach to command and control that does not attempt to impose perfect order, but rather accepts uncertainty and makes allowances for unpredictability.

Mission command helps commanders capitalize on subordinate ingenuity, innovation, and decision making to achieve the commander's intent when conditions change or current orders are no longer relevant. It requires subordinates who seek opportunities and commanders who accept risk for subordinates trying to meet their intent. Subordinate decision making and decentralized execution appropriate to the situation help manage uncertainty and enable necessary tempo at each echelon during operations. Employing the mission command approach during all garrison activities and training events is essential to creating the cultural foundation for its employment in high-risk environments.

Subordinate Decision Making

Successful commanders anticipate future events by developing branches and sequels instead of focusing on details better handled by subordinates during current operations. The higher the echelon, the more time commanders should devote to future operations and the broader the guidance provided to subordinates. Subordinates empowered to make decisions during operations unburden higher commanders from issues that distract from necessary broader perspective and focus on critical issues. Mission command allows those commanders with the best situational understanding to make rapid decisions without waiting for higher echelon commanders to assess the situation and issue orders.

3-4 (Command & Control) Warfighting Function

Commanders delegate appropriate authority to deputies, subordinate commanders, and staff members based upon a judgment of their capabilities and experience. Delegation allows subordinates to decide and act for their commander in specified areas. Delegating decision-making authority reduces the number of decisions made at the higher echelons and reduces response time at lower echelons. In addition to determining the amount of decision-making authority they will delegate, commanders also identify decisions that are their sole responsibility and cannot be delegated to subordinates.

When delegating authority to subordinates, commanders strive to set the necessary conditions for success. They do this by assessing and managing risk. Taking risk is inherent at all levels of command. Commanders and staffs assess hazards and recommend controls to help manage risk, rather than forcing unnecessary risk decisions on subordinates. While commanders can delegate authority, they cannot delegate responsibility. Subordinates are accountable to their commanders for the use of delegated authority, but commanders remain solely responsible and accountable for the actions of their subordinates.

Decentralized Execution

Decentralized execution is the delegation of decision-making authority to subordinates, so they may make and implement decisions and adjust their assigned tasks in fluid and rapidly changing situations.

Subordinate decisions should be ethically based and within the framework of their higher commander's intent. Decentralized execution is essential to seizing, retaining, and exploiting the operational initiative during operations in environments where conditions rapidly change and uncertainty is the norm. Rapidly changing situations and uncertainty are inherent in operations where commanders seek to establish a tempo and intensity that enemy forces cannot match.

Decentralized execution requires disseminating information to the lowest possible level so subordinates can make informed decisions based on a shared understanding of both the situation and their commander's intent. This empowers subordinates operating in rapidly changing conditions to exercise disciplined initiative within their commander's intent. Generally, the more dynamic the circumstances, the greater the need for initiative to make decisions at lower levels. It is the duty of subordinates to exercise initiative to achieve their commander's intent. It is the commander's responsibility to issue appropriate intent and ensure subordinates are prepared in terms of education, training, and experience to exercise initiative.

The commander's intent provides a unifying idea that allows decentralized execution within an overarching framework. It provides guidance within which individuals may exercise initiative to accomplish the desired end state. Understanding the commander's intent two echelons up further enhances unity of effort while providing the basis for decentralized decision making and execution throughout the depth of a formation.

Levels of Control

Determining the appropriate level of control, including delegating decisions and determining how much decentralized execution to employ, is part of the art of command. The level and application of control is constantly evolving and must be continuously assessed and adjusted to ensure the level of control is appropriate to the situation. Commanders should allow subordinates the greatest freedom of action commensurate with the level of acceptable risk in a particular situation. The mission variables (mission, enemy, terrain and weather, troops and support available, time available, and civil considerations) influence how much control to impose on subordinates. Other considerations include—

- Enemy disposition and capabilities
- Level of synchronization and integration required
- Higher echelon headquarters constraints
- Level of risk

(Command & Control) Warfighting Function 3-5

III. Principles of Mission Command

Ref: ADP 6-0, Mission Command (May '12), pp. 1-6 to i-13.

Competence

Tactically and technically competent commanders, subordinates, and teams are the basis of effective mission command. An organization's ability to operate using mission command relates directly to the competence of its Soldiers. Commanders and subordinates achieve the level of competence to perform assigned tasks to standard through training, education, assignment experience, and professional development. Commanders continually assess the competence of their subordinates and their organizations. This assessment informs the degree of trust commanders have in their subordinates' ability to execute mission orders in a decentralized fashion at acceptable levels of risk.

Training and education that occurs in both schools and units provides commanders and subordinates the experiences that allow them to achieve professional competence. Repetitive, realistic, and challenging training creates common experiences that develop the teamwork, trust, and shared understanding that commanders need to exercise mission command and forces need to achieve unity of effort.

Leaders supplement institutional and organizational training and education with continuous self-development. Self-development is particularly important for the skills that rely on the art of command, which is further developed by reading and studying the art of war. These skills can also be developed through coursework, simulations and experience.

Mutual Trust

Mutual trust is shared confidence between commanders, subordinates, and partners that they can be relied on and are competent in performing their assigned tasks. There are few shortcuts to gaining the trust of others. Trust is given by leaders and subordinates, and built over time based on common shared experiences. It is the result of upholding the Army values, exercising leadership consistent with Army leadership principles, and most effectively instilled by the leader's personal example.

Mutual trust is essential to successful mission command, and it must flow throughout the chain of command. Subordinates are more willing to exercise initiative when they believe their commander trusts them. They will also be more willing to exercise initiative if they believe their commander will accept and support the outcome of their decisions. Likewise, commanders delegate greater authority to subordinates who have demonstrated tactical and technical competency and whose judgment they trust.

Shared Understanding

A critical challenge for commanders, staffs, and unified action partners is creating shared understanding of an operational environment, an operation's purpose, problems, and approaches to solving problems. Unified action partners are those military forces, governmental and nongovernmental organizations, and elements of the private sector with whom Army forces plan, coordinate, synchronize, and integrate during the conduct of operations (ADP 3-0).

Shared understanding starts with the Army's doctrine and professional military education that instills a common approach to the conduct of operations, a common professional language, and a common understanding of the principles of mission command. Army professionals understand the most current Army doctrine to ensure a minimum level of shared understanding for the conduct of operations. It is this shared understanding that allows even hastily task-organized units to operate effectively.

Commanders and staffs actively create shared understanding throughout the operations process (planning, preparation, execution, and assessment). They collaboratively frame an operational environment and its problems, and then they visualize approaches to solving those problems.

3-6 (Command & Control) Warfighting Function

Commander's Intent

The commander's intent is a clear and concise expression of the purpose of the operation and the desired military end state that supports mission command, provides focus to the staff, and helps subordinate and supporting commanders act to achieve the commander's desired results without further orders, even when the operation does not unfold as planned (JP 3-0). The higher echelon commander's intent provides the basis for unity of effort throughout the force. Each commander's intent nests within the commander's intent two levels up. During planning, the initial commander's intent drives course of action development. During execution, the commander's intent establishes the limits within which a subordinate may exercise initiative.

Mission Orders

Mission command requires commanders to issue mission orders. Mission orders are directives that emphasize to subordinates the results to be attained, not how they are to achieve them. Mission orders enable subordinates to understand the situation, their commander's mission and intent, and their own tasks. Subordinate commanders decide how to accomplish their own mission. The commander's intent and concept of operations set guidelines that provide unity of effort while allowing subordinate commanders to exercise initiative in planning, preparing, and executing their operations.

A mission order is not a separate type of order; rather, it is a technique for writing orders that allows subordinates maximum freedom of action in accomplishing missions. Mission orders should succinctly state the mission, task organization, commander's intent and concept of operations, tasks to subordinate units, and minimum essential coordinating instructions. Tasks to subordinate units include all the standard elements (who, what, when, where, and why) with particular emphasis on the purpose (why).

Disciplined Initiative

Disciplined initiative refers to the duty individual subordinates have to exercise initiative within the constraints of the commander's intent to achieve the desired end state. Simply put, disciplined initiative is when subordinates have the discipline to follow their orders and adhere to the plan until they realize their orders and the plan are no longer suitable for the situation in which they find themselves. This may occur because the enemy does something unforeseen, there is a new or more serious threat, or a golden opportunity emerges that offers a greater chance of success than the original course of action. The subordinate leader then takes action on their own initiative to adjust to the new situation and achieve their commander's intent, reporting to the commander about the new situation when able to do so.

Leaders and subordinates who exercise disciplined initiative create opportunity by taking action to develop a situation without asking for further guidance. Commanders rely on subordinates to act to meet their intent, not simply adhere to a plan that is no longer working. A subordinate's initiative may be the starting point for seizing, retaining, and exploiting the operational initiative by forcing an enemy to respond to friendly action.

Risk Acceptance

In general terms, risk is the exposure of someone or something valued to danger, harm, or loss. Because risk is part of every operation, it cannot be avoided. Commanders analyze risk in collaboration with subordinates to help determine what level of risk exists and how to mitigate it. When considering how much risk to accept with a course of action, commanders consider risk to the force and risk to the mission against the perceived benefit. They apply judgment with regard to the importance of an objective, time available, and anticipated cost. Commanders need to balance the tension between protecting the force and accepting and managing risks that must be taken to accomplish their mission.

While each situation is different, commanders avoid undue caution or commitment of resources to guard against every perceived threat. An unrealistic expectation of avoiding all risk is detrimental to mission accomplishment.

(Command & Control) Warfighting Function 3-7

IV. Command and Control System

Ref: ADP 6-0, Mission Command (May '12), pp. 1-20 to 1-21 and chap. 4.

Commanders need support to effectively exercise command and control. At every echelon of command, each commander establishes a command and control system—the arrangement of people, processes, networks, and command posts that enable commanders to conduct operations. The command and control system supports the commander's decision making, disseminates the commander's decisions to subordinates, and facilitates controlling forces. Commanders employ their command and control system to enable the people and formations conducting operations to work towards a common purpose. All the equipment and procedures exist to achieve this end. Commanders organize the four components of their command and control system to support decision making and facilitate communication. The most important of these components is people.

People

A commander's command and control system is based on people. The human aspects of operations remain paramount regardless of the technology associated with the system. Therefore, commanders base their command and control systems on human characteristics more than on equipment and processes. Trained personnel are essential to an effective command and control system. Technology cannot support command and control without them.

Processes

Commanders establish and use processes and procedures to organize activities within their headquarters and throughout the force. A process is a series of actions or steps taken to achieve a specific end, such as the military decision-making process. In addition to the major activities of the operations process, commanders and staffs use several integrating processes to synchronize specific functions throughout the operations process.

See pp. 3-20 to 3-21 for discussion of the integrating functions from ADP 3-0.

Procedures are standard, detailed steps that prescribe how to perform specific tasks (CJCSM 5120.01). Processes and procedures can increase organizational competence, for example, by improving a staff's efficiency or by increasing the tempo.

Networks

Generally, a network is a grouping of things that are interconnected for a purpose. Networks enable commanders to communicate information and control forces. Networks enable successful operations. Commanders determine their information requirements and focus their staffs and organizations on using networks to meet these requirements. These capabilities relieve staffs from handling routine data, and they enable extensive information sharing, collaborative planning, execution, and assessment that promote shared understanding. Each network consists of—

- End-user applications
- Information services and data
- Network transport and management

Command Posts

Command posts provide a physical location for the other three components of a command and control system (people, processes, and networks). Command posts vary in size, complexity and focus. Command posts may be comprised of vehicles, containers, and tents, or located in buildings. Commanders systematically arrange platforms, operation centers, signal nodes, and support equipment in ways best suited for a particular operational environment.

See pp. 3-29 to 3-30 for further discussion of command posts.

3-8 (Command & Control) Warfighting Function

Chap 3
I. The Operations Process

Ref: ADP 3-0, Operations (Jul '19), chap. 4.

Commanders establish and use processes and procedures to organize activities within their headquarters and throughout the force.

I. The Operations Process

The Army's framework for organizing and putting command and control into action is the operations process—the major command and control activities performed during operations: planning, preparing, executing, and continuously assessing the operation. Commanders use the operations process to drive the conceptual and detailed planning necessary to understand their operational environment (OE); visualize and describe the operation's end state and operational approach; make and articulate decisions; and direct, lead, and assess operations.

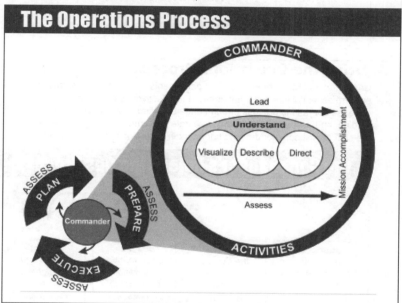

Ref: ADP 5-0, The Operations Process, fig. 1-1, p. 1-4.

Commanders, staffs, and subordinate headquarters employ the operations process to organize efforts, integrate the warfighting functions across multiple domains, and synchronize forces to accomplish missions. This includes integrating numerous processes and activities such as information collection and targeting within the headquarters and with higher, subordinate, supporting, and supported units. The unit's battle rhythm helps to integrate and synchronize the various processes and activities that occur within the operations process.

A goal of the operations process is to make timely and effective decisions and to act faster than the enemy. A tempo advantageous to friendly forces can place the enemy under the pressures of uncertainty and time. Throughout the operations process, making and communicating decisions faster than the enemy can react produces a tempo with which the enemy cannot compete. These decisions include assigning tasks; prioritizing, allocating, and organizing forces and resources; and selecting the critical times and places to act. Decision making during execution includes knowing how and when to adjust previous decisions. The speed and accuracy of a commander's actions to address a changing situation is a key contributor to agility.

II. Principles of the Operations Process

The operations process, while simple in concept, is dynamic in execution. Commanders must organize and train their staffs and subordinates as an integrated team to simultaneously plan, prepare, execute, and assess operations. In addition to the principles of mission command, commanders and staffs consider the following principles for the effective employment of the operations process:

Principles of the Operations Process

- **Drive the operations process**
- **Build and maintain situational understanding**
- **Apply critical and creative thinking**

A. Drive the Operations Process

Commanders are the most important participants in the operations process. While staffs perform essential functions that amplify the effectiveness of operations, commanders drive the operations process through understanding, visualizing, describing, directing, leading, and assessing operations. Accurate and timely running estimates maintained by the staff, assist commanders in understanding situations and making decisions.

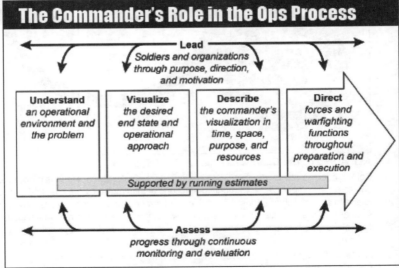

Ref: ADP 5-0, The Operations Process (Jul '19), fig. 1-2, p. 1-8.

See facing page for further discussion.

3-10 (Command & Control) I. The Operations Process

Commanders Drive the Operations Process

Ref: ADP 5-0, The Operations Process (Jul "19), pp. 1-8 to 1-11.

Commanders are the most important participants in the operations process. While staffs perform essential functions that amplify the effectiveness of operations, commanders drive the operations process through understanding, visualizing, describing, directing, leading, and assessing operations. Accurate and timely running estimates maintained by the staff, assist commanders in understanding situations and making decisions.

Understand

Understanding an OE and associated problems is fundamental to establishing a situation's context and visualizing operations. An operational environment is a composite of the conditions, circumstances, and influences that affect the employment of capabilities and bear on the decisions of the commander (JP 3-0). An OE encompasses the air, land, maritime, space, and cyberspace domains; the information environment; the electromagnetic spectrum; and other factors.

Visualize

As commanders build understanding about their OEs, they start to visualize solutions to solve the problems they identify. Collectively, this is known as commander's visualization—the mental process of developing situational understanding, determining a desired end state, and envisioning an operational approach by which the force will achieve that end state (ADP 6-0). Commanders complete their visualization by conceptualizing an operational approach—a broad description of the mission, operational concepts, tasks, and actions required to accomplish the mission (JP 5-0).

Describe

Commanders describe their visualization to their staffs and subordinate commanders to facilitate shared understanding and purpose throughout the force. During planning, commanders ensure subordinates understand their visualization well enough to begin course of action (COA) development. During execution, commanders describe modifications to their visualization in updated planning guidance and directives resulting in fragmentary orders (FRAGORDs) that adjust the original operation order (OPORD).

Commanders describe their visualization in doctrinal terms, refining and clarifying it, as circumstances require. Commanders describe their visualization in terms of commander's intent, planning guidance, commander's critical information requirements (CCIRs), and essential elements of friendly information.

Direct

To direct is implicit in command. Commanders direct action to achieve results and lead forces to mission accomplishment. Commanders make decisions and direct action based on their situational understanding maintained by continuous assessment.

Lead

Leadership is the activity of influencing people by providing purpose, direction, and motivation to accomplish the mission and improve the organization (ADP 6-22). Leadership inspires Soldiers to accomplish things that they otherwise might not. Throughout the operations process, commanders make decisions and provide the purpose and motivation to follow through with the COA they chose. They must also possess the wisdom to know when to modify a COA when situations change.

Assess

Assessment involves deliberately comparing intended forecasted outcomes with actual events to determine the overall effectiveness of force employment. Assessment helps the commander determine progress toward attaining the desired end state, achieving objectives, and completing tasks. Commanders incorporate assessments by the staff, subordinate commanders, and unified action partners into their personal assessment of the situation.

Cmd & Control (ADP 6-0)

(Command & Control) I. The Operations Process 3-11

III. Activities of the Operations Process

Ref: FM 3-0, Operations (Oct. '22), p. 8-7 to 8-8 and ADP 5-0, The Operations Process (Jul "19), pp. 1-4 to 1-7

Commanders employ the operations process to incorporate coalition and joint partners, empower subordinate initiative, and to ensure authorities and risk acceptance are delegated to the appropriate echelon required for the situation. Staffs and subordinate headquarters earn the commander's trust by providing relevant information, anticipating needs, and directing supporting actions.

The major components of the operations process are planning, preparing, executing, and continuously assessing the operation. Planning normally begins upon receipt of orders from a higher echelon headquarters and continues through the execution of the operation. The commander and staff continually assess operations and revise the plan through fragmentary orders. Commanders, assisted by their chiefs of staff or executive officers, drive the preparation for an operation by allocating time, prioritizing resources, and supervising preparation activities such as rehearsals to ensure their forces are ready to execute operations. During execution, commanders and staffs focus their efforts on translating plans into direct action to achieve objectives in accordance with higher commander's intent.

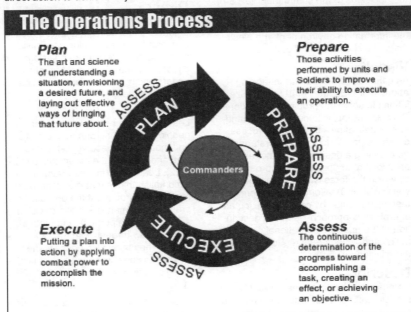

Ref: FM 3-0 (Oct '17), fig. 2-7. The operations process

Army operational planning requires the complete definition of the mission, expression of the commander's intent, completion of the commander and staff estimates, and development of a concept of operations. These form the basis of a plan or order and set the conditions for a successful battle. The initial plan establishes the commander's intent, the concept of operations, and the initial tasks for subordinate units. It allows the greatest possible operational and tactical freedom for subordinate leaders. It is flexible enough to permit leaders to seize opportunities consistent with the commander's intent, thus facilitating quick and accurate decision making during combat operations.

Both commanders and staffs have important roles within the operations process. The commander's role is to drive the operations process through the activities of understanding, visualizing, describing, directing, leading, and assessing operations. The staff's role is to assist commanders with understanding situations, making and implementing decisions, controlling operations, and assessing progress.

The Operations Process

The operations process is a commander-led activity informed by mission command principles. It consists of the major command and control activities performed during operations: planning, preparing, executing, and continuously assessing an operation. These activities may be sequential or simultaneous. They are rarely discrete and often involve a great deal of overlap. Commanders use the operations process to drive the planning necessary to understand, visualize, and describe their unique operational environments; make and articulate decisions; and direct, lead, and assess military operations.

Plan

Planning is the art and science of understanding a situation, envisioning a desired future, and laying out effective ways of bringing that future about (ADP 5-0). Planning consists of two separate but interrelated components: a conceptual component and a detailed component. Successful planning requires the integration of both components. Army leaders employ three methodologies for planning: the Army design methodology, the military decision-making process, and troop leading procedures. Commanders determine how much of each methodology to use based on the scope of the problem, their familiarity with the methodology, the echelon, and the time available.

Prepare

Preparation consists of activities that units perform to improve their ability to execute an operation. Preparation creates conditions that improve friendly forces' opportunities for success. It requires commander, staff, unit, and Soldier actions to ensure the force is trained, equipped, and ready to execute operations. Preparation activities help commanders, staffs, and Soldiers understand a situation and their roles in upcoming operations and set conditions for successful execution.

Execute

Execution is the act of putting a plan into action by applying combat power to accomplish the mission and adjusting operations based on changes in the situation (ADP 5-0). Commanders and staffs use situational understanding to assess progress and make execution and adjustment decisions. In execution, commanders and staffs focus their efforts on translating decisions into actions. They apply combat power to seize, retain, and exploit the initiative to gain and maintain a position of relative advantage. This is the essence of unified land operations.

Assess

Assessment is determination of the progress toward accomplishing a task, creating a condition, or achieving an objective (JP 3-0). Assessment precedes and then occurs during the other activities of the operations process. Assessment involves deliberately comparing forecasted outcomes with actual events to determine the overall effectiveness of force employment. Assessment helps commanders determine progress toward achieving a desired end state, accomplishing objectives, and performing tasks.

Refer to BSS6: The Battle Staff SMARTbook, 6th Ed. (Plan, Prepare, Execute, & Assess Military Operations) for complete discussion of the operations processes. Additional related topics include the three planning methodologies, integrating processes and continuing activities, plans and orders, mission command, rehearsals and after action reviews, and operational terms and military symbols.

(Command & Control) I. The Operations Process 3-13

B. Build and Maintain Situational Understanding

Success in operations demands timely and effective decisions based on applying judgment to available information and knowledge. As such, commanders and staffs seek to build and maintain situational understanding throughout the operations process. Situational understanding is the product of applying analysis and judgment to relevant information to determine the relationships among the operational and mission variables (ADP 6-0). Commanders and staffs continually strive to maintain their situational understanding and work through periods of reduced understanding as a situation evolves. Effective commanders accept that uncertainty can never be eliminated and train their staffs and subordinates to function in uncertain environments.

As commanders build their situational understanding, they share their understanding across the forces and with unified action partners. Creating shared understanding is a principle of mission command and requires communication and information sharing from higher to lower and lower to higher. Higher headquarters ensure subordinates understand the larger situation to include the operation's end state, purpose, and objectives. Staffs from lower echelons share their understanding of their particular situation and provide feedback to the higher headquarters on the operation's progress. Communication and information sharing with adjacent units and unified action partners is also multi-directional. Several tools assist leaders in building situational understanding and creating a shared understanding across the force to include—

- Operational and mission variables *(See p. 1-22.)*
- Running estimates *(See p. 3-16.)*
- Intelligence
- Collaboration
- Liaison

C. Apply Critical and Creative Thinking

Thinking includes awareness, perception, reasoning, and intuition. Thinking is naturally influenced by emotion, experience, and bias. As such, commanders and staffs apply critical and creative thinking throughout the operations process to assist them with understanding situations, making decisions, directing actions, and assessing operations.

Critical thinking is purposeful and reflective thought about what to believe or what to do in response to observations, experiences, verbal or written expressions, or arguments. By thinking critically, individuals formulate judgments about whether the information they encounter is true or false, or if it falls somewhere along a scale of plausibility between true or false. Critical thinking involves questioning information, assumptions, conclusions, and points of view to evaluate evidence, develop understanding, and clarify goals. Critical thinking helps commanders and staffs identify causes of problems, arrive at justifiable conclusions, and make good judgments. Critical thinking helps commanders counter their biases and avoid logic errors.

Creative thinking examines problems from a fresh perspective to develop innovative solutions. Creative thinking creates new and useful ideas, and reevaluates or combines old ideas to solve problems. Leaders face unfamiliar problems that require new or original approaches to solve them. This requires creativity and a willingness to accept change, newness, and a flexible outlook of new ideas and possibilities.

Breaking old habits of thought, questioning the status quo, visualizing a better future, and devising responses to new problems require creative thinking. During operations, leaders routinely face unfamiliar problems or old problems under new conditions. Leaders apply creative thinking to gain new insights, novel approaches, fresh perspectives, and new ways of understanding problems and conceiving ways to solve them.

Refer to ATP 5-0.1 for creative thinking tools and techniques.

3-14 (Command & Control) I. The Operations Process

IV. Army Planning Methodologies

Ref: ADP 3-0, Operations (Jul '19), p. 4-2.

Planning is the art and science of understanding a situation, envisioning a desired future, and laying out effective ways of bringing that future about (ADP 5-0). Planning consists of two separate but interrelated components: a conceptual component and a detailed component. Successful planning requires the integration of both these components. Army leaders employ three methodologies for planning: the Army design methodology, the military decisionmaking process, and troop leading procedures.

A. The Army Design Methodology (ADM)

The Army design methodology is useful as an aid to conceptual thinking about unfamiliar problems. To produce executable plans, commanders integrate the Army design methodology with the detailed planning typically associated with the military decision-making process. Commanders who use the Army design methodology may gain a greater understanding of an operational environment and its problems. Once they have an understanding of the environment, they can better visualize an appropriate operational approach. This greater understanding allows commanders to provide a clear commander's intent and concept of operations.

Army design methodology is iterative, collaborative, and continuous. As the operations process unfolds, the commander, staff, subordinates, and other partners continue collaboration to improve their shared understanding. An improved understanding may lead to modifications to the commander's operational approach or an entirely new approach altogether.

Refer to ATP 5-0.1 for more information on Army design methodology.

B. The Military Decision-Making Process (MDMP)

The military decision-making process is an iterative planning methodology. It integrates activities of the commander, staff, subordinate headquarters, and other partners. This integration enables them to understand the situation and mission; develop, analyze, and compare courses of action; decide on the course of action that best accomplishes the mission; and produce an order for execution. The military decision-making process applies to both conceptual and detailed approaches. It is most closely associated with detailed planning.

For unfamiliar problems, executable solutions typically require integrating the Army design methodology with the military decision-making process. The military decision-making process helps leaders apply thoroughness, clarity, sound judgment, logic, and professional knowledge, so they understand situations, develop options to solve problems, and reach decisions. This process helps commanders, staffs, and others to think critically and creatively while planning.

C. Troop Leading Procedures (TLP)

Troop leading procedures is a dynamic process used by small-unit leaders to analyze a mission, develop a plan, and prepare for an operation. Heavily weighted in favor of familiar problems and short planning periods, troop leading procedures are typically employed by organizations without staffs at the company level and below. Leaders use troop leading procedures to solve tactical problems when working alone or with a small group. For example, a company commander may use the executive officer, first sergeant, fire support officer, supply sergeant, and communications sergeant to help during troop leading procedures.

Refer to BSS6: The Battle Staff SMARTbook, 6th Ed. (Plan, Prepare, Execute, & Assess Military Operations) for complete discussion of the three Army planning methodologies. Additional related topics include the operations process, integrating processes and continuing activities, plans and orders, mission command, rehearsals and after action reviews, and operational terms and military symbols.

V. Imperatives: Command & Control (C2)

Ref: FM 3-0, Operations (Oct. '22), pp. 3-13 to 3-16.

Editor's Note: The following select imperatives of Army operations from FM 3-0 (2022) are highlighted here in greater detail to illustrate application within the command and control warfighting function. See pp. 1-47 to 1-45 for a listing of all imperatives.

Commanders **visualize operational environments** in terms of the factors that are relevant to decision making. Operational environments are dynamic and contain vast amounts of information that can overload C2 systems and impede decision making. Commanders simplify information collection, analysis, and decision making by focusing on how they see themselves, see the enemy, and understand the operational environment. These three categories of factors are interrelated, and leaders must understand how each one relates to the others in the current context.

As part of the **operations process**, Army leaders use different methodologies to understand and weigh options. These methodologies include the Army design methodology, the military decision-making process, and the rapid decision-making and synchronization process. Each methodology provides a process that allows commanders and staffs to see themselves, see the enemy, and understand the operational environment.

See Yourself *(See pp. 3-10 to 3-11.)*

Commanders develop an understanding of their forces relative to mission requirements, enemy capabilities, and impacts from the operational environment. This understanding helps to inform current and potential future advantages relative to enemy forces, allowing staffs to develop and adapt courses of action that exploit advantages and mitigate disadvantages. Commanders and staffs maintain this understanding of their forces through **running estimates**, subordinate commander updates, and friendly forces information requirements (known as FFIRs): A **friendly force information requirement (FFIR)** is information the commander and staff need to understand the status of friendly force and supporting capabilities (JP 3-0). Friendly force information requirements identify the information the commander considers most important to make critical decisions during the execution of operations. The operations officer manages friendly force information requirements for the commander.

Leaders attempt to see themselves from the enemy perspective, in part by understanding essential elements of friendly information. An **essential element of friendly information (EEFI)** is a critical aspect of a friendly operation that, if known by a threat would subsequently compromise, lead to failure, or limit success of the operation and therefore should be protected from enemy detection (ADP 6-0).

Leaders see their formation in relation to their mission and in the broader context of the higher command, adjacent unit, and all domains. Part of "seeing yourself" is understanding how land-based operations enable operations in the other domains, and how capabilities from all domains can enable operations on land.

*See p. 1-29 for a discussion of **joint interdependence**.*

See the Enemy *(See pp. 1-23 to 1-36.)*

Commanders see the enemy in terms of its combat power, advantages, and intentions within the operational environment and broader strategic context. Commanders develop their understanding of enemy forces from their individual knowledge, experience, and judgment honed through self-study, training, and education. From this base of knowledge commanders and staffs build shared understanding of enemy forces and environment through intelligence preparation of the battlefield. Intelligence preparation of the battlefield is the systematic process of analyzing the mission variables of enemy, terrain, weather, and civil considerations in an area of interest to determine their effect on operations (ATP 2-01.3).

3-16 (Command & Control) I. The Operations Process

Intelligence preparation of the battlefield (IPB) provides commanders with aware-ness of information gaps about enemy forces and the operational environment. Staffs translate these gaps into information requirements and assist the commander in determining priority intelligence requirements. A priority intelligence requirement is the intelligence component of commander's critical information requirements used to focus the employment of limited intelligence assets and resources against competing demands for intelligence support (JP 2-0). More importantly, **priority intelligence requirements (PIRs)** identify information about the threat and operational environment that a com-mander considers most important to making decisions in a specific context. Intelligence about civil considerations may be as critical as intelligence about enemy forces in some cases. The intelligence officer, in coordination with the rest of the staff, manages priority intelligence requirements for the commander.

Enemy forces attempt to hide from, deceive, disrupt, and deny friendly collection efforts to prevent friendly forces from perceiving the enemy's true intentions. This requires commanders plan to develop the situation through action and **fight for information**. Information collection operations may require the commander to assume significant risk to determine enemy dispositions and anticipate enemy intentions.

Leaders do not limit their understanding of the enemy to those forces in their assigned area. Enemy forces are capable of employing capabilities from great distances and mul-tiple domains. Leaders must be aware of those capabilities so they can take appropriate action.

See pp. 3-20 and 5-6 for discussion of Intelligence preparation of the battlefield (IPB) .

Understand the Operational Environment *(See pp. 1-17 to 1-22.)*
Leaders view the operational environment in terms of domains, dimensions, operational variables, and mission variables that are relevant to their decisions. The most difficult aspect of an operational environment to understand is how the different factors interact to affect operations.

Understanding is, in the context of decision making, knowledge that has been syn-thesized and had judgment applied to comprehend the situation's inner relationships, enable decision making, and drive action (ADP 6-0). Understanding is judgment applied to knowledge in the context of a particular situation. Understanding is knowing enough about a situation to make an informed decision. Judgment is based on experience, expertise, and intuition—and it informs what decision to make.

Situational Understanding. Successful operations demand timely and effective deci-sions based on the information available. As such, commanders and staffs seek to build and maintain situational understanding throughout an operation. Situational understand-ing is the product of applying analysis and judgment to relevant information to determine the relationships among the operational and mission variables (ADP 6-0). Situational understanding allows commanders to make effective decisions and enables commanders and staffs to assess operations accurately. Commanders and staffs continually strive to maintain their situational understanding and work through periods of uncertainty, accepting that they cannot eliminate them.

Shared Understanding. A critical challenge for commanders, staffs, and unified action partners is creating common understanding of an operational environment, an opera-tion's purpose, its challenges, and the approaches to solving those problems. Shared understanding of the situation, which requires effective flow of information between echelons, forms the basis for unity of effort and subordinate initiative. Effective decentral-ized execution depends on shared understanding.

Common Operational Picture. A common operational picture (COP) is key to achieving and maintaining shared situational understanding in all domains and making effective decisions faster than the threat. The common operational picture is a display of relevant

Continued on next page

Continued on next page

**Cmd & Control
(ADP 6-0)**

(Command & Control) I. The Operations Process 3-17

Imperatives: Command & Control (cont.)

Cmd & Control (ADP 6-0)

Continued from previous page

information within a commander's area of interest tailored to the user's requirements and based on common data and information shared by more than one command (ADP 6-0). Although the COP is ideally a single display, it may include more than one display and information in other forms, such as graphic representations or written reports. The COP facilitates collaborative planning and helps commanders at all echelons achieve shared situational understanding. The COP must account for relevant factors in domains affecting the operation, and it provides and enables a common understanding of the interrelationships between actions and effects through the physical, information, and human dimensions. Shared situational understanding allows commanders to visualize the effects of their decisions on other elements of the force and the overall operation.

Anticipate, Plan, and Execute Transitions *(See p. 1-51.)*

Transitions mark a change of focus in an operation. Leaders plan transitions as part of the initial plan or parts of a **branch or sequel.** They can be unplanned and cause the force to react to unforeseen circumstances. Transitions can be part of progress towards mission accomplishment, or they can reflect a temporary setback. Common transitions are—

- Between competition, crisis, and armed conflict.
- Between operations dominated by offense, defense, and stability.
- Between types of offense or defense.
- Between phases of an operation.
- Between branches and sequels of a campaign or major operation.
- Between command posts during emplacement, movement, and displacement of one or more nodes.
- Shifts of the main effort, supporting effort, and reserve between units.
- Task organization changes.
- Passing responsibility for enemy units crossing unit or echelon boundaries.
- Passing terrain responsibility between units.
- Transferring responsibility for security and governance to legitimate authorities.
- Change in mission from combat operations to reconstitution.
- Moving forces in and out of theater.
- Changes in the environment that cause a reframe of the mission or change in the purpose of the operation.

Place leaders at points of friction. Transitions are critical planning responsibilities for commanders. They anticipate key transitions and issue planning guidance to their staffs. Staffs in turn suggest to their commanders when transitions may be necessary. Staffs monitor current operations and track conditions that require transition. Transitions are typically points of friction or opportunities, and leaders assign subordinate leaders specific responsibilities wherever transitions occur, for example, during passage of lines, at wet gap crossings, at contact points, and along unit boundaries.

Effective transitions require planning and preparation well before their execution, so the force can maintain the momentum and tempo of operations. Risks increase during transitions, so commanders establish clear conditions for their execution. Commanders establish decision points to support successful transitions during operations. The ability of echelons below brigade to execute battle drills mitigates some of the risks higher echelons face during transitions.

A transition occurs for several reasons. An unexpected change in conditions may require commanders to direct an abrupt transition. In such cases, the overall composition of

3-18 (Command & Control) I. The Operations Process

the force remains unchanged despite sudden changes in mission, task organization, and rules of engagement. Typically, task organization evolves to meet changing conditions; however, transition planning must also account for changes in mission. Commanders continuously assess the situation, re-task, re-organize, and cycle their forces in and out of close combat to retain operational initiative. Commanders seek to shift priorities or the main effort without necessitating operational pauses that make friendly forces more vulnerable to enemy action.

Designate, Weight, and Sustain the Main Effort *(See p. 1-62.)*

Commanders frequently face competing demands for limited resources. They resolve these competing demands by establishing priorities. One way in which commanders establish priorities is by designating, weighting, and sustaining the main effort. The **main effort** is a designated subordinate unit whose mission at a given point in time is most critical to overall mission success (ADP 3-0). Commanders provide the main effort with the appropriate resources and support necessary for its success. When designating a main effort, commanders consider augmenting a unit's task organization and giving it priority of resources and support.

The commander designates various priorities of support, such as for air and missile defense (AMD), close air support and other fires, information collection, mobility and countermobility, and sustainment. Commanders and staffs anticipate sustainment requirements of the main effort as it shifts throughout the operation, and they position supplies and capabilities according to the situation. Commanders must balance forward positioning of sustainment assets with the need for freedom of action and operational reach when weighting the main effort.

Commanders shift resources and priorities as circumstances require. While there can be only one main effort at any given time, commanders may **shift the main effort** several times during an operation to increase the endurance of the overall force. They should allow time for the shift of support priorities prior to designating a unit as the main effort, since shifting the main effort may require movement of resources and the positioning of supporting capabilities.

Consolidate Gains Continuously *(See p. 1-16.)*

Leaders add depth to their operations in terms of time and purpose when they consolidate gains. Commanders consolidate gains at the operational and tactical levels as a **strategically informed approach** to current operations with the **desired political outcome of the conflict** in mind. During competition and crisis, commanders expand opportunities created from previous conflicts and activities to sustain enduring U.S. interests, while improving the credibility, readiness, and deterrent effect of Army forces. During large-scale combat operations, commanders consolidate gains continuously or as soon as possible, deciding whether to accept risk with a more moderate tempo during the present mission or in the future as large-scale combat operations conclude.

The multidomain aspects of an operational environment place increased strain on the ability of military forces to create enduring change, particularly in the **human and information dimensions**. The size, scale, and scope of an assigned area of operations (AO) may reduce the duration of effects, just as they dilute the potency of combat power. The speed and pervasiveness of enemy disinformation campaigns is a constant challenge that contests Army forces' ability to change human will and behavior. The need to fix and bypass some enemy forces during operations designed to penetrate or envelop enemy echelons may leave significant enemy threats in rear areas and jeopardize gains made during offensive operations. Commanders therefore continuously assess when and how they will consolidate gains as they develop the situation.

Consolidating gains at every echelon leads to better transitions out of armed conflict and into post-conflict competition. It serves as a preventative against the rise of an insurgency by those wishing to prolong the conflict.

(Command & Control) I. The Operations Process 3-19

VI. Integrating Processes

Ref: ADP 5-0, The Operations Process (Jul '19), pp. 1-15 to 1-17.

Commanders and staffs integrate the warfighting functions and synchronize the force to adapt to changing circumstances throughout the operations process. They use several integrating processes to do this. An integrating process consists of a series of steps that incorporate multiple disciplines to achieve a specific end. For example, during planning, the military decision-making process (MDMP) integrates the commander and staff in a series of steps to produce a plan or order. Key integrating processes that occur throughout the operations process include—

- Intelligence preparation of the battlefield
- Information collection
- Targeting
- Risk management
- Knowledge management

Intelligence Preparation of the Battlefield (IPB) *(See pp. 3-20 & 5-6.)*

Intelligence preparation of the battlefield is the systematic process of analyzing the mission variables of enemy, terrain, weather, and civil considerations in an area of interest to determine their effect on operations (ATP 2-01.3). Led by the intelligence officer, the entire staff participates in IPB to develop and sustain an understanding of the enemy, terrain and weather, and civil considerations. IPB helps identify options available to friendly and threat forces.

IPB consists of four steps. Each step is performed or assessed and refined to ensure that IPB products remain complete and relevant. The four IPB steps are—

- Define the Operational Environment
- Describe Environmental Effects On Operations/*Describe The Effects On Operations*
- Evaluate the Threat/*Adversary*
- Determine Threat/*Adversary* Courses Of Action

IPB begins in planning and continues throughout the operations process. IPB results in intelligence products used to aid in developing friendly COAs and decision points for the commander. Additionally, the conclusions reached and the products created during IPB are critical to planning information collection and targeting. A key aspect of IPB is refinement in preparation and execution.

Refer to ATP 2-01.3 for a detailed discussion of IPB.

Information Collection *(See p. 5-7.)*

Information collection is an activity that synchronizes and integrates the planning and employment of sensors and assets as well as the processing, exploitation, and dissemination systems in direct support of current and future operations (FM 3-55). It integrates the functions of the intelligence and operations staffs that focus on answering CCIRs. Information collection includes acquiring information and providing it to processing elements. It has three steps:

- Collection management
- Task and direct collection
- Execute collection

Information collection helps the commander understand and visualize the operation by identifying gaps in information and aligning reconnaissance, surveillance, security, and intelligence assets to collect information on those gaps. The "decide" and "detect" steps of targeting tie heavily to information collection.

Refer to FM 3-55 for a detailed discussion of information collection to include the relationship between the duties of intelligence and operations staffs.

3-20 (Command & Control) I. The Operations Process

Targeting *(See pp. 6-20 to 6-23.)*

Targeting is the process of selecting and prioritizing targets and matching the appropriate response to them, considering operational requirements and capabilities (JP 3-0). Targeting seeks to create specific desired effects through lethal and nonlethal actions. The emphasis of targeting is on identifying enemy resources (targets) that if destroyed or degraded will contribute to the success of the friendly mission. Targeting begins in planning and continues throughout the operations process. The steps of the Army's targeting process are—

- Decide
- Detect
- Deliver
- Assess

This methodology facilitates engagement of the right target, at the right time, with the most appropriate assets using the commander's targeting guidance. Targeting is a multi-discipline effort that requires coordinated interaction among the commander and several staff sections that together form the targeting working group. The chief of staff (executive officer) or the chief of fires (fire support officer) leads the staff through the targeting process. Based on the commander's targeting guidance and priorities, the staff determines which targets to engage and how, where, and when to engage them. The staff then assigns friendly capabilities best suited to produce the desired effect on each target, while ensuring compliance with the rules of engagement. *Refer to ATP 3-60.*

Risk Management

Risk—the exposure of someone or something valued to danger, harm, or loss—is inherent in all operations. Because risk is part of all military operations, it cannot be avoided. Identifying, mitigating, and accepting risk is a function of command and a key consideration during planning and execution.

Risk management is the process to identify, assess, and control risks and make decisions that balance risk cost with mission benefits (JP 3-0). Commanders and staffs use risk management throughout the operations process to identify and mitigate risks associated with hazards (to include ethical risk and moral hazards) that have the potential to cause friendly and civilian casualties, damage or destroy equipment, or otherwise impact mission effectiveness. Like targeting, risk management begins in planning and continues through preparation and execution. Risk management consists of the following steps:

- Identify hazards
- Assess hazards to determine risks
- Develop controls and make risk decisions
- Implement controls
- Supervise and evaluate

Knowledge Management

Knowledge management is the process of enabling knowledge flow to enhance shared understanding, learning, and decision making (ADP 6-0). It facilitates the transfer of knowledge among commanders, staffs, and forces to build and maintain situational understanding. Knowledge management helps get the right information to the right person at the right time to facilitate decision making. Knowledge management uses a five-step process to create shared understanding. *Refer to ATP 6-01.1.*

Refer to BSS6: The Battle Staff SMARTbook, 6th Ed. (Plan, Prepare, Execute, & Assess Military Operations) for complete discussion of the integrating processes. Additional related topics include the operations process, plans and orders, mission command, rehearsals and after action reviews, and operational terms and military symbols.

(Command & Control) I. The Operations Process 3-21

Both critical and creative thinking must intentionally include ethical reasoning—the deliberate evaluation that decisions and actions conform to accepted standards of conduct. Ethical reasoning within critical and creative thinking helps commanders and staffs anticipate ethical hazards and consider options to prevent or mitigate the hazards within their proposed COAs.

Commanders may form red teams to help the staff think critically and creatively and to avoid groupthink, mirror imaging, cultural missteps, and tunnel vision. Red teaming enables commanders to explore alternative plans and operations in the context of an OE and from the perspective of unified action partners, adversaries, and others. Throughout the operations process, red team members help clarify the problem and explain how others (unified action partners, the population, and the enemy) potentially view the problem. Red team members challenge assumptions and the analysis used to build the plan.

Refer to JP 5-0 for a detailed discussion of red teams and red teaming.

VII. Battle Rhythm

Commanders and staffs must integrate and synchronize numerous activities, meetings, and reports within their headquarters, and with higher, subordinate, supporting, and adjacent units as part of the operations process. They do this by establishing the unit's battle rhythm. Battle rhythm is a deliberate, daily schedule of command, staff, and unit activities intended to maximize use of time and synchronize staff actions (JP 3-33). A unit's battle rhythm provides structure for managing a headquarters' most important internal resource—the time of the commander and staff. A headquarters' battle rhythm consists of a series of meetings, report requirements, and other activities synchronized by time and purpose. These activities may be daily, weekly, monthly, or quarterly depending on the echelon, type of operation, and planning horizon. An effective battle rhythm—

- Facilitates interaction among the commander, staff, and subordinate commanders
- Supports building and maintaining shared understanding throughout the headquarters
- Establishes a routine for staff interaction and coordination

3-22 (Command & Control) I. The Operations Process

Chap 3

II. Command and Support Relationships

Ref: FM 3-0, Operations (Oct. '22), app B.

Command and support relationships provide the basis for unity of command and unity of effort in operations. Command relationships affect Army force generation, force tailoring, and task organization. Commanders use Army support relationships when task-organizing Army forces. All command and support relationships fall within the framework of joint doctrine. *Note: JP 1 discusses joint command relationships and authorities.*

Cmd & Control (ADP 6-0)

I. Chain of Command

The President and Secretary of Defense exercise authority and control of the armed forces through two distinct branches of the chain of command, as described in JP 1, Volume 2: the operational and administrative branches. The operational branch runs from the President, through the Secretary of Defense, to the CCDRs for missions and forces assigned to combatant commands. The administrative branch runs from the President through the Secretary of Defense to the secretaries of the military departments.

The typical operational chain of command extends from the CCDR to a joint task force (JTF) commander, then to a functional component or Service component commander. JTFs comprise forces from more than one Service placed under the operational control (OPCON) of the JTF. Within their commands, CCDRs and JTF commanders establish joint command relationships among forces.

Under joint doctrine, each Unified Command includes a Service component command that provides administrative control (ADCON) for Service forces assigned to that Unified Command. A Service component command consists of the Service Component headquarters and all Service Forces assigned to the Unified Commander. Army doctrine distinguishes between the Army component of a combatant command and Army components of other joint forces.

Army Service Component Command (ASCC)

Under Army doctrine, Army Service component command (ASCC) refers to the Army component assigned to a combatant command. There is only one ASCC within a combatant command's area of responsibility.

ARFORs

The Army components of all other joint forces are called ARFORs. An ARFOR is the Army component and senior Army headquarters of all Army forces assigned or attached to a combatant command, subordinate joint force command, joint functional command, or multinational command. (FM 3-94) It consists of the senior Army headquarters and all Army forces that the CCDR subordinates to the JTF or places under the control of a multinational force commander. The ARFOR becomes the conduit for ADCON functions specified in unit deployment orders.

See FM 3-94 and JP 3-0 for more information on ARFOR.

II. Joint Command Relationships

JP 1, Volume 2 specifies and details four types of joint command relationships:

A. Combatant Command (Command Authority)

COCOM is the command authority over assigned and allocated forces vested only in commanders of combatant commands or as directed by the President or the Secre-

(Command & Control) II. Command & Support Relationships 3-23

tary of Defense in the Unified Command Plan and cannot be delegated or transferred. COCOM only extends to those forces assigned or allocated to the combatant command by the Secretary of Defense. COCOM is established in federal law by Section 164, Title 10, United States Code (USC). Normally, the CCDR exercises this authority through subordinate JFCs, Service components, and functional component commanders. COCOM includes the directive authority for logistics.

B. Operational Control

The authority to exercise OPCON is exclusively derived from COCOM authority. Forces provided by the Services and attached to a combatant command are typically in an OPCON command relationship. OPCON normally includes authority over all aspects of operations and joint training necessary to accomplish missions. It does not include directive authority for logistics or matters of administration, discipline, internal organization, or unit training. OPCON does include the authority to delineate functional responsibilities and operational areas of subordinate JFCs. In two instances, the Secretary of Defense may specify adjustments to accommodate authorities beyond OPCON in an establishing directive: when transferring forces between CCDRs or when transferring members or organizations from the military departments to a combatant command. Adjustments will be coordinated with the participating CCDRs.

C. Tactical Control

TACON is inherent in OPCON. TACON may be delegated to and exercised by commanders at any echelon at or below the level of combatant command. TACON provides sufficient authority for controlling and directing the application of force or tactical use of combat support assets within the assigned mission or task. TACON does not provide organizational authority or authoritative direction for administrative and logistics support; the commander of the parent unit continues to exercise these authorities unless otherwise specified in the establishing directive.

D. Joint Support Relationships

Support is a command authority in joint doctrine. A superior commander establishes a supported and supporting relationship between subordinate commanders when one organization should aid, protect, complement, or sustain another force. Designating supporting relationships is important. It conveys priorities to commanders and staffs planning or executing joint operations. Designating a support relationship does not provide authority to organize and employ commands and forces, nor does it include authoritative direction for administrative and logistic support.

Category	Definition
General support	Support given to the supported force as a whole and not to any particular subdivision thereof (JP 3-09.3).
Mutual support	That support which units render each other against an enemy, because of their assigned tasks, their position relative to each other and to the enemy, and their inherent capabilities (JP 3-31).
Direct support	A mission requiring a force to support another specific force and authorizing it to answer directly to the supported force's request for assistance (JP 3-09.3).
Close support	The action of the supporting force against targets or objectives that are sufficiently near the supported force as to require detailed integration or coordination of the supporting action (JP 3-31).

Ref: FM 3-0 (Oct. '22), table B-1. Joint support categories.

E. Other Authorities

Some authorities exist outside joint command relationships, to include—
- Administrative control.
- Coordinating authority.
- Direct liaison authorized.

3-24 (Command & Control) II. Command & Support Relationships

II. Army Command and Support Relationships

See table B-2 and B-3 (following pages) for Army command and support relationships.

Command relationships define superior and subordinate relationships between units. The type of command relationship often relates to the expected longevity of the relationship between the headquarters involved, and it quickly identifies the degrees of operational and administrative control that the gaining and losing Army commanders provide.

A. Army Command Relationships

Army command relationships are similar but not identical to joint command authorities and relationships. Differences stem from the way Army forces task-organize internally and the need for a system of support relationships between Army forces. Another important difference is the requirement for Army commanders to handle the administrative support requirements that meet the needs of Soldiers. These differences allow for flexible allocation of Army capabilities within various Army echelons.

Organic

Organic forces are those assigned to and forming an essential part of a military organization. For example, a brigade engineer battalion is an organic unit in a brigade combat team (BCT). The Army establishes organic command relationships through organizational documents such as tables of organization and equipment and tables of distribution and allowances. If temporarily task-organized with another headquarters, organic units return to the control of their organic headquarters after completing the mission.

Assigned

Army assigned units remain subordinate to the higher echelon headquarters for extended periods, typically years. Assignment is based on the needs of the Army, and it is formalized by orders rather than organizational documents. Although force tailoring or task-organizing may temporarily detach units, they eventually return to either their headquarters of assignment or their organic headquarters. An Army headquarters is typically responsible for executing the ADCON responsibilities for subordinate Army units under its command unless modified by a higher headquarters.

Attached

Attached units are temporarily subordinated to the gaining headquarters, often for months or longer. They return to their parent headquarters (assigned or organic) when the reason for the attachment ends. The Army headquarters that receives another Army unit through assignment or attachment assumes responsibility for certain ADCON requirements, and particularly sustainment, that normally extend down to that echelon, as specified by directives or orders. For example, when an Army division commander attaches an air defense battery to a BCT, the brigade commander assumes responsibility for unit training, maintenance, resupply, and unit-level reporting for that battalion.

Operational Control (OPCON)

Army commanders normally place a unit OPCON or TACON to a gaining headquarters for a given mission, lasting perhaps a few days. OPCON lets the gaining commander task-organize and direct forces.

Tactical Control (TACON)

TACON does not let the gaining commander task-organize the unit. Neither OPCON nor TACON affects ADCON responsibilities.

(Command & Control) II. Command & Support Relationships 3-25

Army Command & Support Relationships

Ref: FM 3-0, Operations (Oct. '22), pp. B-4 to B-7.

Army command and support relationships are similar but not identical to joint command authorities and relationships. Differences stem from the way Army forces task-organize internally and the need for a system of support relationships between Army forces. Another important difference is the requirement for Army commanders to handle the administrative support requirements that meet the needs of Soldiers

A. Command Relationships

Army command relationships define superior and subordinate relationships between unit commanders. By specifying a chain of command, command relationships unify effort and enable commanders to use subordinate forces with maximum flexibility. Army command relationships identify the degree of control of the gaining Army commander. The type of command relationship often relates to the expected longevity of the relationship between the headquarters involved and quickly identifies the degree of support that the gaining and losing Army commanders provide.

If relation-ship is—	**Then the inherent responsibilities are:**							
	Have command relationship with—	May be task-organized by—	Unless modified, ADCON responsi-bility goes through—	Are assigned position or AO by—	Provide liaison to—	Establish/ maintain communica-tions with—	Have priorities establish-ed by—	Authorities CDR can impose on gaining unit further command or support relationship of—
Organic	Organic HQ	Organic HQ	Organic HQ	Organic HQ	N/A	N/A	Organic HQ	Attached; OPCON; TACON; GS; GSR; R; DS
Assigned	Gaining HQ	Gaining HQ	Gaining HQ	Gaining HQ	N/A	N/A	Gaining HQ	Attached; OPCON; TACON; GS; GSR; R; DS
Attached	Gaining HQ	Gaining HQ	Gaining HQ	Gaining HQ	As required by gaining HQ	Unit to which attached	Gaining HQ	Attached; OPCON; TACON; GS; GSR; R; DS
OPCON	Gaining HQ	Parent unit and gaining unit; gaining unit may pass OPCON to lower HQ	Parent HQ	Gaining HQ	As required by gaining HQ	As required by gaining HQ and parent HQ	Gaining HQ	OPCON; TACON; GS; GSR; R; DS
TACON	Gaining HQ	Parent HQ	Parent HQ	Gaining HQ	As required by gaining HQ	As required by gaining unit and parent HQ	Gaining HQ	TACON; GS GSR; R; DS

ADCON	administrative control		GSR	general support—reinforcing
ASCC	Army Service component command		HQ	headquarters
AO	area of operations		N/A	not applicable
CDR	commander		OPCON	operational control
DS	direct support		R	reinforcing
GS	general support		TACON	tactical control

Ref: FM 3-0 (Oct. '22), table B-2. Army command relationships.

B. Support Relationships

Table B-3 lists Army support relationships. Army support relationships are not a command authority and are more specific than the joint support relationships. Commanders establish support relationships when subordination of one unit to another is inappropriate. If a unit has an established command relationship with a headquarters, a support relationship is unnecessary as the command relationship already grants the gaining commander all the authorities required. Commanders assign a support relationship when—

3-26 (Command & Control) II. Command & Support Relationships

- The support is more effective if a commander with the requisite technical and tactical expertise controls the supporting unit, rather than the supported commander.
- The echelon of the supporting unit is the same as or higher than that of the supported unit. For example, the supporting unit may be a brigade, and the supported unit may be a battalion. It would be inappropriate for the brigade to be subordinated to the battalion, hence the use of an Army support relationship.
- The supporting unit supports several units simultaneously. The requirement to set support priorities to allocate resources to supported units exists. Assigning support relationships is one aspect of command and control (C2).

If relation-ship is—	**Then the inherent responsibilities are:**							
	Have command relation-ship with—	May be task-organiz-ed by—	Receives sustain-ment from—	Are assigned position or an area of operations by—	Provide liaison to—	Establish/ maintain communica-tions with—	Have priorities establish ed by—	Authorities a CDR can impose on gaining unit further command or support relation-ship by—
Direct support	Parent HQ	Parent HQ	Parent HQ	Supported HQ	Supported HQ	Parent HQ; supported HQ	Supported HQ	See note.
Reinforcing	Parent HQ	Parent HQ	Parent HQ	Reinforced HQ	Reinforced HQ	Parent HQ; reinforced HQ	Reinforced HQ; then parent HQ	Not applicable
General support– reinforcing	Parent HQ	Parent HQ	Parent HQ	Parent HQ	Reinforced HQ and as required by parent HQ	Reinforced HQ and as required by parent HQ	Parent HQ; then reinforced HQ	Not applicable
General support	Parent HQ	Parent HQ	Parent HQ	Parent HQ	As required by parent HQ	As required by parent HQ	Parent HQ	Not applicable

Note. Commanders of units in direct support may further assign support relationships between their subordinate units and elements of the supported unit after coordination with the supported commander.

CDR	commander		HQ		headquarters

Ref: FM 3-0 (Oct. '22), table B-3. Army support relationships.

Army support relationships allow supporting commanders to employ their units' capabilities to achieve results required by supported commanders. Support relationships are graduated from an exclusive supported and supporting relationship between two units—as in direct support—to a broad level of support extended to all units under the control of the higher echelon headquarters—as in general support. Support relationships do not alter ADCON. Any transfer of ADCON responsibilities should be specified in the order.

For the Army, **direct support** is a support relationship requiring a force to support another specific force and authorizing it to answer directly to the supported force's request for assistance. A unit assigned a direct support relationship retains its command relationship with its parent unit, but it is positioned by and has priorities of support established by the supported unit. (Joint doctrine considers direct support a mission rather than a support relationship.)

General support is that support given to the supported force as a whole and not to any particular subdivision thereof. Units assigned a general support relationship are positioned and have priorities established by their parent unit.

Reinforcing is a support relationship requiring a force to support another supporting unit. Only like units (for example, artillery to artillery) can be given a reinforcing mission. A unit assigned a reinforcing support relationship retains its command relationship with its parent unit, but it is positioned by the reinforced unit. A unit that is reinforcing has priorities of support established by the reinforced unit, then the parent unit.

General support—reinforcing is a support relationship assigned to a unit to support the force as a whole and to reinforce another similar type unit. A unit assigned a general support—reinforcing support relationship is positioned and has its priorities established by its parent unit and secondly by the reinforced unit.

(Command & Control) II. Command & Support Relationships 3-27

III. Other Relationships

Ref: FM 3-0, Operations (Oct. '22), pp. B-8 to B-9.

Several other relationships established by higher echelon headquarters exist with units that are not in command or support relationships. These relationships are limited or specialized to a greater degree than the command and support relationships, and they may be detailed in a command's implementing directives. These limited relationships are not used when tailoring or task-organizing Army forces. Use of these specialized relationships helps clarify certain aspects of OPCON or ADCON.

Relationship	Operational use	Established by	Authority and limitations
TRO	TRO is an authority exercised by a combatant commander over assigned RC forces not on active duty. Through TRO, CCDRs shape RC training and readiness. Upon mobilization of the RC forces, TRO is no longer applicable.	The CCDR identified in the "Forces for Unified Commands" memorandum. The CCDR normally delegates TRO to the ASCC.	TRO allows the CCDR to provide guidance on operational requirements and training priorities, review readiness reports, and review mobilization plans for RC forces. TRO is not a command relationship. ARNG forces remain under the command and control of their respective State Adjutant Generals until mobilized for Federal service. USAR forces remain under the command and control of the CG, USARC until mobilized.
TRA	TRA is an authority for a designated commander to give direction to an attached unit for leader development, individual and collective training, and unit readiness.	Higher commander.	TRA includes responsibility for all facets of command that enable commanders to accomplish their mission. It does not include those installation command authorities vested in the Army Senior Commander.
DIRLAUTH *Note.* See also paragraph B-16.	Allows planning and direct collaboration between two units assigned to different commands, often based on anticipated tailoring and task organization changes	The parent unit headquarters. This is a coordination relationship, not an authority through which command may be exercised.	Limited to planning and coordination between units.
Aligned	Informal relationship to facilitate planning between a theater army and other Army units identified in operations and exercises in a specific combatant command.	Theater army and parent command.	Normally establishes information channels for coordination between the gaining theater army and Army units that are likely to be committed to that area of responsibility.

ASCC	Army Service component command	RC	Reserve Component
ARNG	Army National Guard	TRO	training and readiness oversight
CCDR	combatant commander	USAR	United States Army Reserve
CG	commanding general	USARC	United States Army Reserve Command
DIRLAUTH	direct liaison authorized		

Ref: FM 3-0 (Oct. '22), table B-4. Other relationships.

Training and readiness oversight (known as TRO) is the authority that CCDRs may exercise over assigned Reserve Component forces when not on active duty or when on active duty for training. **Training and readiness authority** (known as TRA) is the discrete authority, granted by a higher echelon commander, for a designated commander to give direction to an attached unit for leader development, individual and collective training, and unit readiness (to include maintenance, manning, and equipping). **Alignment** is an informal relationship between a theater army and other Army units identified for use in the area of responsibility of a specific geographic combatant command.

3-28 (Command & Control) II. Command & Support Relationships

Chap 3

III. Command Posts

Ref: FM 6-0, Commander and Staff Organization and Operations, (May '22), chap. 7.

I. Command Posts

A command post is a headquarters, or a portion there of, organized for the exercise of command and control. When necessary, commanders control operations from other locations away from the CP. In all cases, the commander alone exercises command when in a CP or elsewhere.

CPs provide a physical location for people, processes, and networks to directly assist commanders as they understand, visualize, describe, direct, lead, and assess operations. CPs can vary in size, complexity and focus, such as the main CP or the tactical CP. CPs may be composed of vehicles, containers, and tents, or located in buildings.

Commanders systematically arrange platforms, operation centers, signal nodes, and support equipment in ways best suited for a particular operational environment. Examples of equipment needed to sustain a CP include vehicles, radio or signal equipment, generators, and lighting. Functions common to all CPs include—

- Conducting knowledge management, information management, and foreign disclosure.
- Building and maintaining situational understanding.
- Controlling operations (by coordinating, synchronizing, and integrating).
- Assessing operations.
- Coordinating with internal and external organizations.
- Performing CP administrative activities.

Effective command and control (C2) requires continuous, and often immediate, close coordination, synchronization, and information sharing across staffs and warfighting functions for directing activities. To promote this, commanders organize their staffs and other components of the C2 system into CPs to assist them in effectively conducting specific operations. Different types of CPs—such as the main CP, the tactical CP, or the rear CP—have specific functions by design.

CPs are arranged by echelon and unit, and they differ based on organization and employment. Commanders systematically arrange platforms, operation centers, signal nodes, and support equipment in ways best suited for a particular operational environment. Depending on the organization, type of unit, and situation, commanders echelon their headquarters into multiple CPs for the conduct of operations. CPs come in many different structures, and they may consist of vehicles, containers, and tents, or located in buildings. CPs provide the physical location for people, processes, and networks to directly assist commanders in understanding, visualizing, describing, directing, leading, and assessing operations.

Refer to BSS6: The Battle Staff SMARTbook, 6th Ed. for further discussion. BSS6 covers the operations process (ADP 5-0); commander's activities; Army planning methodologies; the military decisionmaking process and troop leading procedures (FM 7-0 w/Chg 2); integrating processes (IPB, information collection, targeting, risk management, and knowledge management); plans and orders; mission command, C2 warfighting function tasks, command posts, liaison (ADP 6-0); rehearsals & after action reviews; and operational terms and military symbols (ADP 1-02).

(Command & Control) III. Command Posts 3-29

II. Types of Command Posts

Ref: FM 6-0, Commander and Staff Organization and Operations, (May '22), pp. 7-4 to 7-8.

Main Command Post

A main command post is a portion of a unit headquarters containing the majority of the staff designed to command and control current operations, conduct detailed analysis, and plan future operations. The main CP is the unit's principal CP serving as the primary location for plans, analysis, sustainment coordination, and assessment. It includes representatives of all staff sections, warfighting functions, and information systems to plan, prepare, execute, and assess operations. The main CP is larger in size and in staffing and less mobile than the tactical CP. The COS or XO provides staff supervision of the main CP. All units at battalion and higher echelons have a main CP. General functions of the main CP include—

- Controlling operations.
- Receiving reports for subordinate units and preparing reports required by higher echelon headquarters.
- Planning operations, including branches and sequels.
- Integrating intelligence into current operations and plans.
- Synchronizing the targeting process.
- Planning and synchronizing sustaining operations.
- Assessing the overall progress of operations.

Contingency Command Post

A contingency command post is a portion of a unit headquarters tailored from the theater army headquarters that enables the commander to conduct small-scale operations within the assigned area of operations. Employing the contingency CP for a mission involves a tradeoff between the contingency CP's immediate response capability and its known limitations. These limitations include the scale, scope, complexity, intensity, and duration of operations that it can effectively command without significant augmentation. The contingency CP depends on the main CP for long-range planning and special staff functional support.

Tactical Command Post

A tactical command post is a portion of a unit headquarters designed to command and control operations as directed. Commanders employ the tactical CP as an extension of the main CP to allow the main CP to displace or to control major events while the main CP focuses on other events, such as shaping operations. The tactical CP maintains continuous communications with subordinates, higher echelon headquarters, other CPs, and supporting units. The tactical CP is fully mobile and includes only essential Soldiers and equipment. The tactical CP relies on the main CP for planning, detailed analysis, and coordination. A deputy commander or operations officer (G-3 or S-3) generally leads the tactical CP. While corps through battalion commanders employ a tactical CP as an extension of the main CP, corps and division tactical CPs are resourced differently for functionality. The division tactical CP is resourced to afford the division commander the ability to separate into two functional elements to be used as the commander requires. The corps tactical CP is resourced as a stand-alone capability that nearly mirrors the main CP's functionality. The functions of a tactical CP include but are not limited to—

- Controlling decisive operations or specific shaping operations.
- Controlling a specific task within larger operations such as a gap crossing, passage of lines, relief in place, or air assault operations.
- Controlling the overall unit's operations for a limited time when the main CP is displacing or otherwise not available.
- Performing short-range planning.

3-30 (Command & Control) III. Command Posts

- Providing input to targeting and future operations planning.
- Providing a forward location for issuing orders and conducting rehearsals.
- Forming the headquarters of a task force with subordinate units task-organized under its control.

See FM 3-94 for more details on the corps and division tactical CP.

When the commander does not employ the tactical CP, the staff assigned to it reinforces the main CP. Unit SOPs address the specifics for this reinforcement, including procedures to quickly detach the tactical CP from the main CP.

Rear Command Post

Depending on the situation—including the threat, size of the rear area, and number of units within the support and consolidation areas—division and corps commanders may form a rear CP to assist in controlling operations. The rear CP disperses the C2 signature and enables division and corps commanders to exercise C2 over disparate, functionally-focused elements operating between the close area and the division and corps rear boundary that may exceed the effective span of control of the main CP.

Functions of the rear CP may include planning and directing sustainment, terrain management, movement control, and area security. When augmented by the maneuver enhancement brigade staff, the rear CP may also plan and control combined arms operations with units under division or corps control, coordinate airspace, and employ fires.

Combat Trains Command Post

Combined arms battalions and infantry battalions are also resourced a combat trains CP. The combat trains CP controls and coordinates administrative and logistics support. It consists of members from the S-1 staff section, S-4 staff sections, and aid station. The battalion S-4 leads this CP. The battalion's forward support company normally co-locates with the combat trains CP.

Field Trains Command Post

The Army resources combined arms battalions and infantry battalions with a field trains CP. The battalion field trains and field trains CPs are located in the best positions to facilitate sustainment support. Field trains and field trains CPs are normally located within their battalion's area of operations. Field trains usually include a personnel administration center, elements of the S-4 sustainment staff section, elements of company supply sections, and elements of the forward support company. The brigade headquarters and headquarters company commander leads the field trains CP.

Command Group and Mobile Command Group

A command group and mobile command group allow the commander to maintain command and control when separated from the main, tactical, or rear command post. The mobile command group consists of assigned equipment to support the command group when necessary, and it is directed by the commander. A command group consists of the commander and selected staff members who assist the commander in controlling operations. Command group personnel include staff representation that can immediately affect current operations, such as maneuver, fires (including the air liaison officer), and intelligence. The mission dictates the command group's makeup.

Early-Entry Command Post (EECP)

An early-entry command post is a lead element of a headquarters designed to control operations until the remaining portions of the headquarters are deployed and operational. While not a separate section of the unit's TO&E, the early-entry CP (sometimes referred to as an assault command post) is an ad hoc organization comprised of equipment and personnel from the staff of the tactical, main, and rear CPs. Commanders can establish an early-entry CP to assist them in controlling operations during the deployment phase of operations or to exercise C2 of early-entry forces during a joint forcible entry operation. The early-entry CP performs the functions of the main and tactical CPs until those CPs are deployed and operational.

(Command & Control) III. Command Posts 3-31

III. CP by Echelon and Type of Unit

Ref: FM 6-0, Commander and Staff Organization and Operations, (May '22), pp. 7-4 to 7-5.

Depending on the organization, type of unit, and situation, commanders echelon their headquarters into multiple CPs for the conduct of operations. A theater army is resourced with a main CP and a contingency CP. Corps, divisions, and brigade combat teams can employ a main CP, tactical CP, and a mobile command group. In addition, corps and divisions may operate a rear CP. Combined arms battalions, Stryker battalions, and infantry battalions can employ a main CP, tactical CP, combat trains CP, and a field trains CP. Some multifunctional brigades and battalions operate from a single main CP. Table 7-1 summarizes the various CPs resourced by echelon of command and type of unit. Beyond this publication and the table of organization and equipment (TO&E), units at all levels must establish SOPs to designate the commanders' desired roles and responsibilities of each CP within their organization.

Echelon or Type of Unit	Command Posts
Theater army	• Main command post. • Contingency command post. (See ATP 3-93 for more information on theater army.)
Corps	• Main command post. • Tactical command post. • Mobile command group and early-entry command post (ad hoc with personnel and equipment from the others). • Rear command post. (See FM 3-94 and ATP 3-92 for more information on corps.)
Division	• Main command post. • Tactical command post. • Mobile command group and early-entry command post (ad hoc with personnel and equipment from the others). • Rear command post. (See FM 3-94 and ATP 3-91 for more information on division command posts.)
Brigade combat teams	• Main command post. • Tactical command post. (See FM 3-96 for more information on brigade combat team command posts.)
Multifunctional brigades	• Main command post. • Tactical command post. (These organizations vary extensively. See specific doctrine for each type of multifunctional support brigade.)
Functional brigades and battalions	• Main command post. • Tactical command post. (These organizations vary extensively. See specific doctrine for each type of functional brigade and battalion.)
Maneuver battalions	• Main command post. • Tactical command post. • Combat trains command post. • Field trains command post. (See ATP 3-90.5 for more information on combined arms and infantry battalion command posts.)

Ref: FM 6-0, (May '22), table 7-1. Command posts by echelon and type of unit.

Chap 3

IV. Army Airspace Command & Control

Ref: FM 3-52, Airspace Control (Oct '16), chap. 1.

Airspace is a component of an operational environment critical to successful Army or land operations. Airspace is not owned by individual subordinate organizations in the sense an assigned area of operations confers ownership of the ground. Airspace over an Army area of operations remains under the purview of the joint force commander (JFC). Other military and civilian organizations operating in the joint operations area have airspace requirements over an Army area of operations. These organizations may require airspace to—

- Conduct joint air operations
- Conduct area air defense
- Deliver joint fires
- Conduct civil air operations

See p. 6-17 for discussion of airspace planning and integration.

Cmd & Control (ADP 6-0)

Joint Air Operations

Normally, the JFC designates a joint force air component commander (JFACC) to synchronize the joint air effort. Components retain organic capabilities (sorties) to accomplish missions assigned by the JFC. Components also make capabilities (sorties), either JFC directed or excess, available to the JFC for tasking by the JFACC. Generally, Army capabilities (sorties) are never made available to nor directed by the JFC for tasking by the JFACC. The JFACC plans for and tasks only those joint capabilities (sorties) made available to the JFC for tasking by the JFACC. The JFACC has the authority to direct and employ these joint capabilities (sorties) for a common purpose based on the JFC's concept of operations and air apportionment decisions.

The responsibilities of the JFACC, the area air defense commander (AADC), and airspace control authority (ACA) are interrelated and the JFC normally assigns them to one individual for unity of effort. These responsibilities are normally assigned to the JFACC. Designating one Service component commander as the JFACC, AADC, and ACA often simplifies the coordination required to develop and execute fully integrated air operations.

Area Air Defense (AADC)

The AADC oversees defensive counterair (DCA) operations, which include both air and missile threats. The AADC identifies airspace coordinating measures (ACMs) that support and enhance DCA operations, identifies required airspace management systems, establishes procedures for systems to operate within the airspace, and incorporates them into the airspace control system.

Refer to JP 3-01 for more information on the AADC.

Joint Airspace Control

Competing airspace users balance the demands for and integrate their requirements for airspace. Airspace control is a process used to increase operational effectiveness by promoting the safe, efficient, and flexible use of airspace. To help balance the various airspace user demands, the JFC usually designates an ACA responsible for establishing an airspace control system. An airspace control system is an arrangement of those organizations, personnel, policies, procedures, and facilities required to perform airspace control functions. *Refer to JP 3-52 for more information.*

(Command & Control) IV. Army Airspace Command and Control 3-33

I. Airspace Coordinating Measures (ACMs)

Ref: FM 3-52, Airspace Control (Oct '16), app B (adapted from previous sources).

Methods of airspace control range from positive control of all air assets in an airspace control area to procedural control of air assets, or a combination of both.

Positive Control
Positive control relies on positive identification, tracking, and direction of aircraft within the airspace control area. It uses electronic means such as radar; sensors; identification, friend or foe (IFF) systems; selective identification feature (SIF) capabilities; digital data links; and other elements of the intelligence system and C2 network structures.

Procedural Control
Procedural control relies on a combination of mutually agreed and promulgated orders and procedures. These may include comprehensive AD identification procedures and ROE, aircraft identification maneuvers, fire support coordinating measures (FSCMs), and airspace control measures (ACMs). Service, joint, and multinational capabilities and requirements determine which method, or which elements of each method, that airspace control plans and systems use. Airspace control measures provide a variety of procedural measures of controlling airspace users and airspace.

Airspace Control Measures

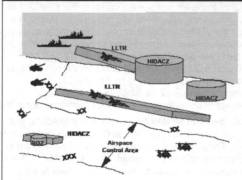

When established, airspace control measures accomplish one or more of the following:
- Reserve airspace for specific airspace users
- Restrict actions of airspace users
- Control actions of specific airspace users
- Require airspace users to accomplish specific actions

1. Air Control Point (ACP)
An easily identifiable point on the terrain or an electronic navigational aid used to provide necessary control during air movement. ACPs are generally designated at each point where the flight route or air corridor makes a definite change in direction and at any other point deemed necessary for timing or control of the operation.

2. Air Corridor
A restricted air route of travel specified for use by friendly aircraft and established to prevent friendly aircraft from being fired on by friendly forces. (Army) — Used to deconflict artillery firing positions with aviation traffic, including unmanned aerial vehicles.

3. Communications Checkpoint (CCP)
An air control point that requires serial leaders to report either to the aviation mission commander or the terminal control facility.

4. Downed Aircrew Pickup Point
A point to where aviators will attempt to evade and escape to be recovered by friendly forces.

5. High Density Airspace Control Zone (HIDACZ)

Airspace designated in an airspace control plan or airspace control order, in which there is a concentrated employment of numerous and varied weapons and airspace users. A high-density airspace control zone has defined dimensions, which usually coincide with geographical features or navigational aids. Access to a high-density airspace control zone is normally controlled by the maneuver commander. The maneuver commander can also direct a more restrictive weapons status within the high-density airspace control zone.

6. Low-Level Transit Route (LLTR)

A temporary corridor of defined dimensions established in the forward area to minimize the risk to friendly aircraft from friendly ADA or ground forces.

7. Minimum-Risk Route (MRR)

A temporary corridor of defined dimensions recommended for use by high-speed, fixed-wing aircraft that presents the minimum known hazards to low-flying aircraft transiting the combat zone.

8. Restricted Operations Zone (ROZ)

A volume of airspace of defined dimensions designated for a specific mission. Entry into that zone is authorized only by the originating headquarters.

9. Standard Use Army Aircraft Flight Route (SAAFR)

Routes established below the coordinating altitude to facilitate the movement of Army aviation assets. Routes are normally located in the corps through brigade rear areas of operation and do not require approval by the airspace control authority.

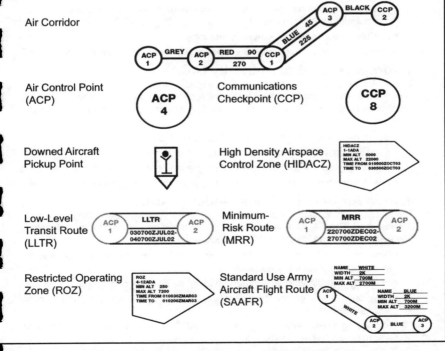

(Command & Control) IV. Army Airspace Command and Control 3-35

II. Key Positions and Responsibilities

There are four key positions critical to planning for and executing airspace control.

1. Joint Force Commander

The joint force commander has many responsibilities, to include the airspace control. For airspace control, the JFC specifically must—

- Include overall responsibility of airspace control and air defense in a joint theater
- Establish airspace control objectives and priorities for the joint force
- Oversee the planning and force integration activities that affect the TAGS
- Resolve matters on which the ACA is unable to obtain agreement
- Possibly retain airspace control responsibilities (or he may appoint an ACA)

The JFC may designate a JFACC as a single component commander for theater- or JOA-wide counterair operations. The JFACC will have the preponderance of air power. He also has the ability to provide C2 and produce and disseminate an ATO and ACO. He is normally appointed as the ACA and AADC. The JFC normally tasks the same person as the ACA, AADC, and JFACC to maintain the flexibility to effectively meet the enemy air threat and manage airspace control. Additional information on the selection and responsibilities of the JFACC can be found in JP 3-56.1.

2. Joint Force Air Component Commander (JFACC)

The JFACC may be sea or land based. JFACC responsibilities include—

- Developing a joint air operations plan to best support force objectives
- Recommending apportionment of the joint air effort to the JFC
- Providing centralized direction for the allocation and tasking capabilities and forces
- Controlling execution of joint operations as specified by the JFC
- Coordinating joint air operations with operations of other component commanders and forces assigned to or supporting the JFC
- Evaluating the results of joint air operations
- Functioning as the supported & supporting commander, as directed by the JFC

3. Airspace Control Authority (ACA)

The ACA is responsible for operating the airspace control system in the airspace control area. Centralized direction by the ACA does not imply assumption of operational or tactical control over any air assets. The ACA has broad responsibilities to include—

- Coordinate, integrate, and regulate the use of the airspace
- Establish broad policies and procedures for airspace control
- Establish airspace control system & integrating host-nation/multinational forces
- Develop airspace control plan (and implement through airspace control order)

4. Area Air Defense Commander (AADC)

The AADC is responsible for planning, coordinating, and integrating the joint area air defense plan. The AADC develops broad policies and procedures for air defense. The AADC has broad responsibilities to include—

- Developing and executing a plan to disseminate timely cueing of information and air and missile early warnings
- Planning, coordinating, and integrating joint air defense operations
- Developing and implementing identification and engagement procedures for air and missile threats
- Appointing a deputy AADC to assist the AADC in planning and coordinating air and missile defense operations

3-36 (Command & Control) IV. Army Airspace Command and Control

Chap 4: Movement & Maneuver Warfighting Function

Ref: ADP 3-0, Operations (Jul '19), p. 5-4 and FM 3-0, Operations (Oct. '22), p. 2-2.

The movement and maneuver warfighting function is the related tasks and systems that move and employ forces to achieve a position of relative advantage over the enemy and other threats. Direct fire and close combat are inherent in maneuver. The movement and maneuver warfighting function includes tasks associated with force projection. Movement is necessary to disperse and displace the force as a whole or in part when maneuvering. Maneuver directly gains or exploits **positions of relative advantage.** Commanders use maneuver for massing effects to achieve surprise, shock, and momentum. Effective maneuver requires close coordination of fires and movement. Both tactical and operational maneuver require sustainment support.

- Movement and maneuver.
- Employ direct fires.
- Occupy an area.
- Conduct mobility and countermobility.
- Conduct reconnaissance and surveillance.
- Employ battlefield obscuration.

Direct fire and close combat are inherent in maneuver. The movement and maneuver warfighting function includes tasks associated with force projection. Movement is necessary to position and disperse the force as a whole or in part when maneuvering. Maneuver directly gains or exploits positions of relative advantage. Commanders use maneuver for massing effects to achieve surprise, shock, and momentum.

Effective maneuver requires some combination of **reconnaissance, surveillance, and security operations** to provide early warning and protect the main body of the formation. Every Soldier on the battlefield is a potential sensor that makes key contributions to information collection and the development of intelligence. Effective maneuver requires close coordination of fires and movement. Movement and maneuver contribute to the development of information advantages through the positioning of units able to employ capabilities in close proximity to the enemy, as well as by physically establishing the facts on the ground that an enemy or adversary cannot refute.

Maneuver requires sustainment. The movement and maneuver warfighting function does not include routine transportation of personnel and materiel that support operations, which falls under the sustainment warfighting function.

For the purposes of The Army Operations & Doctrine SMARTbook, an overview of the following are provided: offense and defense (pp. 4-6 to 4-7), stability operations (pp. 4-8 to 4-9), and deployment/force projection operations (p. 4-10).

Refer to SUTS3: The Small Unit Tactics SMARTbook, 3rd Ed., completely updated with the latest publications for 2019. Chapters and topics include tactical fundamentals, the offense; the defense; train, advise, and assist (stability, peace & counterinsurgency ops); tactical enabling tasks (security, reconnaissance, relief in place, passage of lines, encirclement, and troop movement); special purpose attacks (ambush, raid, etc.); urban and regional environments (urban, fortified areas, desert, cold, mountain, & jungle operations); patrols & patrolling.

(Movement & Maneuver) Warfighting Function 4-1

Imperatives: Movement & Maneuver

Ref: FM 3-0, Operations (Oct. '22), pp. 3-13 to 3-16.

Editor's Note: The following select imperatives of Army operations from FM 3-0 (2022) are highlighted here in greater detail to illustrate application within the movement and maneuver warfighting function. See pp. 1-44 to 1-45 for a listing of all imperatives.

Account for Constant Enemy Observation

Enemy forces possess a wide range of space-, air-, maritime-, and land-based reconnaissance and surveillance capabilities that can detect U.S. forces. To counter these robust and persistent capabilities requires counterintelligence efforts and the disciplined application of operations security.

Enemy forces employ UASs in large numbers and with a diverse array of capabilities. Leaders account for enemy capabilities and likely reconnaissance objectives as they develop their counter-UAS plan. Leaders implement techniques and procedures for countering enemy UASs based on their organic capabilities, attached capabilities, and the mission variables.

Leaders combine multiple measures, including deception, to make it more difficult for enemy forces to detect friendly forces. These measures include—

- Counterreconnaissance, including counter-UAS operations.
- Cover and concealment, both natural and manmade.
- False battle positions and deception obstacles.
- Obscuration.
- Dispersion.
- Noise and light discipline.
- Limited visibility operations, particularly for sustainment functions and large unit movements.
- Electromagnetic emission control and masking, to include social media and personal communication discipline.

Because Army forces employ an increasing number of capabilities that emit electromagnetic radiation that enemies can target, leaders must apply emission control measures, balancing the risks to the force with the risks to the mission. As risk to the force increases, leaders increase their emission control measures. There may be times that the risk of friendly emissions being detected and targeted is assessed as too high, causing Army forces to use methods of communications with no electromagnetic signature.

Implementing Dispersion. Leader efforts to preempt and mitigate enemy detection are essential, but they cannot eliminate the risk of enemy massed and precision fires, including CBRN and weapons of mass destruction. To improve survivability from enemy indirect fires, Army forces maintain dispersion and remain as mobile as possible to avoid presenting themselves as lucrative targets to the enemy's most capable systems. When mission demands require units to remain static for more than short periods of time, those units must dig in to increase survivability.

Refer to ATP 3-37.34 for information on survivability positions.

4-2 (Movement & Maneuver) Warfighting Function

Account for All Forms of Enemy Contact

Leaders consider **nine forms of contact in multiple domains**. They are—

- **Direct**: interactions from line-of-sight weapon systems (including small arms, heavy machine guns, and antitank missiles).
- **Indirect**: interactions from non-line-of-sight weapons systems (including cannon artillery, mortars, and rockets).
- **Non-hostile**: neutral interactions that may degrade or compromise military operations (including civilians on the battlefield).
- **Obstacle**: interactions from friendly, enemy, and natural obstacles (including minefields and rivers).
- **CBRN**: interactions from friendly, enemy, and civilian CBRN effects (including chemical attacks, nuclear attacks, industrial accidents, and toxic or hazardous industrial materials).
- **Aerial**: interactions from air-based combat platforms (including attack helicopters, armed unmanned aircraft systems [UASs], air interdiction, and close air support).
- **Visual**: interactions from acquisition via the human eye, optical, or electro-optical systems (including ground reconnaissance, telescopic, thermal, and infrared sights on weapons and sensor platforms such as unmanned aircraft systems and satellites).
- **Electromagnetic**: interactions via systems used to acquire, degrade, or destroy using select portions of the electromagnetic spectrum (including radar, jamming, cyberspace, space, and electromagnetic systems).
- **Influence***: interactions through the information dimension intended to shape the perceptions, behaviors, and decision making of people relative to a policy or military objective (including through social media, telecommunications, human interaction, and other forms of communication).

* Influence is introduced as a ninth form of contact in this edition of FM 3-0 (2022)

In all contexts, direct, indirect, non-hostile, CBRN, and aerial contacts are sporadic. However, Army forces are typically in continuous visual, electromagnetic, and influence contact with adversaries. Army forces are under persistent visual surveillance by space and other capabilities. Army forces and individuals are in constant electromagnetic contact with adversaries who persistently probe and disrupt individual, group, and Army capabilities dependent on space and cyberspace. Army forces are subject to adversary influence through disinformation campaigns targeting Soldiers and their family and friends through social media and other platforms.

During competition, adversary forces employ multiple methods of collecting on friendly forces to develop an understanding of U.S. capabilities, readiness status, and intentions. They do this in and outside the continental United States. They co-opt civilians and employ space-based surveillance platforms to observe unit training and deployment activities. They also penetrate networks and gain access to individual and group cyberspace personas to create options for future intimidation, coercion, and attack. Soldiers and their families should use telecommunications, the internet, and social media in ways that do not make them or their units vulnerable to adversary surveillance.

During armed conflict, enemy networked land-, maritime-, air-, and space-based capabilities enable threats to detect and rapidly target friendly forces with fires. Forces that are concentrated and static are easy for enemy forces to detect and destroy. Dispersing forces has multiple survivability benefits. It increases opportunities to use cover and concealment to reduce probability of detection. In the event the enemy detects elements of the friendly force, dispersion acts as a form of deception, helping to conceal the intentions of the friendly force. Leaders only concentrate forces when necessary and balance the survivability benefits of dispersion with the negative impacts dispersion has on mis-

Continued on next page

Movement & Maneuver

(Movement & Maneuver) Warfighting Function 4-3

Imperatives: Movement & Maneuver (cont.)

Continued from previous page

sion effectiveness. In addition to dispersion, leaders integrate and synchronize deception, operations security, and other actions to thwart enemy detection efforts.

Command posts are extremely vulnerable to detection from air and space, as well as in
the electromagnetic spectrum. Army forces must ensure their command posts are difficult to detect, dispersed to prevent a single strike from destroying more than one node,
and rapidly displaceable. Once a command post is detected it has only a few minutes to
displace far enough to avoid enemy indirect fire effects. Leaders should focus command
posts on the minimum functions necessary to retain their mobility and do everything
possible to avoid detection. When the risk of enemy fires is high, commanders consider making their operations more decentralized, dispersing command post nodes into
smaller component nodes, and greater dispersion of electromagnetic signatures. Use of
existing hardened structures and restrictive terrain to conceal headquarters equipment
and vehicles, instead of tents organized in standard configurations, are options commanders have to improve command post survivability.

Create and Exploit Relative Advantages

The employment of lethal force is based on the premise that destruction and other physical consequences compels enemy forces to change their decision making and behavior,
ultimately accepting defeat. The type, amount, and ways in which lethal force compels
enemy forces varies, and this depends heavily on enemy forces, their capabilities, goals,
and the will of relevant populations. Understanding the relationship between physical,
information, and human factors enables leaders to take advantage of every opportunity
and limit the negative effects of undesirable and unintended consequences.

Dimensions. Actions taken focused on one dimension can create advantages in the other dimensions. The physical dimension dominates tactical actions and the employment
of destructive force to compel an outcome. Physical actions, particularly the employment of violence, usually generate cognitive effects in the human dimension. Information
dimension factors inform and reflect the interaction between human and physical factors.
The information dimension deals with how relevant actors and populations communicate
what is happening in the physical and human dimensions. The human dimension is
where perceptions, decision making, and behavior is determined, and is therefore the
dimension that ultimately determines human will. Commanders combine, reinforce, and
exploit advantages through all the dimensions, expanding them as they accrue over
time.

Decision Dominance. Successful military operations often depend on a commander's
ability to gain and maintain the operational initiative by achieving decision dominance—a
desired state in which a force generates decisions, counters threat information warfare
capabilities, strengthens friendly morale and will, and affects threat decision making
more effectively than the opponent. Decision dominance requires developing a variety of
information advantages relative to that of the threat and then exploiting those advantages to achieve objectives. Commanders employ relevant military capabilities from all
warfighting functions to create and exploit decision dominance.

Make Initial Contact with the Smallest Element Possible

Army forces are extremely vulnerable when they do not sufficiently understand the disposition of enemy forces and become decisively engaged on terms favorable to enemy
forces. To avoid being surprised and incurring heavy losses, leaders must set conditions
for making enemy contact on terms favorable to the friendly force. They anticipate when
and where to make enemy contact, the probability and impact of making enemy contact,
and actions to take on contact. Quickly applying multiple capabilities against enemy
forces while preventing the bulk of the friendly force from being engaged itself requires
an understanding of the advantageous way. Judicious employment of all available recon-

Continued from previous page

4-4 (Movement & Maneuver) Warfighting Function

naissance and security capabilities is the most effective way to make direct contact with the smallest possible friendly force. Friendly forces should attempt to make contact with sensors and unmanned systems first, incorporating them into their movement techniques. Employment of UAS and other platforms activates enemy systems and enables their detection without creating risks to manned friendly reconnaissance and maneuver forces. After detecting an enemy capability, Army forces cue intelligence platforms from other domains to improve their understanding of enemy force dispositions and engage those forces on advantageous terms.

Using capabilities from multiple domains, such as air and ground, commanders cause threat systems to activate or emit electromagnetic signals that reveal their capabilities and the locations of their critical nodes, such as sensors, shooters, and command posts. During competition, commanders and staffs use this information to improve understanding, update target lists, and refine plans for attacking threat vulnerabilities. By doing this, commanders and staffs set conditions for success during armed conflict.

There are situations in which it is not advisable to make contact with the smallest possible element. When commanders are confident they have superior forces, have the element of surprise, and know the enemy's disposition and course of action, they make contact with as much combat power as possible to maximize surprise and shock effect against enemy forces.

Impose Multiple Dilemmas on the Enemy

Imposing multiple dilemmas on enemy forces complicates their decision making and forces them to prioritize among competing options. It is a way of seizing the initiative and making enemy forces react to friendly operations. Simultaneous operations encompassing multiple domains—conducted in depth and supported by deception—present enemy forces with multiple dilemmas. Employing capabilities from multiple domains degrades enemy freedom of action, reduces enemy flexibility and endurance, and disrupts enemy plans and coordination. The application of capabilities in complementary and reinforcing ways creates more problems than an enemy commander can solve, which erodes both enemy effectiveness and the will to fight.

Deception contributes to creating multiple dilemmas, achieving operational surprise, and maintaining the initiative. Deception efforts by tactical formations seek to delay enemy decision making until it is too late to matter, or to cause an enemy commander to make the wrong decision. Deception requires an understanding of how to surprise enemy forces; time to plan, prepare, execute, and assess a deception operation; and the ability to properly resource the deception effort.

Forcible entry operations and envelopments into locations offset from how enemy defenses are oriented can create multiple dilemmas by dislocating enemy forces' prepared operational approach or exceeding their capability to respond. The capability to project power across operational distances presents enemy forces with difficult decisions about how to array their forces in time and space. Rapid tactical maneuver to exploit a penetration or envelopment defeats enemy attempts to reposition integrated fires networks or integrated air defense systems, which in turn are typically less effective when moving.

Creating multiple dilemmas requires recognizing exploitable opportunities. Understanding enemy dispositions, systems, and vulnerabilities, and the characteristics of the terrain and population, informs situational understanding and course of action development. Employing mutually supporting forces along different axes to strike from unexpected directions creates dilemmas, particularly when Army and joint forces simultaneously create effects against enemy forces in multiple domains. Commanders seek every opportunity to make enemy forces operate in different directions at the time and locations of their choosing. Commanders are not limited to destructive means for imposing multiple dilemmas on the enemy. For example, they can employ psychological operations and civil affairs capabilities to influence and garner the support of civilian populations. This creates a dilemma for enemy forces who must react and divert resources to counter passive or active resistance.

(Movement & Maneuver) Warfighting Function 4-5

I. Offense and Defense
Ref: ADP 3-90, Offense and Defense (Jul '19).

ADP 3-90, Offense and Defense, articulates how Army forces conduct the offense and defense. It contains the fundamental tactics related to the execution of these elements of decisive action. Tactics employs, orders arrangement of, and directs actions of forces in relation to each other. Commanders select tactics that place their forces in positions of relative advantage. The selected tactics support the attainment of goals. Tactics create multiple dilemmas for an enemy allowing the friendly commander to defeat the enemy in detail. Successful tactics require synchronizing all the elements of combat power.

Tactics is the employment, ordered arrangement, and directed actions of forces in relation to each other. Tactics always require judgment and adaptation to a situation's unique circumstances. Techniques and procedures are established patterns or processes that can be applied repeatedly with little judgment to various circumstances. Together, tactics, techniques, and procedures (TTP) provide commanders and staffs with the fundamentals to develop solutions to tactical problems. The solution to any specific problem is a unique combination of these fundamentals, current TTP, and the creation of new TTP based on an evaluation of the situation. Commanders determine acceptable solutions by mastering doctrine and current TTP. They gain this mastery through experiences in education, training, and operations.

The Tactical Level of War *(See p. 4-11.)*
The tactical level of warfare is the level of warfare at which battles and engagements are planned and executed to achieve military objectives assigned to tactical units or task forces (JP 3-0). Activities at this level focus on achieving assigned objectives through the ordered arrangement, movement, and maneuver of combat elements in relation to each other and to enemy forces. The strategic and operational levels of warfare provide the context for tactical operations. Without this context, tactical operations become disconnected from operational end states and strategic goals.

The Offense *(See pp. 1-118 to 1-125.)*
The offense is the decisive form of war. The offense is the ultimate means commanders have of imposing their will on enemy forces. Army forces conduct the offense to defeat and destroy enemy forces as well as gain control of terrain, resources, and population centers. Commanders may also conduct the offense to deceive or divert an enemy force, develop intelligence, or hold an enemy force in position. Commanders seize, retain, and exploit the initiative when conducting the offense. Specific operations may orient on an enemy force or terrain objective to achieve a position of relative advantage. Taking the initiative from an enemy force requires the conduct of the offense, even in the defense.

The main purposes of the offense are to defeat enemy forces, destroy enemy forces, and gain control of terrain, resources, and population centers. Additionally, commanders conduct the offense to secure decisive terrain, to deprive the enemy of resources, to gain information, to deceive and divert the enemy, to hold the enemy in position, to disrupt his attack, and to set the conditions for future successful operations.

The Defense *(See pp. 1-110 to 1-117.)*
While the offense is more decisive, the defense is usually stronger. However, the conduct of the defense alone normally cannot determine the outcome of battles. Army forces generally conduct the defense to create conditions favorable for the offense.

The purpose of the defense is to create conditions for the offense that allows Army forces to regain the initiative. Other reasons for conducting the defense include retaining decisive terrain or denying a vital area to an enemy, attriting or fixing an enemy as a prelude to the offense, countering enemy action, and increasing an enemy's vulnerability by forcing an enemy commander to concentrate subordinate forces.

A defensive operation is an operation to defeat an enemy attack, gain time, economize forces, and develop conditions favorable for offensive or stability operations (ADP 3-0).

4-6 (Movement & Maneuver) Warfighting Function

The inherent strengths of the defense are the defender's ability to occupy positions before an attack and use the available time to improve those defenses. A defending force stops improving its defensive preparations only when it retrogrades or begins to engage enemy forces.

Enabling Operations *(See pp. 1-108 to 1-109.)*

Commanders perform enabling operations to help in the planning, preparation, and execution of any of the four elements of decisive action. Enabling operations are never decisive operations. Enabling operatons discussed in ADP 3-90 include reconnaissance, security, troop movement, relief in place, passage of lines, encirclement operations, and urban operations. Other publications discuss other enabling operations. For example, FM 3-13 discusses information operations, ATP 3-90.4 discusses mobility operations, and ATP3-90.8 discusses countermobility operations.

Elements of Decisive Action

Offensive operations
- Movement to Contact
    Search and Attack
    Cordon and Search
- Attack
    Ambush
    Counterattack
    Demonstration
    Feint
    Raid
    Spoiling attack
- Exploitation
- Pursuit
    Frontal
    Combination

Defensive operations
- Area Defense
- Mobile Defense
- Retrograde
    Delay
    Withdraw
    Retirement

Stability operations tasks
- Establish civil security
- Support to civil control
- Restore essential services
- Support to governance
- Support to economic and infrastructure development
- Conduct security cooperation

Defensive support of civil authorities tasks
- Provide support for Domestic disasters
- Provide support for domestic chemical, biological, radiological, and nuclear incidents
- Provide support for domestic civilian law enforcement agencies
- Provide other designated domestic support

Enabling operations

- **Reconnaissance**
    Area
    Reconnaissance in force
    Route
    Special
    Zone
- **Passage of lines**
    Forward
    Rearward
- **Troop movement**
    Administrative movement
    Approach march
    Tactical road march
- **Relief in place**
    Sequential
    Simultaneous
    Staggered
- **Security**
    Screen
    Guard
    Cover
    Area

Tactical Mission Tasks

- Ambush
- Attack by fire
- Block
- Breach
- Bypass
- Canalize
- Clear
- Contain
- Control
- Counterreconnaisance
- Destroy
- Defeat
- Disengagement
- Disrupt
- Exfiltrate
- Fix
- Follow and assume
- Follow and support
- Interdict
- Isolate
- Neutralize
- Occupy
- Reduce
- Retain
- Secure
- Seize
- Support by fire
- Suppress
- Turn

Forms of Maneuver and Forms of the Defense

- Envelopment
- Frontal assault
- Infiltration
- Penetration
- Turning movement

- Defense of a linear obstacle
- Perimeter defense
- Reverse slope defense

Ref: ADP 3-90, Offense and Defense (Jul '19), Figure 2-1. Taxonomy of Army tactics.

Refer to SUTS3: The Small Unit Tactics SMARTbook (Planning & Conducting Tactical Operations) for complete discussion of offensive and defensive operations. Related topics include tactical mission fundamentals, stability & counterinsurgency operations, tactical enabling operations, special purpose attacks, urban operations & fortifications, and patrols & patrolling.

(Movement & Maneuver) Warfighting Function 4-7

II. Stability Operations

Ref: ADP 3-07, Stability (Jul '19).

Ultimately, stability is the set of conditions in which a local populace regards its governance institutions as legitimate and its living situation as acceptable and predictable. Actions to maintain or reestablish stability first aim to lessen the level of violence. These actions also aim to enable the functioning of governmental, economic, and societal institutions. Lastly, these actions encourage the general adherence to local laws, rules, and norms of behavior.

A stability operation is an operation conducted outside the United States in coordination with other instruments of national power to establish or maintain a secure environment and provide essential governmental services, emergency infrastructure reconstruction, and humanitarian relief (ADP 3-0). A stability operation occurs as part of decisive action in a joint operation or as an activity (often in peacetime). Stability operations be performed as tasks (specified or implied) in an operation focused on combat, or be performed as activities (often in peacetime).

Stabilization is a process in which personnel identify and mitigate underlying sources of instability to establish the conditions for long-term stability. Stabilization also includes efforts to counter an adversary's attempts to consolidate its gains in a region or to reassert its influence. While long-term development requires stability, stability does not require long-term development. Therefore, stability tasks focus on identifying and targeting the root causes of instability and by building the capacity of local institutions.

Primary Army Stability Tasks

Six Army stability operations tasks correspond to the stability sectors adopted by the DOS.

1. Establish Civil Security. Establishing civil security involves providing for the safety of the host nation and its population, including protection from internal and external threats. Establishing civil security provides needed space for host-nation and civil agencies and organizations to work toward sustained peace.

2. Establish Civil Control. Establishing civil control supports efforts to institute rule of law and stable, effective governance. Civil control relates to public order—the domain of the police and other law enforcement agencies, courts, prosecution services, and prisons (known as the Rule of Law sector).

3. Restore Essential Services. The restoration of essential services in a fragile environment is essential toward achieving stability. The basic functions of local governance stop during conflict and other disasters. Initially, military forces lead efforts to establish or restore the most basic civil services: the essential food, water, shelter, and medical support necessary to sustain the population until forces restore local civil services. Military forces follow the lead of other USG agencies, particularly United States Agency for International Development, in the long restoration of essential services.

4. Support to Governance. When a legitimate and functional host-nation government exists, military forces operating to support the state have a limited role. However, if the host-nation government cannot adequately perform its basic civil functions—whatever the reason—some degree of military support to governance may be necessary. Military efforts to support governance focus on restoring public administration and resuming public services.

5. Support to Economic and Infrastructure Development. Military efforts to support the economic sector are critical to sustainable economic development. The economic viability of a host nation often exhibits stress and ultimately fractures as conflict, disaster, and internal strife overwhelms the government.

6. Conduct Security Cooperation. Security cooperation is all DoD interactions with foreign security establishments to build security relationships that promote specific United States security interests, develop allied and partner nation military and security capabilities for self-defense and multinational operations, and provide U.S. forces with peacetime and contingency access to allied and partner nations (JP 3-20).

4-8 (Movement & Maneuver) Warfighting Function

Stability Underlying Logic

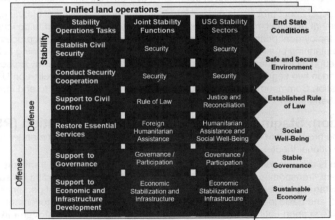

Ref: ADP 3-07, Stability (Jul '19), Introductory figure-1. Stability underlying logic.

Refer to TAA2: Military Engagement, Security Cooperation & Stability SMARTbook (Foreign Train, Advise, & Assist) for further discussion. Topics include the Range of Military Operations (JP 3-0), Security Cooperation & Security Assistance (Train, Advise, & Assist), Stability Operations, Peace Operations, Counterinsurgency Operations, Civil-Military Operations, and more!

(Movement & Maneuver) Warfighting Function 4-9

III. Force Projection/Deployment Operations

Ref: ATP 3-35 (FM 3-35), Army Deployment and Redeployment (Mar '15).

Note: The movement and maneuver warfighting function includes tasks associated with force projection, but does not include administrative movements. Administrative movements fall under the sustainment warfighting function.

ATP 3-35 is the Army's authoritative doctrine for planning, organizing, executing, and supporting deployment and redeployment.

Force Projection (See pp. 1-83 to 1-85.)

Force projection is the ability to project the military instrument of national power from the United States or another theater, in response to requirements for military operations. (JP 3-0) It is a demonstrated ability to alert, mobilize, rapidly deploy, and operate effectively anywhere in the world. The Army, as a key member of the joint team, must be ready for global force projection with an appropriate mix of combat forces together with support and sustainment units.

Force projection encompasses a range of processes including mobilization, deployment, employment, sustainment, and redeployment. These processes have overlapping timelines, are continuous, and can repeat throughout an operation. Force projection operations are inherently joint and require detailed planning and synchronization.

Deployment Operations (See p. 1-84.)

Deployment is composed of activities required to prepare and move forces, supplies, and equipment to a theater. This involves the force as it task organizes, tailors itself for movement based on the mission, concept of operations, available lift, and other resources.

The Joint deployment process is divided into four phases -- deployment planning; predeployment activities; movement; and JRSOI. The terminology used to describe the Army deployment phases is in synch with the Joint process. The Joint process includes a planning phase at the outset whereas the Army considers planning to be woven through all the phases. Moreover, the movement phase in the Army process is discussed in two segments -- fort to port and port to port. The Army relies on U.S. Transportation Command (USTRANSCOM) to provide the strategic lift to and from the port of embarkation (POE).

Reception, Staging, Onward Movement, & Integration (RSOI)

RSOI is the process that delivers combat power to the Joint Force Commander (JFC) in the operational theater.

Reception is the unloading of personnel and equipment from strategic transport, marshaling them, transporting them to staging areas, and if required, providing life support services. **Staging** is the assembling, holding, and organizing arriving of personnel, equipment, and basic loads into units; preparing the units for onward movement; and providing life support until the unit becomes self-sustaining. **Onward Movement** is moving units from reception facilities and staging areas to TAAs or other theater destinations; placing arriving nonunit personnel to gaining commands; and providing sustainment to distribution sites. **Integration** is the synchronized transfer of authority of units to a designated component or functional commander for employment in the theater of operations. (See p. 1-85.)

Refer to SMFLS5: The Sustainment & Multifunctional Logistics SMARTbook (Guide to Operational & Tactical Level Sustainment), chapter nine for 44 pages on deployment operations from ATP 3-35 and JP 3-35. Topics include Predeployment, Movement; (RSO&I) Reception, Staging, Onward movement, Integration; and redepolyment.

Chap 4
I. Tactics and Tactical Mission Tasks

Ref: ADP 3-90, Offense & Defense (Jul '19), chap. 1 and ADP 1-02, Terms and Military Symbols (Aug '18), chap. 9.

I. The Tactical Level of War

Tactics is the employment, ordered arrangement, and directed actions of forces in relation to each other. Tactics always require judgment and adaptation to a situation's unique circumstances. Techniques and procedures are established patterns or processes that can be applied repeatedly with little judgment to various circumstances. Together, tactics, techniques, and procedures (TTP) provide commanders and staffs with the fundamentals to develop solutions to tactical problems. The solution to any specific problem is a unique combination of these fundamentals, current TTP, and the creation of new TTP based on an evaluation of the situation. Commanders determine acceptable solutions by mastering doctrine and current TTP. They gain this mastery through experiences in education, training, and operations.

The tactical level of warfare is the level of warfare at which battles and engagements are planned and executed to achieve military objectives assigned to tactical units or task forces (JP 3-0). Activities at this level focus on achieving assigned objectives through the ordered arrangement, movement, and maneuver of combat elements in relation to each other and to enemy forces. The strategic and operational levels of warfare provide the context for tactical operations. (See JP 3-0 and ADP 3-0 for more discussions on strategic and operational levels of warfare.) Without this context, tactical operations become disconnected from operational end states and strategic goals.

See pp. 3-2 to 3-3 for related discussion and an overview of the offense and defense.

An **engagement** is a tactical conflict, usually between opposing lower echelon maneuver forces (JP 3-0). Brigades and lower echelon units generally conduct engagements. Engagements result from deliberate closure with or chance encounters between two opponents.

A battle is a set of related engagements that lasts longer and involves larger forces than an engagement. Battles affect the course of a campaign or major operation, as they determine the outcome of a division or corps echelon achieving one or more significant objectives. The outcomes of battles determine strategic and operational success and contribute to the overall operation or campaign achieving a strategic purpose. The outcomes of engagements determine tactical success and contribute to friendly forces winning a battle.

Echelons of command, sizes of units, types of equipment, or components do not define the strategic, operational, or tactical levels of warfare. Instead, the level of warfare is determined by what level objective is achieved by the action. National assets, including space-based and cyberspace capabilities previously considered principally strategic, provide important support to tactical operations.

Refer to SUTS3: The Small Unit Tactics SMARTbook, 3rd Ed., completely updated with the latest publications for 2019. Chapters and topics include tactical fundamentals, the offense; the defense; train, advise, and assist (stability, peace & counterinsurgency ops); tactical enabling tasks (security, reconnaissance, relief in place, passage of lines, encirclement, and troop movement); special purpose attacks (ambush, raid, etc.); urban and regional environments (urban, fortified areas, desert, cold, mountain, & jungle operations); patrols & patrolling.

(Movement & Maneuver) I. Tactics & Tactical Mission Tasks 4-11

II. The Art and Science of Tactics

Army leaders at all echelons master the art and science of tactics—two distinct yet inseparable concepts—to solve the problems they will face on the battlefield. A tactical problem occurs when the mission variables—mission, enemy, terrain and weather, troops and support available, time available, and civil considerations (known as METT-TC)—of the desired tactical situation differ from the current situation.

A. The Art

The art of tactics is three interrelated aspects: the creative and flexible array of means to accomplish missions, decision making under conditions of uncertainty when faced with a thinking and adaptive enemy, and the understanding of the effects of combat on Soldiers. An art, as opposed to a science, requires exercising intuition based on operational experiences and cannot be learned solely by study. Leaders exercise the art of tactics by balancing study with a variety of relevant and practical experiences. Repetitive practice under a variety of realistic conditions increases an individual's mastery of the art of tactics.

Leaders apply the art of tactics to solve tactical problems within their commander's intent by choosing from interrelated options, including—

- The types of operations, forms of maneuver, and tactical mission tasks.
- Task organization of available forces and allocation of resources.
- The arrangement and choice of control measures.
- Controlling the tempo of the operation.
- The level of necessary risk.

Combat is a lethal clash of opposing wills and a violent struggle between thinking and adaptive commanders with opposing goals. Commanders strive to defeat their enemies. Defeat is to render a force incapable of achieving its objectives (ADP 3-0). Commanders seek to accomplish missions that support operational or strategic purposes while preventing their enemies from doing the same.

B. The Science

The science of tactics is the understanding of those military aspects of tactics—capabilities, techniques, and procedures—that can be measured and codified. The science of tactics includes the physical capabilities of friendly and enemy organizations and systems. It also includes techniques and procedures used to accomplish specific tasks. The science of tactics is straightforward. Much of what subordinate doctrine publications contain are the science of tactics—techniques and procedures for employing the various elements of the combined arms team. A combined arms team is a team that uses combined arms—the synchronized and simultaneous application of arms to achieve an effect greater than if each element was used separately or sequentially (ADP 3-0).

III. Tactical Mission Tasks

A tactical mission task is a specific activity performed by a unit while executing a form of tactical operation or form of maneuver. A tactical mission task may be expressed as either an action by a **friendly force** or **effects on an enemy force** (FM 7-15). The tactical mission tasks describe the results or effects the commander wants to achieve.

Not all tactical mission tasks have symbols. Some tactical mission task symbols will include unit symbols, and the tactical mission task "delay until a specified time" will use an amplifier. However, no modifiers are used with tactical mission task symbols. Tactical mission task symbols are used in course of action sketches, synchronization matrixes, and maneuver sketches. They do not replace any part of the operation order.

See following pages (pp. 4-20 to 4-21) for tactical mission tasks.

4-12 (Movement & Maneuver) I. Tactics & Tactical Mission Tasks

Tactical Doctrinal Taxonomy

Ref: Adapted from ADP 3-90, Offense and Defense (Jul '19), fig. 2-1, p. 2-3.

The following shows the Army's tactical doctrinal taxonomy for the four elements of decisive action (in accordance with ADP 3-0) and their subordinate tasks. The commander conducts tactical enabling tasks to assist the planning, preparation, and execution of any of the four elements of decisive action. Tactical enabling tasks are never decisive operations in the context of the conduct of offensive and defensive tasks. (They are also never decisive during the conduct of stability tasks.) The commander uses tactical shaping tasks to assist in conducting combat operations with reduced risk.

Elements of Decisive Action (and subordinate tasks)

Offensive Operations

Movement to Contact
Search and attack
Cordon and search

Attack
Ambush*
Counterattack*
Demonstration*
Spoiling attack*
Feint*
Raid*
*Also known as special purpose attacks

Exploitation

Pursuit
Frontal
Combination

Forms of Maneuver
Envelopment
Frontal attack
Infiltration
Penetration
Turning Movement

Defensive Operations

Area Defense

Mobile Defense

Retrograde
Delay
Withdraw
Retirement

Forms of the Defense
Defense of linear obstacle
Perimeter defense
Reverse slope defense

Stability Operations
Civil security
Civil control
Restore essential services
Support to governance
Support to economic and
infrastructure development
Conduct security cooperation

Defense Support to Civil Authorities
Provide support for domestic
disasters
Provide support for domestic
CBRN incidents
Provide support for domestic
law enforcement agencies
Provide other designated
support

Enabling Operations

Reconnaissance Operations
Area
Reconnaissance in force
Route
Special
Zone

Security
Screen
Guard
Cover
Area

Passage of Lines
Forward
Rearward

Troop Movement
Administrative movement
Approach march
Road march

Encirclement Operations

Relief in Place
Sequential
Simultaneous
Staggered

Other Enabling Operations (Examples)
Information Operations
(FM 3-13)
Mobility Operations
(ATP 3-90.4)
Countermobility Operations
(ATP 3-90.8)

Tactical Mission Tasks

Actions by Friendly Forces
Attack-by-Fire
Breach
Bypass
Clear
Control
Counterreconnaissance
Disengage
Exfiltrate
Follow and Assume
Follow and Support

Occupy
Reduce
Retain
Secure
Seize
Support-by-Fire

Effects on Enemy Force
Block
Canalize
Contain
Defeat
Destroy
Disrupt
Fix
Interdict
Isolate
Neutralize
Suppress
Turn

A. Actions by Friendly Forces

Attack by Fire		*Attack-by-fire* is a tactical mission task in which a commander uses direct fires, supported by indirect fires, to engage an enemy without closing with him to destroy, suppress, fix, or deceive him.
Breach		*Breach* is a tactical mission task in which the unit employs all available means to break through or secure a passage through an enemy defense, obstacle, minefield, or fortification.
Bypass		Bypass is a tactical mission task in which the commander directs his unit to maneuver around an obstacle, position, or enemy force to maintain the momentum of the operation while deliberately avoiding combat with an enemy force.
Clear		*Clear* is a tactical mission task that requires the commander to remove all enemy forces and eliminate organized resistance within an assigned area.
Control	*No graphic*	*Control* is a tactical mission task that requires the commander to maintain physical influence over a specified area to prevent its use by an enemy or to create conditions for successful friendly operations.
Counterrecon	*No graphic*	*Counterreconnaissance* is a tactical mission task that encompasses all measures taken by a commander to counter enemy reconnaissance and surveillance efforts.
Disengage	*No graphic*	*Disengage* is a tactical mission task where a commander has his unit break contact with the enemy to allow the conduct of another mission or to avoid decisive engagement.
Exfiltrate	*No graphic*	*Exfiltrate* is a tactical mission task where a commander removes soldiers or units from areas under enemy control by stealth, deception, surprise, or clandestine means.
Follow and Assume		*Follow and assume* is a tactical mission task in which a second committed force follows a force conducting an offensive operation and is prepared to continue the mission if the lead force is fixed, attritted, or unable to continue. The follow-and-assume force is not a reserve but is committed to accomplish specific tasks.
Follow and Support		*Follow and support* is a tactical mission task in which a committed force follows and supports a lead force conducting an offensive operation. The follow-and-support force is not a reserve but is a force committed to specific tasks.
Occupy		*Occupy* is a tactical mission task that involves moving a friendly force into an area so that it can control that area. Both the force's movement to and occupation of the area occur without enemy opposition.
Reduce	*No graphic*	*Reduce* is a tactical mission task that involves the destruction of an encircled or bypassed enemy force.
Retain		*Retain* is a tactical mission task in which the cdr ensures that a terrain feature controlled by a friendly force remains free of enemy occupation or use. The commander assigning this task must specify the area to retain and the duration of the retention, which is time- or event-driven.
Secure		*Secure* is a tactical mission task that involves preventing a unit, facility, or geographical location from being damaged or destroyed as a result of enemy action. This task normally involves conducting area security operations.
Seize		*Seize* is a tactical mission task that involves taking possession of a designated area by using overwhelming force. An enemy force can no longer place direct fire on an objective that has been seized.
Support by Fire		*Support-by-fire* is a tactical mission task in which a maneuver force moves to a position where it can engage the enemy by direct fire in support of another maneuvering force. The primary objective of the support force is normally to fix and suppress the enemy so he cannot effectively fire on the maneuvering force.

4-14 (Movement & Maneuver) I. Tactics & Tactical Mission Tasks

B. Effect on Enemy Forces

Block		*Block* is a tactical mission task that denies the enemy access to an area or prevents his advance in a direction or along an avenue of approach.
		Block is also an engineer obstacle effect that integrates fire planning and obstacle effort to stop an attacker along a specific avenue of approach or prevent him from passing through an engagement area.
Canalize		*Canalize* is a tactical mission task in which the commander restricts enemy movement to a narrow zone by exploiting terrain coupled with the use of obstacles, fires, or friendly maneuver.
Contain		*Contain* is a tactical mission task that requires the commander to stop, hold, or surround enemy forces or to cause them to center their activity on a given front and prevent them from withdrawing any part of their forces for use elsewhere.
Defeat	*No graphic*	*Defeat* occurs when an enemy has temporarily or permanently lost the physical means or the will to fight. The defeated force is unwilling or unable to pursue his COA, and can no longer interfere to a significant degree. Results from the use of force or the threat of its use.
Destroy		*Destroy* is a tactical mission task that physically renders an enemy force combat-ineffective until it is reconstituted. Alternatively, to destroy a combat system is to damage it so badly that it cannot perform any function or be restored to a usable condition without being entirely rebuilt.
Disrupt		*Disrupt* is a tactical mission task in which a commander integrates direct and indirect fires, terrain, and obstacles to upset an enemy's formation or tempo, interrupt his timetable, or cause his forces to commit prematurely or attack in a piecemeal fashion.
		Disrupt is also an engineer obstacle effect that focuses fire planning and obstacle effort to cause the enemy to break up his formation and tempo, interrupt his timetable, commit breaching assets prematurely, and attack in a piecemeal effort.
Fix		*Fix* is a tactical mission task where a commander prevents the enemy from moving any part of his force from a specific location for a specific period. Fixing an enemy force does not mean destroying it. The friendly force has to prevent the enemy from moving in any direction.
		Fix is also an engineer obstacle effect that focuses fire planning and obstacle effort to slow an attacker's movement within a specified area, normally an engagement area.
Isolate		*Isolate* is a tactical mission task that requires a unit to seal off-both physically and psychologically-an enemy from his sources of support, deny him freedom of movement, and prevent him from having contact with other enemy forces.
Neutralize		*Neutralize* is a tactical mission task that results in rendering enemy personnel or materiel incapable of interfering with a particular operation.
Suppress		*Suppress* is a tactical mission task that results in the temporary degradation of the performance of a force or weapon system below the level needed to accomplish its mission.
Turn		*Turn* is a tactical mission task that involves forcing an enemy element from one avenue of approach or movement corridor to another.
		Turn is also a tactical obstacle effect that integrates fire planning and obstacle effort to divert an enemy formation from one avenue of approach to an adjacent avenue of approach or into an engagement area.

(Movement & Maneuver) I. Tactics & Tactical Mission Tasks 4-15

C. Mission Symbols

Counterattack (dashed axis)	CATK	A form of attack by part or all of a defending force against an enemy attacking force, with the general objective of denying the enemy his goal in attacking (FM 3-0).
Cover	C □ C	A form of security operation whose primary task is to protect the main body by fighting to gain time while also observing and reporting information and preventing enemy ground observation of and direct fire against the main body.
Delay	D	A form of retrograde in which a force under pressure trades space for time by slowing down the enemy's momentum and inflicting maximum damage on the enemy without, in principle, becoming decisively engaged (JP 1-02, see delaying operation).
Guard	G □ G	A form of security operations whose primary task is to protect the main body by fighting to gain time while also observing and reporting information and preventing enemy ground observation of and direct fire against the main body. Units conducting a guard mission cannot operate independently because they rely upon fires and combat support assets of the main body.
Penetrate		A form of maneuver in which an attacking force seeks to rupture enemy defenses on a narrow front to disrupt the defensive system (FM 3-0).
Relief in Place	RIP	A tactical enabling operation in which, by the direction of higher authority, all or part of a unit is replaced in an area by the incoming unit.
Retirement	R	A form of retrograde [JP 1-02 uses *operation*] in which a force out of contact with the enemy moves away from the enemy (JP 1-02).
Screen	S □ S	A form of security operations that primarily provides early warning to the protected force.
Withdraw	W	A planned operation in which a force in contact disengages from an enemy force (JP 1-02) [The Army considers it a form of retrograde.]

4-16 (Movement & Maneuver) I. Tactics & Tactical Mission Tasks

Chap 4 — Movement & Maneuver

II. Reconnaissance

Ref: ADP 3-90, Offense and Defense (Jul '19), chap 5, pp. 5-1 to 5-2 and FM 3-90-1, Reconnaissance, Security, and Tactical Enabling Tasks, Vol. 2 (Mar '13), chap. 1.

Reconnaissance is a mission undertaken to obtain, by visual observation or other detection methods, information about the activities and resources of an enemy or adversary, or to secure data concerning the meteorological, hydrographical, or geographical characteristics of a particular area (JP 2-0). Reconnaissance accomplished by small units primarily relies on the human dynamic rather than technical means. Reconnaissance is a focused collection effort. It is performed before, during, and after operations to provide commanders and staffs information used in the intelligence preparation of the battlefield (IPB) process so they can formulate, confirm, or modify courses of action (COAs).

Reconnaissance is a process of gathering information to help the commander shape his understanding of the battlespace. Reconnaissance uses many techniques and technologies to collect this information, but it is still largely a human endeavor. (Dept. of Army photo.)

Reconnaissance Objective

Commanders orient their reconnaissance assets by identifying a reconnaissance objective within an area of operations (AO). The reconnaissance objective is a terrain feature, geographic area, enemy force, adversary, or other mission or operational variable about which the commander wants to obtain additional information. Every reconnaissance mission specifies a reconnaissance objective that clarifies the intent of the effort, and prioritizes those efforts, by specifying the most important information to obtain. Commanders assign reconnaissance objectives based on priority information requirements resulting from the IPB process and the reconnaissance asset's capabilities and limitations. A reconnaissance objective can be information about a specific geographical location, such as the cross country trafficability of a specific area, to confirm or deny a specific activity of a threat, or to specify a specific location of a threat.

(Movement & Maneuver) II. Reconnaissance 4-17

Reconnaissance

Ref: ADP 3-90, Offense and Defense (Jul '19), pp. 5-1 to 5-2.

Reconnaissance is a mission undertaken to obtain, by visual observation or other detection methods, information about the activities and resources of an enemy or adversary, or to secure data concerning the meteorological, hydrographical, or geographical characteristics of a particular area (JP 2-0). Reconnaissance accomplished by small units primarily relies on the human dynamic rather than technical means. Reconnaissance is a focused collection effort. It is performed before, during, and after operations to provide commanders and staffs information used in the intelligence preparation of the battlefield (IPB) process so they can formulate, confirm, or modify courses of action (COAs).

Types of Reconnaissance

The five types of reconnaissance: route, zone, area, reconnaissance in force, and special.

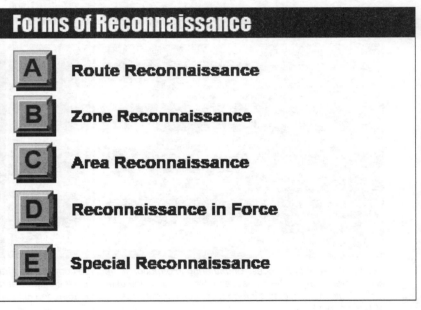

The responsibility for accomplishing reconnaissance does not reside solely with reconnaissance units. Every unit has an implied mission to report information about the terrain, civilian activities, and friendly and enemy dispositions. Troops in contact with an enemy and reconnaissance patrols of maneuver units, at all echelons, collect information on enemy units and activities. In echelon support and consolidation areas, reserve maneuver forces, functional and multifunctional support and sustainment elements, other governmental agencies, and multinational forces observe and report civilian, adversary, and enemy activity and significant changes in terrain trafficability. Although all units conduct reconnaissance, ground cavalry, aviation attack reconnaissance units, scouts, and special forces are specifically trained to conduct reconnaissance operations. Some branches, such as the Corps of Engineers and Chemical Corps, conduct specific reconnaissance operations that complement the force's overall reconnaissance effort. However, BCT, division, and corps commanders primarily use their organic or attached reconnaissance—ground or air—and intelligence elements to accomplish reconnaissance.

4-18 (Movement & Maneuver) II. Reconnaissance

A. Route Reconnaissance

Route reconnaissance is a type of reconnaissance operation to obtain detailed information of a specified route and all terrain from which the enemy could influence movement along that route. Route reconnaissance provides new or updated information on route conditions, such as obstacles and bridge classifications, and enemy, adversary, and civilian activity along the route.

B. Zone Reconnaissance

Zone reconnaissance is a type of reconnaissance operation that involves a directed effort to obtain detailed information on all routes, obstacles, terrain, and enemy forces within a zone defined by boundaries. Obstacles include existing, reinforcing, and areas with CBRN contamination. Commanders assign a zone reconnaissance mission when they need additional information on a zone before committing other forces. Zone reconnaissance is the most time-and resource-intensive form of reconnaissance

C. Area Reconnaissance

Area reconnaissance is a type of reconnaissance operation that focuses on obtaining detailed information about the terrain or enemy activity within a prescribed area. Commanders assign an area reconnaissance when information on the enemy situation is limited or when focused reconnaissance will yield specific information on the area in question. An area reconnaissance differs from a zone reconnaissance in that the unit conducting an area reconnaissance starts from an LD.

D. Reconnaissance in Force

A reconnaissance in force is a type of reconnaissance operation designed to discover or test the enemy's strength, dispositions, and reactions or to obtain other information. A commander assigns a reconnaissance in force when an enemy force is operating within an area and the commander cannot obtain adequate intelligence by other means. The unit commander plans for both the retrograde or reinforcement of the friendly force (in case it encounters superior enemy forces) and for the exploitation of its success.

E. Special Reconnaissance

Special reconnaissance is reconnaissance and surveillance actions conducted as a special operation in hostile, denied, or diplomatically and/or politically sensitive environments to collect or verify information of strategic or operational significance, employing military capabilities not normally found in conventional forces (JP 3-05). Special reconnaissance provides an additional capability for commanders and supplements other conventional reconnaissance and surveillance actions.

Reconnaissance is characterized as either stealthy or aggressive. A key factor in executing reconnaissance is the time available to conduct the mission. The commander recognizes the increased risk to both the reconnaissance element and the main body when accelerating the pace of reconnaissance. This risk can be somewhat offset by employing air reconnaissance and technical means to cover open terrain or areas of lower threat. *(FM 3-90-2, pp. 1-13 to 1-14.)*

Refer to SUTS3: The Small Unit Tactics SMARTbook, 3rd Ed., completely updated with the latest publications for 2019. Chapters and topics include tactical fundamentals, the offense; the defense; train, advise, and assist (stability, peace & counterinsurgency ops); tactical enabling tasks (security, reconnaissance, relief in place, passage of lines, encirclement, and troop movement); special purpose attacks (ambush, raid, etc.); urban and regional environments (urban, fortified areas, desert, cold, mountain, & jungle operations); patrols & patrolling.

(Movement & Maneuver) II. Reconnaissance 4-19

II. Reconnaissance Fundamentals

1. Ensure Continuous Reconnaissance

Effective reconnaissance is continuous. The commander conducts reconnaissance before, during, and after all operations. Before an operation, reconnaissance focuses on filling gaps in information about the enemy and the terrain. During an operation, reconnaissance focuses on providing the commander with updated information that verifies the enemy's composition, dispositions, and intentions as the battle progresses.

2. Do Not Keep Reconnaissance Assets in Reserve

Reconnaissance assets, like artillery assets, are never kept in reserve. When committed, reconnaissance assets use all of their resources to accomplish the mission. This does not mean that all assets are committed all the time. The commander uses his reconnaissance assets based on their capabilities and METT-TC to achieve the maximum coverage needed to answer the commander's critical information requirements (CCIR). At times, this requires the commander to withhold or position reconnaissance assets to ensure that they are available at critical times and places.

3. Orient on the Reconnaissance Objective

The commander uses the reconnaissance objective to focus his unit's efforts. Commanders of subordinate reconnaissance elements remain focused on achieving this objective, regardless of what their forces encounter during the mission. When time, limitations of unit capabilities, or enemy action prevents a unit from accomplishing all the tasks normally associated with a particular form of reconnaissance, the unit uses the reconnaissance objective to focus the reconnaissance effort.

4. Report Information Rapidly and Accurately

Reconnaissance assets must acquire and report accurate and timely information on the enemy, civil considerations, and the terrain over which operations are to be conducted. Information may quickly lose its value. Reconnaissance units report exactly what they see and, if appropriate, what they do not see. Seemingly unimportant information may be extremely important when combined with other information.

5. Retain Freedom of Maneuver

Reconnaissance assets must retain battlefield mobility to successfully complete their missions. If these assets are decisively engaged, reconnaissance stops and a battle for survival begins. Reconnaissance assets must have clear engagement criteria that support the maneuver commander's intent. They must employ proper movement and reconnaissance techniques, use overwatching fires, and standing operating procedures.

6. Gain and Maintain Enemy Contact

Once a unit conducting reconnaissance gains contact with the enemy, it maintains that contact unless the commander directing the reconnaissance orders otherwise or the survival of the unit is at risk. This does not mean that individual scout and reconnaissance teams cannot break contact with the enemy. The commander of the unit conducting reconnaissance is responsible for maintaining contact using all available resources. That contact can range from surveillance to close combat.

7. Develop the Situation Rapidly

When a reconnaissance asset encounters an enemy force or an obstacle, it must quickly determine the threat it faces. For an enemy force, it must determine the enemy's composition, dispositions, activities, and movements and assess the implications of that information. For an obstacle, it must determine the type and extent of the obstacle and whether it is covered by fire. Obstacles can provide the attacker with information concerning the location of enemy forces, weapon capabilities, and organization of fires. In most cases, the reconnaissance unit developing the situation uses actions on contact.

4-20 (Movement & Maneuver) II. Reconnaissance

Chap 4

Movement & Maneuver

III. Security Operations

Ref: ADP 3-90, Offense and Defense (Jul '19), chap 5, pp. 5-3 to 5-4 and FM 3-90-2, Reconnaissance, Security, and Tactical Enabling Tasks, Vol. 2 (Mar '13), chap. 2.

The main difference between conducting security operations and reconnaissance is that security operations orient on the force or facility being protected while reconnaissance orients on the enemy and terrain. Security operations aim to protect a force from surprise and reduce the unknowns in any situation. Commanders conduct security operations to the front, flanks, or rear of a friendly force. Security operations are shaping operations. As a shaping operation, economy of force is often a consideration when planning. (ADP 3-90)

I. Security Operations Tasks

The four types of security operations are—screen, guard, cover, and area security.

A. Screen

Screen is a type of security operation that primarily provides early warning to the protected force. (ADP 3-90) A unit performing a screen observes, identifies, and reports enemy actions. Generally, a screening force engages and destroys enemy reconnaissance elements in its capabilities— augmented by indirect fires—but otherwise fights only in self-defense. The screen has the minimum combat power necessary to provide the desired early warning, which allows the commander to retain the bulk of the main body's combat power for commitment at the decisive place and time. A screen provides the least amount of protection of any security mission; it does not have the combat power to develop the situation.

A screen is appropriate to cover gaps between forces, exposed flanks, or the rear of stationary and moving forces. The commander can place a screen in front of a stationary formation when the likelihood of enemy action is small, the expected enemy force is small, or the main body needs only limited time, once it is warned, to react effectively. Designed to provide minimum security with minimum forces, a screen is usually an economy-of-force operation based on prudent risk. If a significant enemy force is expected or a significant amount of time and space is needed to provide the required degree of protection, the commander assigns and resources a guard or cover mission instead of a screen. The security element forward of a moving force must conduct a guard or cover because a screen lacks the combat power to defeat or contain the lead elements of an enemy force.

Critical Tasks for a Screen

Unless the commander orders otherwise, a security force conducting a screen performs certain tasks within the limits of its capabilities. A unit can normally screen

Refer to SUTS3: The Small Unit Tactics SMARTbook, 3rd Ed., completely updated with the latest publications for 2019. Chapters and topics include tactical fundamentals, the offense; the defense; train, advise, and assist (stability, peace & counterinsurgency ops); tactical enabling tasks (security, reconnaissance, relief in place, passage of lines, encirclement, and troop movement); special purpose attacks (ambush, raid, etc.); urban and regional environments (urban, fortified areas, desert, cold, mountain, & jungle operations); patrols & patrolling.

(Movement & Maneuver) III. Security Operations 4-21

an avenue of approach two echelons larger than itself. If a security force does not have the time or other resources to complete all of these tasks, the security force commander must inform the commander assigning the mission of the shortfall and request guidance on which tasks must be completed and their priority. After starting the screen, if the security unit commander determines that he cannot complete an assigned task, such as maintain continuous surveillance on all avenues of approach into an AO, he reports and awaits further instructions. Normally, the main force commander does not place a time limit on the duration of the screen, as doing so may force the screening force to accept decisive engagement. Screen tasks are to:

- Allow no enemy ground element to pass through the screen undetected and unreported
- Maintain continuous surveillance of all avenues of approach larger than a designated size into the area under all visibility conditions
- Destroy or repel all enemy reconnaissance patrols within its capabilities.
- Locate the lead elements of each enemy advance guard and determine its direction of movement in a defensive screen
- Maintain contact with enemy forces and report any activity in the AO
- Maintain contact with the main body and any security forces operating on its flanks
- Impede and harass the enemy within its capabilities while displacing

B. Guard

Guard is a type of security operation done to protect the main body by fighting to gain time while preventing enemy ground observation of and direct fire against the main body. Units performing a guard cannot operate independently. They rely upon fires, functional support, and multifunctional support assets of the main body. (ADP 3-90)

A guard differs from a screen in that a guard force contains sufficient combat power to defeat, cause the withdrawal of, or fix the lead elements of an enemy ground force before it can engage the main body with direct fire. A guard force routinely engages enemy forces with direct and indirect fires. A screening force, however, primarily uses indirect fires or close air support to destroy enemy reconnaissance elements and slow the movement of other enemy forces. A guard force uses all means at its disposal, including decisive engagement, to prevent the enemy from penetrating to a position were it could observe and engage the main body. It operates within the range of the main body's fire support weapons, deploying over a narrower front than a comparable-size screening force to permit concentrating combat power.

The three types of guard operations are advance, flank, and rear guard. A commander can assign a guard mission to protect either a stationary or a moving force.

Guard tasks:

- Destroy the enemy advance guard
- Maintain contact with enemy forces and report activity in the AO
- Maintain continuous surveillance of avenues of approach into the AO under all visibility conditions
- Impede and harass the enemy within its capabilities while displacing
- Cause the enemy main body to deploy, and then report its direction of travel
- Allow no enemy ground element to pass through the security area undetected and unreported
- Destroy or cause the withdrawal of all enemy reconnaissance patrols
- Maintain contact with its main body and any other security forces operating on its flanks

4-22 (Movement & Maneuver) III. Security Operations

II. Fundamentals of Security Ops

Ref: FM 3-90-2, Recon, Security, & Tactical Enabling Tasks, Vol. 2 (Mar '13), pp. 2-2 to 2-3.

1. Provide Early and Accurate Warning

The security force provides early warning by detecting the enemy force quickly and reporting information accurately to the main body commander. The security force operates at varying distances from the main body based on the factors of METT-TC. As a minimum, it should operate far enough from the main body to prevent enemy ground forces from observing or engaging the main body with direct fires. The earlier the security force detects the enemy, the more time the main body has to assess the changing situation and react. The commander positions ground security and aeroscouts to provide long-range observation of expected enemy avenues of approach, and he reinforces and integrates them with available intelligence collection systems to maximize warning time.

2. Provide Reaction Time and Maneuver Space

The security force provides the main body with enough reaction time and maneuver space to effectively respond to likely enemy actions by operating at a distance from the main body and by offering resistance to enemy forces. The commander determines the amount of time and space required to effectively respond from information provided by the intelligence preparation of the battlefield (IPB) process and the main body commander's guidance regarding time to react to enemy courses of action (COA) based on the factors of METT-TC. The security force that operates farthest from the main body and offers more resistance provides more time and space to the main body. It attempts to hinder the enemy's advance by acting within its capabilities and mission constraints.

3. Orient on the Force or Facility to Be Secured

The security force focuses all its actions on protecting and providing early warning to the secured force or facility. It operates between the main body and known or suspected enemy units. The security force must move as the main body moves and orient on its movement. The security force commander must know the main body's scheme of maneuver to maneuver his force to remain between the main body and the enemy. The value of terrain occupied by the security force hinges on the protection it provides to the main body commander.

4. Perform Continuous Reconnaissance

The security force aggressively and continuously seeks the enemy and reconnoiters key terrain. It conducts active area or zone reconnaissance to detect enemy movement or enemy preparations for action and to learn as much as possible about the terrain. The ultimate goal is to determine the enemy's COA and assist the main body in countering it. Terrain information focuses on its possible use by the enemy or the friendly force, either for offensive or defensive operations. Stationary security forces use combinations of OPs, aviation, patrols, intelligence collection assets, and battle positions (BPs) to perform reconnaissance. Moving security forces perform zone, area, or route reconnaissance along with using OPs and BPs, to accomplish this fundamental.

5. Maintain Enemy Contact

Once the security force makes enemy contact, it does not break contact unless specifically directed by the main force commander. The security asset that first makes contact does not have to maintain that contact if the entire security force maintains contact with the enemy. The security force commander ensures that his subordinate security assets hand off contact with the enemy from one security asset to another in this case. The security force must continuously collect information on the enemy's activities to assist the main body in determining potential and actual enemy COAs and to prevent the enemy from surprising the main body. This requires continuous visual contact, the ability to use direct and indirect fires, freedom to maneuver, and depth in space and time.

(Movement & Maneuver) III. Security Operations 4-23

C. Cover

Cover is a type of security operation done independent of the main body to protect them by fighting to gain time while preventing enemy ground observation of and direct fire against the main body. Commanders use the cover task offensively and defensively. (Cover as a doctrinal term also has other definitions.) (ADP 3-90)

The covering force's distance forward of the main body depends on the intentions and instructions of the main body commander, the terrain, the location and strength of the enemy, and the rates of march of both the main body and the covering force. The width of the covering force area is the same as the AO of the main body.

In Army doctrine, a *covering force* is a self-contained force capable of operating independently of the main body, unlike a screening or guard force. A covering force, or portions of it, often becomes decisively engaged with enemy forces. Therefore, the covering force must have substantial combat power to engage the enemy and accomplish its mission. A covering force develops the situation earlier than a screen or a guard force. It fights longer and more often and defeats larger enemy forces.

While a covering force provides more security than a screen or guard force, it also requires more resources. Before assigning a cover mission, the main body commander must ensure that he has sufficient combat power to resource a covering force and the decisive operation. When the commander lacks the resources to support both, he must assign his security force a less resource-intensive security mission, either a screen or a guard.

A covering force accomplishes all the tasks of screening and guard forces. A covering force for a stationary force performs a defensive mission, while a covering force for a moving force generally conducts offensive actions. A covering force normally operates forward of the main body in the offense or defense, or to the rear for a retrograde operation. Unusual circumstances could dictate a flank covering force, but this is normally a screen or guard mission.

D. Area Security

Area security is a type of security operation conducted to protect friendly forces, lines of communications, and activities within a specific area. The security force may be protecting the civilian population, civil institutions, and civilian infrastructure with the unit's AO. (ADP 3-90)

Area security operations occur regardless of which element of operations is currently dominant. They focus on the protected force, installation, route, or area. Protected forces range from echelon headquarters through artillery and echelon reserves to the sustaining base. Protected installations can also be part of the sustaining base, or they can constitute part of the area's infrastructure. Areas to secure range from specific points (bridges and defiles) and terrain features (ridge lines and hills) to large civilian population centers and their adjacent areas. Population-centric area security missions are common across the range of military operations, but are almost a fixture during irregular warfare. These population-centric area security operations typically combine aspects of the area defense and offensive tasks to eliminate the efficacy of internal defense threats.

Operations in noncontiguous AOs require commanders to emphasize area security. During offensive and retrograde operations, the speed at which the main body moves provides some measure of security. Rapidly moving units in open terrain rely on technical assets to provide advance warning of enemy forces. In restrictive terrain, security forces focus on key terrain such as potential choke points.

Chap 4

IV. Mobility and Countermobility

Ref: Adapted from ADP 3-90, Offense and Defense (Jul '19), pp. 3-11 to 3-13 and FM 3-34, Engineer Operations (Apr '14).

I. Mobility

Mobility tasks are those combined arms activities that mitigate the effects of natural and man-made obstacles to enable freedom of movement and maneuver (ATP 3-90.4).

Mobility Tasks

- **Conduct Breaching**
- **Conduct Clearing (Areas and Routes)**
- **Conduct Gap Crossing**
- **Construct and Maintain Combat Roads and Trails**
- **Construct and Maintain Forward Airfields and Landing Zones**
- **Conduct Traffic Operations and Enforcement**

Offensive Considerations

Mobility is necessary for successful offensive actions. Its major focus is to enable friendly forces to move and maneuver freely on the battlefield. The commander seeks the capability to move, exploit, and pursue the enemy across a wide front. When attacking, the commander concentrates the effects of combat power at selected locations. This may require the unit to improve or construct combat trails through areas where routes do not exist. The surprise achieved by attacking through an area believed to be impassable may justify the effort expended in constructing these trails. The force bypasses existing obstacles and minefields identified before starting the offensive operation instead of breaching them whenever possible. Units mark bypassed minefields whenever the mission variables of METT-TC allow.

Maintaining the momentum of the offense requires the attacking force to quickly pass through obstacles as it encounters them. There is a deliberate effort to capture bridges, beach and port exits, and other enemy reserved obstacles intact. The preferred method of fighting through a defended obstacle is employing a hasty (in-stride) breach, because it avoids the loss of time and momentum associated with conducting a deliberate breach.

Rivers and other gaps remain major obstacles despite advances in high-mobility weapon systems and extensive aviation support. Wet gap crossings are among the most critical, complex, and vulnerable combined arms operations. A crossing is conducted as a hasty crossing and as a continuation of the attack whenever possible because the time needed to prepare for a gap crossing allows the enemy more time to strengthen the defense. The size of the gap, as well as the enemy and friendly situations, will dictate the specific tactics, techniques, and procedures used in conducting the crossing. Functional engineer brigades contain the majority of tactical bridging assets. Military police and CBRN assets may also be required.

(Movement & Maneuver) IV. Mobility & Countermobility 4-25

Assured Mobility

Ref: Adapted from ADRP 3-90, Offense and Defense (Aug '12), pp. 3-12 to 3-13.

Assured mobility is a framework of -- processes, actions, and capabilities -- that assure the ability of a force to deploy, move, and maneuver where and when desired, without interruption or delay, to achieve the mission. The assured mobility fundamentals predict, detect, prevent, avoid, neutralize, and protect support the assured mobility framework. This framework is one means of enabling a force to achieve the commander's intent. Assured mobility emphasizes the conduct of proactive mobility, countermobility, and protection tasks in an integrated manner so as to increase the probability of mission accomplishment. While focused primarily on movement and maneuver, the assured mobility concept links to each warfighting function and both enables and is enabled by those functions.

Refer to ATP 3-90.4, Combined Arms Mobility Operations for further discussion.

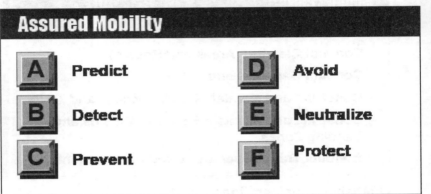

A. Predict

Commanders and staffs must accurately predict potential obstacles to force mobility by analyzing the enemy's capabilities and tactics, techniques, and procedures. This involves understanding how the enemy will evolve in reaction to friendly force countermeasures. It also involves understanding how the effects of terrain and the effects of the population, such as vehicular traffic and dislocated civilians, will impact force mobility. This helps build the mobility portion of the common operational picture and facilitates decisionmaking.

B. Detect

Commanders and staffs use intelligence products and information collection assets to identify the location of natural and manmade obstacles and potential means the enemy can use to create obstacles. Commanders employ available information collection assets to detect enemy obstacle preparations and also identify areas where there are no or only limited obstacles to ground movement and maneuver. This knowledge can be obtained through sustained surveillance of an area. Commanders identify both actual and potential obstacles and propose solutions and alternate COAs to minimize or eliminate their potential impact.

C. Prevent

Commanders and staffs apply this fundamental by preventing civilian interference with operations and denying the enemy's ability to influence friendly mobility. This is accomplished by forces acting proactively to elicit local populace support, or at least non-interference, and to eliminate enemy countermobility capabilities before those capabilities can emplace or activate obstacles, and by mitigating the factors that result in natural obstacles to friendly force movement and maneuver. This may include the employment of information-related capabilities to decrease uncertainty among the population to build support for or acceptance of operations.

Prevention may also consist of aggressive action to destroy enemy assets and capabilities before they can be used to create obstacles. In recent operations this included disrupting terrorist bomb-making cells by all available means, such as cutting off their funding, eliminating safe houses where bombs can be constructed, jamming frequencies to prevent remote detonators from being triggered, and either capturing or killing members of these cells. Forces also apply this fundamental by conducting countermobility operations to shape enemy movement and maneuver that may affect friendly movement and maneuver. This includes denying the enemy the ability and opportunity to attack critical infrastructure that supports mobility, such as airfields, roads, and bridges; or that could result in an obstacle; or have an obstacle effect if destroyed, such as dams and industrial chemical production and storage facilities.

D. Avoid

If prevention fails, the commander will move or maneuver forces to avoid impediments to mobility, if this is viable within the scheme of maneuver. If detection efforts can tell the commander where the enemy has not been, this frees up the unit to maneuver rapidly through those areas, even if they are not the most favorable movement routes.

E. Neutralize

Commanders and staffs plan to neutralize, reduce, or overcome obstacles and impediments as soon as possible to allow unrestricted movement of forces. The specific tactics, techniques, and procedures employed will depend on the mission variables of METT-TC, the rules of engagement, and where along the range of military operations the unit finds itself. For example, a small unit involved in major operations encountering surface-laid mines on a road in an urban area might attempt to destroy the mines in place using organic methods, such as aimed rifle or machinegun fire, after only minimal checks to reduce the danger to local civilians and accepting collateral damage to civilian buildings before proceeding on with its mission. That same unit encountering the same situation during the conduct of a peace-keeping operation would more likely secure the site, evacuate civilians from the area, and call for an explosive ordnance disposal team to disarm the mines in place to preclude any collateral damage.

F. Protect

Commanders and staffs plan and implement survivability and other protection measures that will prevent observation of the maneuvering force and thereby reduce the enemy's ability to engage or otherwise interfere with that force. This includes the use of combat formations and movement techniques. It may involve the use of electronic warfare systems—such as counter-radio controlled improvised explosive device electronic warfare systems, mine plows and rollers, and modifications to the rules of engagement. This may also include the conduct of countermobility missions to deny the enemy the capability to maneuver in certain directions and thereby provide additional protection to friendly maneuvering forces. It can also be as simple as altering patrol routes.

While engineers are the principal staff integrators for assured mobility, other staff sections play critical roles in ensuring the effective application and integration of mobility, countermobility, and protection tasks. In the case of amphibious operations, this would include naval forces that are responsible for assured mobility from amphibious shipping to beach and landing zone exits. These critical roles include providing information on threats to the routes. The senior engineer staff officer's role within assured mobility is similar to the role of the assistant chief of staff, intelligence (G-2) or the intelligence staff officer's (S-2s) integrating role with intelligence preparation of the battlefield. Ultimately, assured mobility is the commander's responsibility.

(Movement & Maneuver) IV. Mobility & Countermobility 4-27

Aspects of Mobility

Ref: FM 3-34, Engineer Operations (Apr '14), pp. 3-5 to 3-7.

Mobility operations are intended to maintain freedom of tactical maneuver and operational movement through the following five functional areas:

- Countermine activities detect, neutralize (through a combined arms breach or bypass), mark, and proof mined areas
- Countering employs tactics and equipment to breach or bypass and ultimately reduce obstacles other than mines
- Gap crossing fills gaps in the terrain to allow personnel and equipment to pass
- Constructing combat roads and trails expediently prepares or repairs routes of travel for personnel and equipment. This includes temporary bypasses for damaged roads and bridges.
- FACE prepares or repairs expedient landing zones (LZs), forward arming and refueling points (FARPs), landing strips, or other aviation support sites in the forward combat area

1. Countermine Activities

Countermine operations are all efforts taken to counter an enemy's mine effort. Countermine operations are difficult because detection systems are imperfect and mine neutralization systems are only partially effective. Normally, countermine operations using explosive systems are conducted under enemy observation and fire. Countermine operations include mine detection, enemy minefield reconnaissance, combined arms breaching, and enemy mine operations prevention.

- **Mine Detection.** The detection of mines is linked to proper reconnaissance techniques to include geospatial engineering assistance. A proper analysis of enemy techniques and devices lays the groundwork for effective reconnaissance. FM 20-32, Mine/Countermine Operations, is the primary reference for a discussion of mine detection.

- **Enemy Minefield Reconnaissance.** Engineer reconnaissance and the specific considerations of minefield reconnaissance are discussed in FM 5-170. Proper reconnaissance creates the conditions necessary for successful obstacle breaching.

- **Combined Arms Breaching.** As engineers plan for mobility operations, they may realize that they will have to conduct breaching operations. Enemy obstacles that disrupt, fix, turn, or block the force can affect the timing and force of the operation. Most obstacles can and will be observed by the enemy and will be protected with fires. They should be bypassed if possible. For those that must be breached, constant coordination and integration of all elements of the TF are vital for success. Combat engineers are key to the orchestration of the operation and are responsible for employing the tactics and techniques necessary to penetrate obstacles in the path of the force. Combined arms breaching operations are some of the most complex of modern warfare but are not an end in themselves. They exist only as a part of the maneuver forces' operation focused on the objective. The goal of breaching operations is the continued, uninterrupted momentum of ground forces to the objective; therefore, these operations should be planned and executed in support of the ground forces' needs to ensure that actions at the objective are supported by actions at the breach. Fundamentals of combined arms breaching operations have evolved in concert with the fundamentals of ground combat and provide a logical and time-proven set of rules. These fundamentals include suppress, obscure, secure, reduce, and assault (SOSRA).

4-28 (Movement & Maneuver) IV. Mobility & Countermobility

- **Enemy Mine Operations Prevention.** The most effective means of countering a mine threat is to prevent mine laying. Proactive countermine operations destroy enemy mine manufacturing and storage facilities or mine-laying capabilities before the mines are laid. Planners must consider enemy mine storage and mine production facilities and assets for inclusion on the target lists. In addition to destroying mine manufacturing and storage facilities, units must consider targeting enemy engineers and equipment capable of laying mines.

2. Countering an Obstacle

Many issues encountered in counter-obstacle operations apply to non-mine obstacles. Engineer reconnaissance should focus on collection efforts to detect the presence of enemy obstacles, determine their types, and provide the necessary information to plan appropriate combined arms breaching or bypass operations to negate the impact on the friendly scheme of maneuver. Reconnaissance also allows friendly forces to anticipate when and where the enemy may employ obstacles that could impede operations. It is prudent to incorporate plans whenever possible to deny the enemy the opportunity to establish effective obstacles.

3. Gap Crossing

Engineers focus on projecting combat power over gaps by tailoring the appropriate resources for the specific mission set. Engineer planners task-organize the appropriate bridge units to support gap crossing operations. Combat engineers can assist with an assault gap crossing using organic armored vehicle-launched bridges (AVLBs) and Wolverines or their heavy equipment to modify the existing gap or by using expedient bridging (rope bridges, small nonstandard bridging using local materials). Operational-level engineers resource subordinate units that do not possess the necessary organic standard bridging equipment. River crossing is a unique gap crossing mission that requires specific and dedicated assets.

4. Combat Roads and Trails

The ability to move personnel and equipment is essential to maneuver warfare. This ability provides the commander with the means to increase tempo and speed, as well as concentrate mass at crucial times and places. The construction and maintenance of trails and roads are normally considered general engineering tasks and are, therefore, performed by engineering support units. However, locations near the forward line of own troops (FLOT) or time restrictions may require forward combat engineer units to perform these functions in an expedient manner or for short durations until support engineers are available. The two most likely scenarios that would involve this requirement would be bypass operations or support FACE operations. It is important for the engineer commander and staff to perform this function only in support of the maneuver plan. They should not allow engineering assets to be dissipated, rendering them unable to perform their primary role of supporting the commander's operational scheme of maneuver.

5. Forward Aviation Combat Engineering

With the advent of airpower and its associated support requirements, engineers have acquired a mission to support aviation assets. This frontline support will normally take the form of creating LZs for helicopters and vertical and/or short take-off and landing aircraft or parachute drop zones for personnel, equipment or supplies. Engineers should always strive to take full advantage of existing infrastructure and natural terrain features when constructing expeditionary landing and/or drop zones. The use of forward landing and/or drop zones can increase the speed and tempo of operations. For example, a closer proximity decreases turnaround time for aircraft and helicopters (for example, FARP sites); decreases personnel, equipment, and supply travel times from rear areas to the forward combat area; and decreases the response times of close air and helicopter support missions.

Clearing operations are conducted to eliminate the enemy's obstacle effort or residual obstacles within an assigned area or along a specified route. A clearing operation is a mobility operation, and, as with most mobility operations, it is typically performed by a combined arms force built around an engineer-based clearing force. A clearing operation could be conducted as a single mission to open or reopen a route or area, or it may be conducted on a recurring basis in support of efforts to defeat a sustained threat to a critical route.

Defensive Considerations

During the defense, mobility tasks include maintaining routes, coordinating gaps in existing obstacles, and supporting counterattacks. Engineers also open helicopter landing zones and tactical landing strips for fixed-wing aircraft. Maintaining and improving routes and creating bypass or alternate routes at critical points are major engineering tasks because movement routes are subjected to fires from enemy artillery and aircraft systems. These enemy fires may necessitate deploying engineer equipment, such as assault bridging and bulldozers, forward. The commander can also evacuate dislocated civilians or restrict their movements to routes not required by friendly forces to avoid detracting from the mobility of the defending force. The commander can do this provided the action is coordinated with the host nation or the appropriate civil-military operations units and fulfills the commander's responsibilities to dislocated civilians under the law of armed conflict.

The commander's priority of mobility support is first to routes used by counterattacking forces, then to routes used by main body forces displacing to subsequent positions. This mainly involves reducing obstacles and improving or constructing combat roads and trails to allow tactical support vehicles to accompany moving combat vehicles. The commander coordinates carefully to ensure that units leave lanes or gaps in their obstacles that allow for the repositioning of main body units and the commitment of the counterattack force. CBRN reconnaissance systems also contribute to the force's mobility in a contaminated environment.

II. Countermobility

Countermobility operations are those combined arms activities that use or enhance the effects of natural and manmade obstacles to deny an adversary freedom of movement and maneuver (ATP 3-90.8). Countermobility operations help isolate the battlefield and protect attacking forces from enemy counterattack, even though force mobility in offensive actions normally has first priority. Obstacles provide security for friendly forces as the fight progresses into the depth of the enemy's defenses. They provide flank protection and deny the enemy counterattack routes. They assist friendly forces in defeating the enemy in detail and can be vital in reducing the amount of forces required to secure a given area. Further, they can permit the concentration of forces by allowing a relatively small force to defend a large AO. The commander ensures the use of obstacles is integrated with fires and fully synchronized with the concept of operations to avoid hindering the attacking force's mobility.

During visualization, the commander identifies avenues of approach that offer natural flank protection to an attacking force, such as rivers or ridgelines. Staff running estimates support this process. Flanks are protected by destroying bridges, emplacing minefields, and by using scatterable munitions to interdict roads and trails. Swamps, canals, lakes, forests, and escarpments are natural terrain features that can be quickly reinforced for flank security.

Offensive Considerations

Countermobility operations during the offense must stress rapid emplacement and flexibility. Engineer support must keep pace with advancing maneuver forces and be prepared to emplace obstacles alongside them. Obstacles are employed to maximize the effects of restrictive terrain, such as choke points, or deny the usefulness of key terrain, since time and resources will not permit developing the terrain's full

4-30 (Movement & Maneuver) IV. Mobility & Countermobility

defensive potential. The commander first considers likely enemy reactions and then plans how to block enemy avenues of approach or withdrawal with obstacles. The commander also plans the use of obstacles to contain bypassed enemy elements and prevent the enemy from withdrawing. The plan includes obstacles to use upon identification of the enemy's counterattack. Speed and interdiction capabilities are vital characteristics of the obstacles employed. The commander directs the planning for air- and artillery-delivered munitions on enemy counterattack routes. The fire support system delivers these munitions in front of or on top of enemy lead elements once they commit to one of the routes. Rapid cratering devices and surface mine-fields provide other excellent capabilities.

Control of minefields and obstacles and accurate reporting to all units are vital. Obstacles will hinder both friendly and enemy maneuver. Control of obstacle initiation is necessary to prevent the premature activation of minefields and obstacles.

Refer to FM 90-7 and FM 5-102 for information on obstacle integration and FM 3-34.210 for information on mine warfare.

Defensive Considerations

Countermobility operations help isolate the battlefield and protect friendly forces from enemy attacks. The commander normally concentrates engineer efforts on countering the enemy's mobility. A defending force typically requires large quantities of Class IV and V materiel and specialized equipment to construct fighting and survivability positions and obstacles. With limited assets, the commander must establish priorities among countermobility, mobility, and survivability efforts. The commander ensures that the unit staff synchronizes these efforts with the unit's sustainment plans.

The commander may plan to canalize the enemy force into a salient. In this case, the commander takes advantage of the enemy force's forward orientation by fixing the enemy and then delivering a blow to the enemy's flank or rear. As the enemy's attacking force assumes a defensive posture, the defending commander rapidly coordinates and concentrates all defending fires against unprepared and unsupported segments of the attacking enemy force. The unit may deliver these fires simultaneously or sequentially.

When planning obstacles, commanders and staffs consider not only current operations but also future operations. The commander should design obstacles for current operations so they do not hinder planned future operations. Any commander authorized to employ obstacles can designate certain obstacles to shape the battlefield as high-priority reserve obstacles. The commander assigns responsibility for preparation to a subordinate unit but retains authority for ordering their completion. One example of a reserve obstacle is a highway bridge over a major river. Such obstacles receive the highest priority in preparation and, if ordered, execution by the designated subordinate unit.

A commander integrates reinforcing obstacles with existing obstacles to improve the natural restrictive nature of the terrain to halt or slow enemy movement, canalize enemy movement into engagement areas, and protect friendly positions and maneuver. The commander may choose to employ scatterable mines, if allowed by the rules of engagement. Obstacles must be integrated with fires to be effective. This requires the ability to deliver effective fires well beyond the obstacle's location. When possible, units conceal obstacles from hostile observation. They coordinate obstacle plans with adjacent units and conform to the obstacle zone or belts of superior echelons.

Effective obstacles block, turn, fix, disrupt, or force the enemy to attempt to breach them. The defender tries to predict enemy points of breach based on terrain and probable enemy objectives. The defending force develops means to counter enemy breach attempts, such as pre-coordinated fires. The attacker will try to conceal the time and location of the breach. The defending commander's plan addresses how to counter such a breach attempt, to include reestablishing the obstacle by using scatterable mines and other techniques.

(Movement & Maneuver) IV. Mobility & Countermobility 4-31

Obstacle Planning

Ref: Adapted from FM 5-102, Countermobility, chap. 2 and BCBL.

Obstacle planning begins with understanding the fundamentals of the obstacle framework. Precise use of these terms creates a common language and prevents confusion during planning and execution.

A. Obstacle Classification

Obstacle classification consists of two types of obstacles: existing obstacles and reinforcing obstacles.

- **Existing Obstacles.** Existing obstacles are obstacles that are present in the battlespace as inherent aspects of the terrain. The two types of existing obstacles are: natural (terrain features) and cultural (man-made terrain features).

- **Reinforcing Obstacles**. Reinforcing obstacles are obstacles that military forces specifically construct, emplace, or detonate to reduce threat mobility. The two types of reinforcing obstacles are tactical and protective.

B. Obstacle Intent

Obstacle intent is how the commander wants to use tactical obstacles to support his scheme of maneuver. Obstacle intent consists of:

- **Target.** Target is defined as the threat force's size and type along an avenue of approach.

- **Obstacle Effect.** Obstacle effect is achieved through the integration of both fires and obstacles to manipulate the threat's movement in support of the commander's scheme of maneuver. All tactical obstacles should disrupt, turn, fix, or block the enemy.

- **Relative location.** Relative location is where the commander wants the desired effect on the target.

C. Principles of Obstacle Emplacement

Regardless of the type defense employed by the tactical commander, there are five basic employment principles for reinforcing obstacles:

- **Reinforcing obstacles support the maneuver commander's plan.** Reinforcing obstacles must be planned and emplaced to support the tactical plan. Obstacles other than mines emplaced outside the range of friendly weapons are of little use. Reinforcing obstacles that do not accomplish one or more of the basic purposes are also of little value.

- **Reinforcing obstacles are integrated with observed fires.** Obstacles are used to develop engagement areas in which enemy maneuver is restricted and slowed, thereby increasing the hit probability of friendly direct and indirect fires.

- **Reinforcing obstacles are integrated with existing obstacles and with other reinforcing obstacles.** Reinforcing obstacles are sited to take the maximum advantage of existing obstacles. They are placed where they can close the gaps or openings between existing obstacles and/or close any passages through them.

- **Reinforcing obstacles are employed in depth.** A series of simple obstacles arranged one behind the other along a probable axis of enemy advance is far more effective than one large, elaborate obstacle. Restricting the design of obstacles to correspond with the strength of the existing obstacle helps to conserve effort and direct it toward executing obstacles in depth.

- **Reinforcing obstacles are employed for surprise.** Using obstacles in order to obtain surprise is one means available to the commander to retain a degree of initiative even when defending.

4-32 (Movement & Maneuver) IV. Mobility & Countermobility

D. Obstacle Protection

Obstacle protection is protecting the integrity of obstacles. This protection can be achieved through counterreconnaissance, breach asset destruction, obstacle repair, and using phony obstacles. Counterreconnaissance prevents the threat from gathering information on friendly preparation. The reconnaissance and surveillance plan includes obstacle protection.. Early breach asset destruction will reduce the threat's ability to maneuver and ensure maximum effectiveness of the obstacles. Obstacle repair must occur when the threat has attempted to breach tactical obstacles and during lulls in the battle, between echelons.

E. Obstacle C2

Obstacle C2 focuses on obstacle emplacement authority and obstacle control.

- **Obstacle Emplacement Authority.** The authority that a unit commander has to emplace reinforcing obstacles.

- **Obstacle Control.** The commander uses control measures, specific guidance, and orders to maintain obstacle control.

Obstacle Control Measures

Obstacle control measures are specific control measures that simplify granting obstacle emplacement authority and providing obstacle control *(FM 3-0, Operations, w/Chg. 1)*.

Obstacle control measure	Emplacement authority From	Emplacement authority To	Graphic	Example
Zone	CORPS	DIV	Unit designation / Effect graphic	52ID / 2/52
	DIV	BDE		
Belt	CORPS	DIV[1]	Unit designation / Effect graphic	3/52 / 3/52
	DIV	BDE[ii]		
	BDE	TF		
Group	BDE[2]	TF[2]	Unit designation / Effect symbol is the graphic	Obstacle groups in a belt / 3-68AR 2-13IN
	TF	CO TRP		
Obstacle restricted area	Any		52ID 120900-162400ZSEP22	

1 - Rarely done by corps and divisions, but possible.
2 - Done only when directed and integrated with corps or division fire plans.

Graphic effects symbols				
	Disrupt	Fix	Turn	Block

Ref: FM 3-0, Operations (Oct '17), fig. 6-8, p. 6-11.

(Movement & Maneuver) IV. Mobility & Countermobility 4-33

Improvement to defensive positions is continuous. Given time and resources, the defending force constructs additional obstacle systems in-depth, paying special attention to its assailable flanks and rear. The rear is especially vulnerable if there are noncontiguous areas of operations or nontraditional threats. Obstacle systems can provide additional protection from enemy attacks by forcing the enemy to spend time and resources to breach or bypass them. This gives the defending force more time to engage enemy forces attempting to execute a breach or bypass.

The commander designates the unit responsible for establishing and securing each obstacle. The commander may retain execution authority for some obstacles or restrict the use of some types of obstacles to allow other battlefield activities to occur.

Tactical and protective obstacles are constructed primarily at company level and below. Small-unit commanders ensure that observation and fires cover all obstacles to hinder breaching. Deliberate protective obstacles are common around fixed sites. Protective obstacles are a key enabler of survivability operations. They are tied in with FPFs and provide the friendly force with close-in protection.

III. Engineer Support

Army engineer support to operations encompasses a wide range of tasks that require many capabilities. Commanders use engineers throughout unified land operations across the range of military operations. They use them primarily to assure mobility, enhance protection, enable force projection and logistics, and build partner capacity and develop infrastructure.

Engineer Support to Warfighting Functions

Unified land operations require the continuous generation and application of combat power, often for protracted periods.

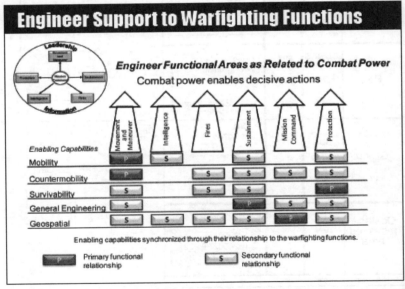

Ref: FM 3-34 (2016), fig. 2-1. Engineer application of combat power.

The three engineer disciplines -- combat, general and geospatial engineering -- encompass tasks along the lines of engineer support. The combat engineering discipline, due to its support to maneuver forces in close combat, is primarily focused on tasks that assure mobility and enhance protection. The general engineering and geospatial engineering disciplines performs tasks along all four lines of engineer support.

4-34 (Movement & Maneuver) IV. Mobility & Countermobility

Chap 5

Intelligence Warfighting Function

Ref: ADP 2-0, Intelligence (Jul '19), chap. 2; ADP 3-0, Operations (Jul '19), pp. 5-4 to 5-5; and FM 3-0, Operations (Oct '22), p. 2-2.

The intelligence warfighting function is the related tasks and systems that facilitate understanding the enemy, terrain, weather, civil considerations, and other significant aspects of the operational environment (ADP 3-0). Intelligence involves analyzing information from all sources, which includes the other warfighting functions, and conducting operations to collect information. The integration of intelligence into operations facilitates understanding of an operational environment and assists in determining when and where to employ capabilities against adversaries and enemies. Intelligence likewise facilitates responses by Army forces to other situations, such as public health crises and events precipitating noncombatant evacuation. The intelligence warfighting function provides support to force generation, situational understanding, targeting and information operations, and information collection. The intelligence warfighting function fuses the information collected through reconnaissance, surveillance, security operations, and intelligence operations. Commanders drive intelligence and intelligence drives operations. Army forces execute intelligence, surveillance, and reconnaissance (ISR) through the operations and intelligence processes, with an emphasis on intelligence analysis and information collection.

Timely, accurate, relevant, and predictive intelligence enables decision making, tempo, and agility during operations. Due to the fog and friction of warfare, commanders must fight for intelligence and share it with adjacent units and across echelons.

The intelligence warfighting function includes the following tasks:

- Provide support to force generation.
- Provide support to situational understanding.
- Conduct information collection
- Provide intelligence support to targeting and information operations.

Fighting for Intelligence

Information collection begins immediately following receipt of mission. Units must be prepared to fight for intelligence against a range of threats, enemy formations, and unknowns. These challenges include integrated air defense systems (IADSs) and long range fires, counter reconnaissance, cyberspace and EW operations, deception operations, and camouflage. It may be necessary for commanders to allocate maneuver, fires, and other capabilities to conduct combat operations to enable information collection.

Priority intelligence requirements, information requirements, and targeting requirements inform the integrated information collection plan. All units (maneuver, fires, maneuver support, and sustainment units) are part of the information collection effort. Commanders and staffs integrate and synchronize all activities that provide useful information as a part of the information collection effort, including Soldier and leader engagements, patrols, observation posts and listening posts, convoys, and checkpoints.

During planning, combat information and intelligence is especially useful in determining the viability of potential courses of action. During the execution phase of the operations process, a layered and continuous information collection effort ensures detection of any enemy formations, lethal fires capabilities, or specialized capabilities that provide the enemy advantage.

Intelligence (ADP 2-0)

(Intelligence) Warfighting Function 5-1

I. Intelligence Overview

Ref: ADP 2-0, Intelligence (Jul '19), preface and pp. 1 to 2.

Operations and intelligence are closely linked. The intelligence process is continuous and directly drives and supports the operations process. This principle will remain true well into the future. Intelligence will continue to be a critical part of the conduct-planning, preparing, executing, and assessing--of operations. Future operations will be difficult. They will occur in complex operational environments against capable peer threats, who most likely will start from positions of relative advantage. U.S. forces will require effective intelligence to prevail during these operations.

Intelligence supports joint and Army operations across unified action, the Army's strategic roles, unified land operations, and decisive action at each echelon-from the geographic combatant command down to the battalion level. Specifically, intelligence supports commanders and staffs by facilitating situational understanding across all domains and the information environment. Commanders and staffs use situational understanding to identify and exploit multi-domain windows of opportunity and to achieve and exploit positions of relative advantage.

Intelligence is inherently joint, interagency, intergovernmental, and multinational. Every aspect of intelligence is synchronized, networked, and collaborative across all unified action partners. This synchronization occurs through national to tactical intelligence support. The Army both benefits from and contributes to national to tactical intelligence and focuses the Army intelligence effort through the intelligence warfighting function, which is larger than military intelligence. Critical participants within the function include commanders and staffs, decision makers, collection managers, and intelligence leaders.

Despite a thorough understanding of intelligence fundamentals and a proficient staff, an effective intelligence effort is not assured. Large-scale ground combat operations are characterized by complexity, chaos, fear, violence, fatigue, and uncertainty. The fluid and chaotic nature of large-scale ground combat operations causes the greatest degree of fog, friction, and stress on the intelligence warfighting function. Threat forces will attempt to counter friendly collection capabilities by using integrated air defense systems, long-range fires, counterreconnaissance, cyberspace and electronic warfare operations, camouflage and concealment, and deception.

Ensuring an effective intelligence effort is a challenge described as **fighting for intelligence.**

Intelligence

Intelligence is (1) the product resulting from the collection, processing, integration, evaluation, analysis, and interpretation of available information concerning foreign nations, hostile or potentially hostile forces or elements, or areas of actual or potential operations; (2) the activities that result in the product; and (3) the organizations engaged in such activities (JP 2-0).

Throughout modern history, intelligence has been and remains an inherent part of military operations. From national and Department of Defense (DOD) levels down to the Army battalion level, intelligence is an activity that never stops. Army forces are globally engaged, always executing operations and preparing for future operations as part of a joint team. A key part of global engagement is the continuous use of intelligence, the collection and analysis of information, and the production of intelligence. This constant activity, referred to as intelligence, is never at rest. To understand Army intelligence, it is important to understand intelligence within the larger context of joint large-scale combat operations, including Army large-scale ground combat operations; the operational environment; unified action; the Army strategic roles; and unified land operations.

5-2 (Intelligence) Warfighting Function

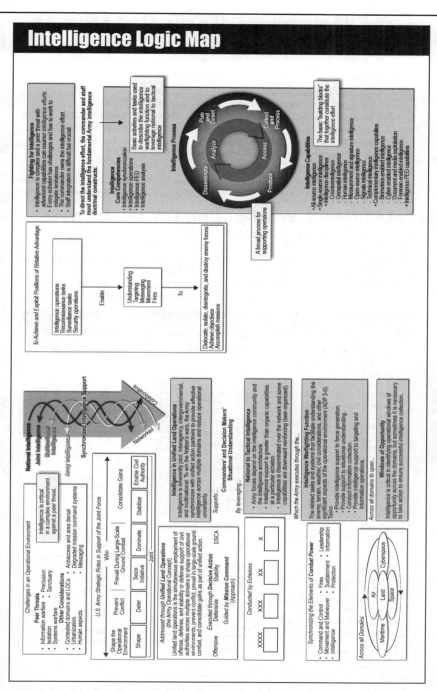

Ref: ADP 2-0, Intelligence (Jul '19), introductory figure. ADP 2-0 logic chart.

(Intelligence) Warfighting Function 5-3

Ensuring an effective intelligence effort is a challenge described as fighting for intelligence. The following aspects of fighting for intelligence are critical:

- Effective intelligence requires developing an effective intelligence architecture well before large-scale combat operations.
- The commander must own the intelligence effort.
- The commander and staff-
  - Must forge an effective relationship and excel in staff integration.
  - Must understand intelligence limitations, especially collection gaps, at their echelon overcome or mitigate those limitations through effective information collection.
  - At times, may have to conduct combat operations or find creative solutions to enable information collection.
- The unit must adjust the information collection plan, adapt to threat counter-collection measures, and maintain a layered and aggressive information collection effort.

II. Intelligence Warfighting Function Tasks

The intelligence warfighting function is the Army's contribution to the joint intelligence effort. The intelligence warfighting function is the related tasks and systems that facilitate understanding the enemy, terrain, weather, civil considerations, and other significant aspects of the operational environment (ADP 3-0). Specifically, other significant aspects of the operational environment include threats, adversaries, the operational variables, and can include other aspects depending on the nature of operations.

Intelligence Warfighting Function Tasks

Intelligence tasks ▶	Commander's focus ▶	Commander's decisions
Provide intelligence support to force generation: • Provide intelligence readiness. • Establish an intelligence architecture. • Provide intelligence overwatch. • Generate intelligence knowledge. • Tailor the intelligence force.	Orient on contingencies.	• Should the unit's level of readiness be increased? • Should the operation plan be implemented?
Provide support to situational understanding: • Perform IPB. • Perform situation development. • Provide intelligence support to protection. • Provide tactical intelligence overwatch. • Conduct police intelligence operations. • Provide intelligence support to civil affairs operations.	• Plan an operation. • Prepare. • Execute. • Assess. • Secure the force. • Determine 2d and 3d order effects on operations and the populace.	• Which COA will be implemented? • Which enemy actions are expected? • What mitigation strategies should be developed and implemented to reduce the potential impact of operations on the population?
Conduct information collection: • Collection management. • Direct information collection. • Execute collection. • Conduct intelligence-related missions and operations.	• Plan information collection for an operation, including PED requirements. • Prepare. • Execute. • Assess.	• Which DPs, HPTs, and HVTs are linked to the threat's actions? • Are the assets available and in position to collect on the DPs, HPTs, and HVTs? • Have the assets been repositioned for branches or sequels?
Provide intelligence support to targeting and information operations: • Provide intelligence support to targeting. • Provide intelligence support to information operations. • Provide intelligence support to combat assessment.	• Create lethal or nonlethal effects against targets. • Destroy, suppress, disrupt, or neutralize targets. • Reposition intelligence or attack assets.	• Are the unit's lethal and nonlethal actions and maneuver effective? • Which targets should be re-engaged? • Are the unit's information operations effective?

Ref: ADP 2-0 (Jul '19), table 2-1. Overview of intelligence warfighting function tasks.

III. Intelligence Support to Commanders and Decisionmakers

Ref: Adapted from ADP 2-0, Intelligence (Aug '12), pp. 2 to 3.

Commanders provide guidance and resources to support unique requirements of the staffs and subordinate commanders. Although commanders drive operations, as the principal decisionmakers, their relationship with their staffs must be one of close interaction and trust. This relationship must encourage initiative within the scope of the commander's intent. Independent thought and timely actions by staffs are vital to mission command.

Commanders provide guidance and continuous feedback throughout operations by—
- Providing direction
- Stating clear, concise commander's critical information requirements (CCIRs)
- Synchronizing the intelligence warfighting function
- Participating in planning
- Collaborating with the G-2/S-2 during the execution of operations

Teamwork within and between staffs produces integration essential to effective mission command and synchronized operations. While all staff sections have clearly defined functional responsibilities, they cannot work efficiently without complete cooperation and coordination among all sections and cells. Key staff synchronization and integration occur during—

- **Intelligence preparation of the battlefield (IPB).** The G-2/S-2 leads the IPB effort with the entire staff's participation during planning.
- **Army design methodology, the military decisionmaking process, and the rapid decisionmaking and synchronization process.** Intelligence provides important input that helps frame operational problems and drives decisionmaking processes.
- **Information collection.** The G-2/S-2 staff provides the analysis, supporting products, and draft plan necessary for the G-3/S-3 to task the information collection plan.
- **Targeting.** Intelligence is an inherent part of the targeting process and facilitates the execution of the decide, detect, deliver, and assess functions.
- **Assessments.** The G-2/S-2 staff collaborates closely with the rest of the staff to ensure timely and accurate assessments occur throughout operations.

The staff performs many different activities as a part of the intelligence warfighting function. This effort is extremely intensive during planning and execution. After the commander establishes CCIRs, the staff focuses the intelligence warfighting function on priority intelligence requirements and other requirements. The staff assesses the situation and refines or adds new requirements, as needed, and quickly retasks units and assets. It is critical for the staff to plan for and use well-developed procedures and flexible planning to track emerging targets, adapt to changing operational requirements, and meet the requirement for combat assessment.

Refer to BSS6: The Battle Staff SMARTbook, 6th Ed. (Plan, Prepare, Execute, & Assess Military Operations) for further discussion of the intelligence warfighting function as it relates to the operations process -- to include warfighting function tasks, intelligence core competencies, the intelligence process, and types of intelligence products.

(Intelligence) Warfighting Function 5-5

IV. Intelligence Integrating Functions

Ref: ATP 2-01.3, Intelligence Preparation of the Battlefield (Mar '19), p. xi.

Commanders and staffs integrate the warfighting functions and synchronize the force to adapt to changing circumstances throughout the operations process. They use several integrating processes to do this. (See pp. 3-20 to 3-21.)

Intelligence Preparation of the Battlefield (IPB)

Intelligence preparation of the battlefield (IPB) is a systematic, continuous process of analyzing the threat and other aspects of an operational environment within a specific geographic area. The figure below lists relevant IPB products. The four IPB steps are—

- Define the Operational Environment
- Describe Environmental Effects on Operations
- Evaluate the Threat
- Determine Threat Courses of Action

See p. 3-20 for discussion of IPB as one of the integrating processes.

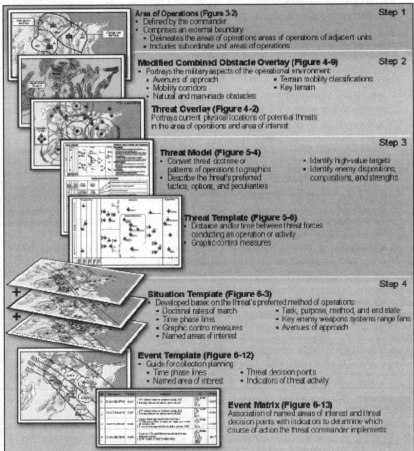

Ref: Introductory figure. Products of the IPB process.

5-6 (Intelligence) Warfighting Function

Information Collection

Ref: FM 3-55, Information Collection (May '13).

Knowledge is the precursor to effective action in the informational or physical domains. Knowledge about an operational environment requires aggressive and continuous operations to acquire information. Information collected from multiple sources and analyzed becomes intelligence that provides answers to commander's critical information requirements (CCIRs). Commanders use reconnaissance and surveillance to provide intelligence to reduce the inherent uncertainty of war.

Information collection is an activity that synchronizes and integrates the planning and employment of sensors and assets as well as the processing, exploitation, and dissemination systems in direct support of current and future operations. *(See p. 3-20.)*

Ref: FM 3-55, fig. 1-1. Information collection activities.

Refer to BSS6: The Battle Staff SMARTbook, 6th Ed. (Plan, Prepare, Execute, & Assess Military Operations) for complete discussion of intelligence preparation of the battlefield (50 pgs) from ATP 2-01.3 (Mar '19) and information collection from FM 3-55 (May '13).

(Intelligence) Warfighting Function 5-7

V. Types of Intelligence Products

The G-2/S-2 staff produces and maintains a broad variety of products tailored to its consumers. These products are developed and maintained in accordance with the commander's guidance. For all of these products, the primary focus of the G-2/S-2 staff's analysis is presenting predictive intelligence to support operations. The intelligence products include the—

A. Intelligence Estimate

An intelligence estimate is the appraisal, expressed in writing or orally, of available intelligence relating to a specific situation or condition with a view to determining the courses of action open to the threat and the order of probability of their adoption. The G-2/S-2 staff develops and maintains the intelligence estimate. The primary purpose of the intelligence estimate is to—

- Determine the full set of COAs open to the threat and the probable order of their adoption
- Disseminate information and intelligence
- Determine requirements concerning threats and other relevant aspects of the operational environment

B. Intelligence Summary

INTSUMs provide the context for commander's situational understanding. The INTSUM reflects the G-2's/S-2's interpretation and conclusions regarding threats, terrain and weather, and civil considerations over a designated period of time. This period will vary with the desires of the commander and the requirements of the situation. The INTSUM provides a summary of the threat situation, threat capabilities, the characteristics of terrain and weather and civil considerations, and COAs. The INTSUM can be presented in written, graphic, or oral format, as directed by the commander.

C. Intelligence Running Estimate

Effective plans and successful execution hinge on accurate and current running estimates. A running estimate is the continuous assessment of the current situation used to determine if the current operation is proceeding according to the commander's intent and if the planned future operations are supportable (ADP 5-0). Failure to maintain accurate running estimates may lead to errors or omissions that result in flawed plans or bad decisions during execution.

Running estimates are principal knowledge management tools used by the commander and staff throughout the operations process. In their running estimates, the commander and each staff section continuously consider the effect of new information and update the following:

- Facts
- Assumptions
- Friendly force status
- Threat activities and capabilities
- Civil considerations
- Recommendations and conclusions

D. Common Operational Picture (COP)

A common operational picture is a single display of relevant information within a commander's area of interest tailored to the user's requirements and based on common data and information shared by more than one command (ADRP 6-0). The COP is the primary tool for supporting the commander's situational understanding. All staff sections provide input from their area of expertise to the COP.

5-8 (Intelligence) Warfighting Function

I. The Intelligence Process

Chap 5

Ref: ADP 2-0, Intelligence (Jul '19), chap. 3.

Commanders use the operations process to drive the planning necessary to understand, visualize, and describe their operational environment; make and articulate decisions; and direct, lead, and assess military operations. Commanders successfully accomplish the operations process by using information and intelligence. The design and structure of the intelligence process support commanders by providing intelligence needed to support mission command and the commander's situational understanding. The commander provides guidance and focus by defining operational priorities and establishing decision points and CCIRs.

The Joint Intelligence Process
The joint intelligence process provides the basis for common intelligence terminology and procedures. (JP 2-0.) It consists of six interrelated categories of intelligence operations:

- Planning and direction
- Collection
- Processing and exploitation
- Analysis and production
- Dissemination and integration
- Evaluation and feedback

The Army Intelligence Process
Due to the unique characteristics of Army operations, the Army intelligence process differs from the joint process in a few subtle ways while accounting for each category of the joint intelligence process. The Army intelligence process consists of four steps (plan and direct, collect, produce, and disseminate) and two continuing activities (analyze and assess).

Commander's guidance drives the intelligence process. The process generates information, products, and knowledge about threats, terrain and weather, and civil considerations for the commander and staff. The intelligence process supports all of the activities of the operations process (plan, prepare, execute, and assess). The intelligence process can be conducted multiple times to support each activity of the operations process. Although the intelligence process includes unique aspects and activities, it is designed similarly to the operations process:

- The *plan and direct step* of the intelligence process closely corresponds with the plan activity of the operations process
- The *collect, produce, and disseminate steps* of the intelligence process together correspond to the execute activity of the operations process
- *Assess*, which is continuous, is part of the overall assessment activity of the operations process

Intelligence support to operations requires leveraging national to tactical intelligence. This support is coordinated through the intelligence staff at each echelon by using the intelligence process.

Intelligence (ADP 2-0)

(Intelligence) I. The Intelligence Process 5-9

The G-2/S-2 produces intelligence for the commander as part of a collaborative process. The commander drives the G-2's/S-2's intelligence production effort by establishing intelligence and information requirements with clearly defined goals and criteria. Differing unit missions and operational environments dictate numerous and varied production requirements to the G-2/S-2 and staff.

The G-2/S-2 and staff provide intelligence products that enable the commander to—
- Plan operations and employ maneuver forces effectively
- Recognize potential COAs
- Conduct mission preparation
- Employ effective tactics, techniques, and procedures
- Take appropriate security measures
- Focus information collection
- Conduct effective targeting

Commander's Guidance

Commanders drive the intelligence process by both providing commander's guidance and approving priority intelligence requirements (PIRs). While issuing their guidance, commanders should limit the number of PIRs so the staff can focus its efforts and allocate sufficient resources. Each commander dictates which intelligence products are required, when they are required, and in what format.

I. Intelligence Process

Just as the activities of the operations process overlap and recur as the mission demands, so do the steps of the intelligence process.

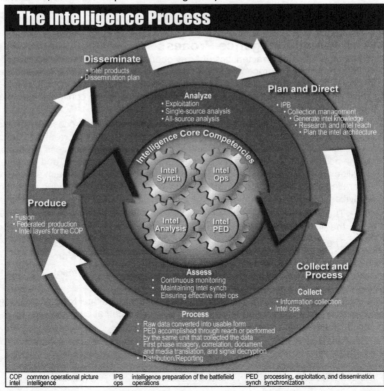

II. Requirements Management
Ref: ADP 2-0, Intelligence (Jul '19), p. 3-5.

For collection managers, there are three types of requirements resulting from collection management. The following three types of validated information requirements are prioritized for purposes of assigning information collection tasks: priority intelligence requirements (PIRs), intelligence requirements, and information requirements.

Refer to FM 3-55 and ATP 2-01 for more details on requirements and indicators.

Collection managers must understand how collection and PED assets are distributed as they develop and validate requirements. They must also understand that some requirements can be answered through intelligence reach. The following shows the process of developing requirements and integrating them into the information collection process.

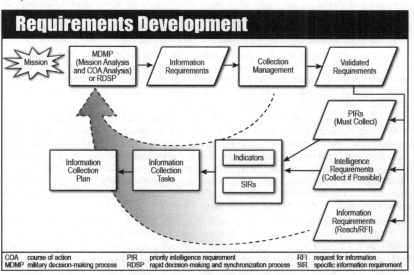

Ref: ADP 2-0, Intelligence (Jul '19), fig. 3-2. Requirements Development.

Priority Intelligence Requirement (PIR)
An intelligence requirement, stated as a priority for intelligence support, that the commander and staff need to understand the adversary or the operational environment. Also called PIR. *Refer to FM 2-01.3.*

Intelligence Requirements
A type of information requirement developed by subordinate commanders and the staff (including subordinate staffs) that requires dedicated information collection for the elements of threat, terrain and weather, and civil considerations.

Information Requirements
In intelligence usage, those items of information regarding the adversary and other relevant aspects of the operational environment that need to be collected and processed in order to meet the intelligence requirements of a commander. Any information elements the commander and staff require to successfully conduct operations.

Intelligence Process Steps

Ref: ADP 2-0, Intelligence (Jul '19), pp. 3-3 to 3-8.

A. Plan and Direct

Each staff element must conduct analysis before operational planning can begin. Planning consists of two separate but closely related components--conceptual and detailed planning. Conceptual planning involves understanding the operational environment and the problem, determining the operation's end state, and visualizing an operational approach. Detailed planning translates the broad operational approach into a complete and practical plan.

The initial generation of intelligence knowledge about the operational environment occurs far in advance of detailed planning and orders production. This intelligence assists in focusing information collection once a mission is received or in anticipation of a mission. Commanders and staffs often begin planning in the absence of a complete and approved higher headquarters' operation plan or operation order. In these instances, the headquarters begins a new planning effort based on a warning order and other directives, such as a planning order or an alert order from their higher headquarters.

Intelligence planning is also an inherent part of the Army design methodology and the military decision-making process. Intelligence analysts must prepare detailed planning products for the commander and staff for orders production and the conduct of operations. Through thorough and accurate planning, the staff allows the commander to focus the unit's combat power to achieve mission success.

The plan and direct step also includes activities that identify key information requirements and develops the means for satisfying those requirements. The intelligence staff collaborates with the operations and signal staffs to plan the intelligence architecture. Collaboration facilitates parallel planning and enhances all aspects of the intelligence process by enriching analysis, incorporating different points of view and broadening situational understanding. The staffs produce a synchronized and integrated information collection plan focused on answering PIRs and other requirements. PIRs and other requirements drive the information collection effort.

Commanders must employ organic collection assets as well as plan, coordinate, and articulate requirements to leverage DOD intelligence capabilities. Commanders and staffs cannot expect higher echelons to automatically provide all of the information and intelligence they need. While intelligence each is a valuable tool, the push of intelligence products from higher echelons does not relieve subordinate staffs from developing specific and detailed requirements. Commanders and staffs must focus requests for intelligence support by clearly articulating requirements.

The staff focuses information collection plans on answering CCIRs and other requirements and enables the quick retasking of units and assets as the situation changes. Collection management includes continually identifying intelligence gaps. This ensures the developing threat situation and civil considerations-not only the operation order-drive information collection. Specifically, G-2/S-2s-

- Evaluate information collection assets for suitability (availability, capability, vulnerability, and performance history) to execute information collection tasks and make appropriate recommendations on asset tasking to the primary operations staff officers.

- Assess information collection against CCIRs and other requirements to determine the effectiveness of the information collection plan. They maintain awareness to identify gaps in coverage and identify the need to cue or recommend redirecting information collection assets to the primary operations staff officers.

- Update the collection management tools as requirements are satisfied, added, modified, or deleted. They remove satisfied requirements and recommend new requirements as necessary.

5-12 (Intelligence) I. The Intelligence Process

B. Collect and Process

Collection is synchronized to provide critical information at key times throughout the phases of an operation and during the transition from one operation to another operation. A successful information collection effort results in the timely collection and reporting of relevant and accurate information, which supports the production of intelligence. Collection consists of collecting, processing, and reporting information in response to information collection tasks. Different units and systems collect information and data about threats, terrain and weather, and civil considerations. Collected information is used in intelligence databases, intelligence production, and the G-2's/S-2's awareness—and ultimately supports the commander's situational understanding. Information collection activities transition as requirements change, the unit mission changes, the unit proceeds through the phases of an operation, and the unit prepares for future operations.

C. Produce

Production is the development of intelligence through the analysis of collected information and existing intelligence. Analysts create intelligence products, conclusions, or projections regarding threats and relevant aspects of the operational environment to answer known or anticipated requirements in an effective format. The G-2/S-2 staff processes and analyzes information from single or multiple sources, disciplines, and complementary intelligence capabilities, and integrates the information with existing intelligence to create finished intelligence products.

Intelligence products must be timely, relevant, accurate, predictive, and tailored to facilitate situational understanding and support decisionmaking. The accuracy and detail of intelligence products have a direct effect on operational success. Due to time constraints, analysts sometimes develop intelligence products that are not as detailed as they prefer.

The G-2/S-2 staff prioritizes and synchronizes the unit's information processing and intelligence production efforts. The G-2/S-2 staff addresses numerous and varied production requirements based on PIRs and other requirements; diverse missions, environments, and situations; and user-format requirements. Through analysis, collaboration, and intelligence reach, the G-2/S-2 and staff use the intelligence capability of higher, lateral, and subordinate echelons to meet processing and production requirements.

Processing is often an important production activity. The G-2/S-2 staff processes information collected by the unit's assets as well as information received from higher, subordinate, and lateral echelons and other organizations. Processing includes sorting through large amounts of collected information and intelligence and converting relevant information into a form suitable for analysis, production, or immediate use.

Analysis occurs to ensure the information is relevant, to isolate significant elements of information, and to integrate the information into an intelligence product. Additionally, analysis of information and intelligence is important to ensure the focus, prioritization, and synchronization of the unit's intelligence production effort is in accordance with the PIRs and other requirements.

D. Disseminate

Commanders must receive combat information and intelligence products in time and in an appropriate format to facilitate situational understanding and support decisionmaking. Timely dissemination of intelligence is critical to the success of operations. Dissemination is deliberate and ensures consumers receive intelligence to support operations.

This step does not include the normal reporting and technical channels otherwise conducted by intelligence warfighting function organizations and units during the intelligence process. Each echelon with access to information may perform analysis on that information. Then each echelon ensures that resulting intelligence products are properly disseminated. Determining the product format and selecting the means to deliver it are key aspects of dissemination.

(Intelligence) I. The Intelligence Process 5-13

III. Intelligence Process Continuing Activities

Analyze and assess are two continuing activities that shape the intelligence process. They occur continually throughout the intelligence process.

A. Analyze

Analysis assists commanders, staffs, and intelligence leaders in framing the problem, stating the problem, and solving it. Leaders at all levels conduct analysis to assist in making many types of decisions. Analysis occurs at various stages throughout the intelligence process and is inherent throughout intelligence support to situational understanding and decisionmaking. Collectors perform initial analysis before reporting. For example, a HUMINT collector analyzes an intelligence requirement to determine the best possible collection strategy to use against a specific source.

Analysis in requirements management is critical to ensuring information requirements receive the appropriate priority for collection. The G-2/S-2 staff analyzes each requirement to determine—

- The requirement's feasibility and whether it supports the commander's guidance
- The best method of satisfying the requirement (for example, what unit or capability and where to position that capability)
- If the collected information satisfies the requirement

Analysis is used in situation development to determine the significance of collected information and its significance relative to predicted threat COAs and PIRs and other requirements. Through predictive analysis, the staff attempts to identify threat activity or trends that present opportunities or risks to the friendly force. They often use indicators developed for each threat COA as the basis for their analysis and conclusions.

B. Assess

Assess is part of the overall assessment activity of the operations process. For intelligence purposes, assessment is the continuous monitoring and evaluation of the current situation, particularly significant threat activities and changes in the operational environment. Assessing the situation begins upon receipt of the mission and continues throughout the intelligence process. This assessment allows commanders, staffs, and intelligence leaders to ensure intelligence synchronization. Friendly actions, threat actions, civil considerations, and events in the area of interest interact to form a dynamic operational environment. Continuous assessment of the effects of each element on the others, especially the overall effect of threat actions on friendly operations, is essential to situational understanding.

The G-2/S-2 staff continuously produces assessments based on operations, the information collection effort, the threat situation, and the status of relevant aspects of the operational environment. These assessments are critical to—

- Ensure PIRs are answered
- Ensure intelligence requirements are met
- Redirect collection assets to support changing requirements
- Ensure operations run effectively and efficiently
- Ensure proper use of information and intelligence
- Identify threat efforts at deception and denial

The G-2/S-2 staff continuously assesses the effectiveness of the information collection effort. This type of assessment requires sound judgment and a thorough knowledge of friendly military operations, characteristics of the area of interest, and the threat situation, doctrine, patterns, and projected COAs.

5-14 (Intelligence) I. The Intelligence Process

Chap 5

II. Army Intelligence Capabilities

Ref: ADP 2-0, Intelligence (Jul '19), chap. 4.

The intelligence warfighting function executes the intelligence process by employing intelligence capabilities. All-source intelligence and single-source intelligence are the building blocks by which the intelligence warfighting function facilitates situational understanding and supports decisionmaking. The intelligence warfighting function receives information from a broad variety of sources. Some of these sources are commonly referred to as single-source capabilities. Single-source capabilities are employed through intelligence operations with the other means of information collection (reconnaissance, surveillance, and security operations). The intelligence produced based on all of those sources is called all-source intelligence.

I. All-Source Intelligence

Army forces conduct operations based on all-source intelligence assessments and products developed by the intelligence staff. All-source intelligence is the integration of intelligence and information from all relevant sources in order to analyze situations or conditions that impact operations. In joint doctrine, all-source intelligence is intelligence products and/or organizations and activities that incorporate all sources of information in the production of finished intelligence (JP 2-0).

All-Source Analysis

The fundamentals of all-source intelligence analysis comprise intelligence analysis techniques an the all-source analytical tasks: situation development, generating intelligence knowledge, IPB, and support to targeting and information operations.

Through the receipt and processing of incoming reports and messages, the intelligence staff determines the significance and reliability of incoming information, integrates incoming information with current intelligence holdings, and through analysis and evaluation determines changes in threat capabilities, vulnerabilities, and probable COAs. The intelligence staff supports the integrating processes (IPB, targeting, risk management, information collection, and knowledge management) by providing all-source analysis of threats, terrain and weather, and civil considerations.

All-source intelligence is used to develop the intelligence products necessary to aid situational understanding, support the development of plans and orders, and answer information requirements. Although all-source intelligence normally takes longer to produce, it is more reliable and less susceptible to deception than single-source intelligence.

All-Source Production

Fusion facilitates all-source production. For Army purposes, fusion is consolidating, combining and correlating information together. Fusion occurs as an iterative activity to refine information as an integral part of all-source analysis.

All-source intelligence production is continuous and occurs throughout the intelligence and operations processes. Most of the products from all-source intelligence are initially developed during planning and updated, as needed, throughout preparation and execution based on information gained from continuous assessment.

Intelligence (ADP 2-0)

(Intelligence) II. Army Intel Capabilities 5-15

The Intelligence Disciplines

Ref: ADP 2-0, Intelligence (Jul '19), pp. 4-3 to 4-10.

In joint operations, the intelligence enterprise is commonly organized around the intelligence disciplines. The intelligence disciplines are—

Intelligence Disciplines

- **Counterintelligence (CI)**
- **Geospatial Intelligence**
- **Human Intelligence (HUMINT)**
- **Measurement and Signature Intelligence (MASINT)**
- **Open-Source Intelligence (OSINT)**
- **Signals Intelligence (SIGINT)**
- **Technical Intelligence (TECHINT)**

The intelligence disciplines are integrated to ensure a multidiscipline approach to intelligence analysis, and ultimately all-source intelligence facilitates situational understanding and supports decisionmaking. Each discipline applies unique aspects of support and guidance through technical channels.

Refer to JP 2-0.1.

Counterintelligence (CI)

CI counters or neutralizes intelligence collection efforts through collection, CI investigations, operations, analysis, production, and technical services and support. CI includes all actions taken to detect, identify, track, exploit, and neutralize multidiscipline intelligence activities of foreign intelligence and security services (FISS), international terrorist organizations, and adversaries, and is the key intelligence community contributor to protect U.S. interests and equities.

The mission of Army CI is to conduct aggressive, comprehensive, and coordinated investigations, operations, collection, analysis and production, and technical services. These functions are conducted worldwide to detect, identify, assess, counter, exploit, or neutralize the FISS, international terrorist organization, and adversary collection threat.

Refer to ATP 2-22.2-1.

Geospatial Intelligence

Geospatial intelligence is the exploitation and analysis of imagery and geospatial information to describe, assess, and visually depict physical features and geographically referenced activities on the Earth. Geospatial intelligence consists of imagery, imagery intelligence, and geospatial information (JP 2-03). (Section 467, Title 10, USC [10 USC 467], establishes GEOINT.) Note. GEOINT consists of any one or any combination of the following components: imagery, IMINT, and geospatial information and services.

For more information on GEOINT, refer to ATP 2-22.7, ATP 3-34.80, and AR 115-11.

Human Intelligence (HUMINT)

Human intelligence is the collection by a trained human intelligence collector of foreign information from people and multimedia to identify elements, intentions, composition, strength, dispositions, tactics, equipment, and capabilities (FM 2-0).

A HUMINT source is a person from whom foreign information is collected for the purpose of producing intelligence. HUMINT sources can include friendly, neutral, or hostile personnel. The source may either possess first- or second-hand knowledge normally obtained through sight or hearing. Categories of HUMINT sources include but are not limited to detainees, enemy prisoners of war, refugees, displaced persons, local inhabitants, friendly forces, and members of foreign governmental and nongovernmental organizations.

For more information on HUMINT, refer to ATP 2-22.31.

Measurement and Signature Intelligence (MASINT)
Measurement and signature intelligence is intelligence obtained by quantitative and qualitative analysis of data (metric, angle, spatial, wavelength, time dependence, modulation, plasma, and hydromagnetic) derived from specific technical sensors for the purpose of identifying any distinctive features associated with the emitter or sender, and to facilitate subsequent identification and/or measurement of the same. The detected feature may be either reflected or emitted (JP 2-0).

For more information on MASINT, refer to JP 2-0 and ATP 2-22.8 (classified).

Open-Source Intelligence (OSINT)
Open-source intelligence is information of potential intelligence value that is available to the general public (JP 2-0). For the Army, OSINT is the discipline that pertains to intelligence produced from publicly available information that is collected, exploited, and disseminated in a timely manner to an appropriate audience for the purpose of addressing a specific intelligence requirement. OSINT operations are integral to Army intelligence operations.

For more information on OSINT, refer to ATP 2-22.9.

Signals Intelligence (SIGINT)
Signals intelligence is intelligence derived from communications, electronic, and foreign instrumentation signals (JP 2-0). SIGINT provides unique intelligence information, complements intelligence derived from other sources, and is often used for cueing other sensors to potential targets of interest. For example, SIGINT, which identifies activities of interest, may be used to cue GEOINT to confirm that activity. Conversely, changes detected by GEOINT can cue SIGINT collection against new targets. The discipline is subdivided into three subcategories:

- Communications intelligence (COMINT)
- Electronic intelligence (ELINT)
- Foreign instrumentation signals intelligence (FISINT)

Technical Intelligence (TECHINT)
Technical intelligence is intelligence derived from the collection, processing, analysis, and exploitation of data and information pertaining to foreign equipment and materiel for the purposes of preventing technological surprise, assessing foreign scientific and technical capabilities, and developing countermeasures designed to neutralize an adversary's technological advantages (JP 2-0). The role of TECHINT is to ensure Soldiers understand the threat's full technological capabilities. With this understanding, U.S. forces can adopt appropriate countermeasures, operations, and tactics, techniques, and procedures.

Every TECHINT mission supports tactical through strategic requirements by the timely collection and processing of materiel and information, follow-on analysis and resulting production of intelligence, and dissemination to a wide range of consumers. Commanders rely on TECHINT to provide them with tactical and technological advantages to successfully synchronize and execute operations. TECHINT combines information to identify specific individuals, groups, and nation states, matching them to events, places, devices, weapons, equipment, or contraband that associates their involvement in hostile or criminal activity.

For more information on TECHINT, refer to ATP 2-22.4.

(Intelligence) II. Army Intel Capabilities 5-17

II. Single-Source Intelligence

Single-source intelligence includes the joint intelligence disciplines and complementary intelligence capabilities. One important aspect within single-source intelligence is processing, exploitation, and dissemination (PED) activities.

A. The Intelligence Disciplines

In joint operations, the intelligence enterprise is commonly organized around the intelligence disciplines. The intelligence disciplines are—

- CI
- Geospatial intelligence (GEOINT)
- HUMINT
- MASINT
- Open-source intelligence (OSINT)
- SIGINT
- Technical intelligence (TECHINT)

The intelligence disciplines are integrated to ensure a multidiscipline approach to intelligence analysis, and ultimately all-source intelligence facilitates situational understanding and supports decisionmaking. Each discipline applies unique aspects of support and guidance through technical channels.

See previous pages (pp. 5-16 to 5-17) for further discussion.

B. Complementary Intelligence Capabilities

Complementary intelligence capabilities contribute valuable information for all-source intelligence to facilitate the conduct of operations. The complementary intelligence capabilities are specific to the unit and circumstances at each echelon and can vary across the intelligence enterprise. These capabilities include but are not limited to—

- Biometrics-enabled intelligence (BEI)
- Cyber-enabled intelligence
- Document and media exploitation (DOMEX)
- Forensic-enabled intelligence (FEI)

C. Processing, Exploitation, and Dissemination (PED)

Processing and exploitation in intelligence usage, is the conversion of collected information into forms suitable to the production of intelligence (JP 2-01). Dissemination and integration, in intelligence usage, is the delivery of intelligence to users in a suitable form and the application of the intelligence to appropriate missions, tasks, and functions (JP 2-01). These two definitions are routinely combined into the acronym PED. PED is exclusive to single-source intelligence and fits within the larger intelligence process.

In joint doctrine, PED is a general concept that facilitates the allocation of assets to support intelligence operations. Under the PED concept, planners examine all collection assets and then determine if allocation of additional personnel and systems is required to support the exploitation of the collected information. Accounting for PED facilitates processing collected information into usable and relevant information for subsequent all-source production in a timely manner. Beyond doctrine, PED plays an important role within larger DOD intelligence programmatics.

There are many enablers that support PED activities. PED enablers are the specialized intelligence and communications systems, advanced technologies, and the associated personnel that conduct intelligence processing as well as single-source analysis within intelligence units.

5-18 (Intelligence) II. Army Intel Capabilities 4-18

Chap 6

Fires Warfighting Function

Ref: ADP 3-19, Fires (Jul '19), chap. 1, ADP 3-0, Operations (Jul '19), p. 5-5 and FM 3-0, Operations (Oct. '22), p. 2-2.

I. The Fires Warfighting Function

The fires warfighting function is the related tasks and systems that create and converge effects in all domains against the threat to enable actions across the range of military operations (ADP 3-0). These tasks and systems create lethal and nonlethal effects delivered from both Army and Joint forces, as well as other unified action partners. The fires warfighting function does not wholly encompass, nor is it wholly encompassed by, any particular branch or function.

Many of the capabilities that contribute to fires also contribute to other warfighting functions, often simultaneously. For example, an aviation unit may simultaneously execute missions that contribute to the movement and maneuver, fires, intelligence, sustainment, protection, and command and control warfighting functions. Additionally, air defense artillery (ADA) units conduct air and missile defense (AMD) operations in support of both fires and protection warfighting functions.

Space and cyberspace capabilities can provide commanders with options to defeat, destroy, disrupt, deny, or manipulate enemy networks, information, and decision making.

Commanders must execute and integrate fires, in combination with the other elements of combat power, to create and converge effects and achieve the desired end state. Fires tasks are those necessary actions that must be conducted to create and converge effects in all domains to meet the commander's objectives. The tasks of the fires warfighting function are:

Integrate Army, multinational, and joint fires through:

- Targeting.
- Operations process.
- Fire support.
- Airspace planning and management.
- Electromagnetic spectrum management.
- Multinational integration.
- Rehearsals.
- Air and missile defense planning and integration.

Execute fires across all domains and in the information environment, employing:

- Surface-to-surface fires.
- Air-to-surface fires.
- Surface-to-air fires.
- Cyberspace operations and EW.
- Space operations.
- Multinational fires.
- Special operations.
- Information operations.

Fires (ADP 3-19)

(Fires) Warfighting Function 6-1

II. Fires Overview

Ref: ADP 3-19, Fires (Jul '19).

Success in large-scale combat operations is dependent on the Army's ability to employ fires. Fires enable maneuver. Over the past two decades, potential peer threats have invested heavily in long-range fires and integrated air defense systems, making it even more critical that the U.S. Army possess the ability to maneuver and deliver fires in depth and across domains.

Fires in Support of Unified Land Operations

The Army operational concept for conducting operations as part of a joint team is unified land operations. Unified land operations is the simultaneous execution of offense, defense, stability, and defense support of civil authorities across multiple domains to shape operational environments, prevent conflict, prevail in large-scale ground combat, and consolidate gains as part of unified action (ADP 3-0). The goal of unified land operations is to achieve the JFC's end state by applying landpower as part of unified action. Commanders employ fires to set conditions for the successful employment of other elements of combat power to conduct unified land operations. The targeting process can help commanders and staffs to prioritize and integrate assets to create effects that allow for achievement of the commander's objectives within unified land operations.

The Army's primary mission is to organize, train, and equip its forces to conduct prompt and sustained land combat to defeat enemy ground forces and seize, occupy, and defend land areas. During the conduct of unified land operations, Army forces support the joint force through four strategic roles:

- Shape OEs.
- Prevent conflict.
- Prevail during large-scale ground combat.
- Consolidate gains.

Fires in Support of Large-Scale Combat Operations

The Army, as part of the joint force, conducts large-scale combat operations. The preponderance of large-scale combat operations will consist of offensive and defensive operations initially, although some stability operations will occur simultaneously as part of consolidating gains. Commanders employ fires as part of large-scale combat operations by creating effects to enable joint force freedom of action.

Commanders use Army and joint targeting to select and prioritize targets, integrating lethal and nonlethal effects from different capabilities in support of large-scale combat operations. Commanders may converge effects from multiple systems, either simultaneously or in close succession, to create an even greater effect than would have been achieved if each effect was created individually. Convergence is the massing of capabilities from multiple domains to create effects in a single domain. Convergence overwhelms the enemy, giving them too many dilemmas to address simultaneously, which creates gaps for exploitation by the joint force. The convergence of multiple effects within an area requires careful synchronization prior to execution to ensure effects don't interfere with one another or pose a risk to the force.

To effectively enable joint force freedom of action during large-scale combat operations, commanders must synchronize the effects created with fires with the actions of the rest of the joint force. This synchronization initially takes place during planning, where commanders and their staffs determine the timing of the creation of the effect and link that timing to a clearly defined, conditions-based trigger. Commanders must also plan for assessment of the effects and determine alternate courses of action if the effects are not created as planned.

6-2 (Fires) Warfighting Function

Fires Logic Diagram

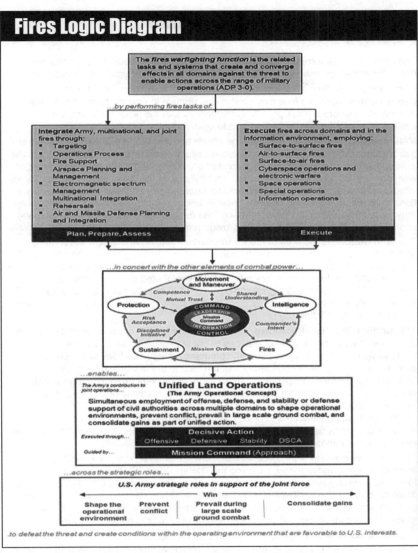

Ref: ADP 3-19, Fires (Jul '19), introductory figure, ADP 3-19 Logic chart.

During large-scale combat operations, multiple Army echelons must synchronize and deconflict their activities, including the creation of effects. The use of deep and close areas can help with dividing responsibilities among echelons within an AO. The close area is the portion of the commander's area of operations where the majority of subordinate maneuver forces conduct close combat (ADP 3-0). The deep area is where the commander sets conditions for future success in close combat (ADP 3-0). Commanders may focus the effects of fires in their deep area to execute shaping operations against enemy forces not in contact with friendly forces in the close area, while subordinate units are responsible for creation of effects in the close area the commander has assigned to them.

(Fires) Warfighting Function 6-3

III. Fires Across the Domains

Ref: ADP 3-19, Fires (Jul '19), pp. 1-4 to 1-5.

The Army operates within all **domains**: land, air, maritime, space, and cyberspace (including the electromagnetic spectrum) as well as in the information environment.

Commanders use fires to create effects in support of Army and joint operations. **Cross-domain fires** are fires executed in one domain to create effects in a different domain. Cross-domain fires provide commanders with the flexibility to find the best system to create the required effect and to build redundancy into their plan.

Multi-domain fires are fires that converge effects from two or more domains against a target. Multi-domain fires may produce synergistic effects that are greater than the sum of the individual effects that would have been created separately. Surface-based fires converged with other effects across domains creates multiple dilemmas, taxing the enemy's ability to effectively respond. For example, a commander may employ offensive cyberspace operations to attack an enemy air defense network while surface-to-surface fires destroy enemy air defense radars and air-to-surface fires destroy the air defense command and control nodes. The converged effects provide reduced risk to allied operational aircraft.

The **land domain** is the area of the Earth's surface ending at the high water mark and overlapping with the maritime domain in the landward segment of the littorals (JP 3-31). The joint force land component commander (JFLCC) is the supported commander within the land area of operations (AO) designated by the joint force commander (JFC). Within the designated AO, the JFLCC has the authority to designate target priority, effects, and timing of fires in order to integrate and synchronize maneuver, fires, and interdiction.

The **air domain** is the atmosphere, beginning at the Earth's surface, extending to the altitude where its effects upon operations become negligible (JP 3-30). The JFC normally assigns joint force air component commander responsibilities to the component commander having the preponderance of forces and the ability to effectively plan, task, and control joint air operations. In addition, as all component commands will need to utilize the air domain to some extent, the JFC normally designates the joint force air component commander as the airspace control authority to promulgate airspace coordinating measures to deconflict the multiple users on behalf of the JFC. The Army air-ground system is the Army's system to synchronize, coordinate, and integrate air-ground operations, joint air support, and airspace.

The **maritime domain** is the oceans, seas, bays, estuaries, islands, coastal areas, and the airspace above these, including the littorals (JP 3-32). Naval and maritime forces operate on (surface), under (subsurface), or above the sea (air). Fires from the maritime domain support the land scheme of fires with traditional naval surface fires, and joint fires to include cruise missile and anti-ship missiles, as well as protecting global shipping lanes and friendly maritime assets to maintain freedom of maneuver.

The **space domain** is the space environment, space assets, and terrestrial resources required to access and operate in, to, or through the space environment (FM 3-14). Space is a physical domain where military operations are conducted. Space capabilities include the ability to access information collection; environmental monitoring; early warning, satellite based sensors and communications; and positioning, navigation, and timing.

Cyberspace is a global domain within the information environment consisting of the interdependent networks of information technology infrastructures and resident data, including the Internet, telecommunications networks, computer systems, and embedded processors and controllers (JP 3-12.) Commanders will generally create effects in the cyberspace domain through offensive and defensive cyberspace operations. However, they may also create effects on the physical network layer of cyberspace.

See pp. 6-5 to 6-14 for further discussion of fires across the domains.

6-4 (Fires) Warfighting Function

Chap 6

I. Execute Fires Across the Domains

Ref: ADP 3-19, Fires (Jul '19), chap. 3. (See also p. 6-4.)

The commander is responsible for the integration of fires within the AO. The commander consults the fire support coordinator, chief of fires, air liaison officer, fire support officer, and experts on AMD, cyberspace, EW, space, special operations, and information operations for advice on the allocation, integration, and use of available fires resources. Fires in all domains require detailed coordination and planning to support the commander's objectives. Employment of these systems requires the use of common terminology and coordination measures across the joint force. It includes surface-to-surface fires, air-to-surface fires, and nonlethal means that the commander uses to support the concept of the operation.

I. Surface-to-Surface Fires

Army surface-to-surface indirect fires includes cannon, rocket, and missile systems as well as mortars organic to maneuver elements. Field Artillery is the equipment, supplies, ammunition, and personnel involved in the use of cannon, rocket, or surface-to-surface missile launchers (JP 3-09). The role of the field artillery (FA) is to destroy, neutralize, or suppress the enemy by cannon, rocket, and missile fire and to integrate and synchronize all fire support assets into operations. Fire support is fires that directly support land, maritime, amphibious, and special operations forces to engage enemy forces, combat formations, and facilities in pursuit of tactical and operational objectives (JP 3-09).

At times, the FA must enable fires in other domains through the employment of surface-to-surface fires converged with effects in other domains. For example, surface-to-surface capabilities contribute significantly to counterair operations, targeting and destroying enemy air and missile weapons, command and control elements, and supporting infrastructure, ultimately reducing the threats that Army, joint, and multinational air-to-surface assets must face.

FA cannon, rocket, and missile systems organic, assigned, attached, or operational control to the FA battalions of brigade combat teams, division artillery, and field artillery brigades provide continuously available fires under all weather conditions and in all types of terrain. FA can shift and mass fires rapidly from dispersed locations and displace rapidly to new position areas. FA units are positioned to provide continuous fires.

Army surface-to-surface fires are applied to deliver effects in concert with all other fires capabilities. Fires are integrated through the targeting process, fire support planning, unit airspace plan (UAP), and military decision making process (MDMP). Surface-to-surface fires are integrated with other airspace users to facilitate massing of effects.

The integration of surface fires is a critical factor in the success of operations. The commander is responsible for the integration of fires within the AO. The fire support coordinator, chief of fires, and fire support officer advise the commander on the allocation and use of available indirect fires and fire support resources. The chief of fires is the senior fires staff officer at echelons above corps who advises the commander on the best use of available fires resources and provides input to the necessary orders.

Commanders integrate fire support into the concept of operations during planning. FA commanders assisted by fire support personnel and organizations at all echelons

Fires (ADP 3-19)

(Fires) I. Execute Fires Across the Domains 6-5

integrate Army, joint, interagency and multinational fires capabilities during the operational process for use at the designated place and time. Fires are critical to accomplishing offensive and defensive tasks. However, nonlethal effects are also important contributors to decisive action, regardless of which element dominates. Accomplishing the mission by creating an appropriate mix of effects remains an important consideration for every commander.

Naval surface fire support provides fire support by naval surface gun, missile, and EW systems in support of a unit or units tasked with achieving the commander's objectives. Naval assets can provide support in a unique manner and should be considered as one source of fire support along with other components and weapon systems.

II. Air-to-Surface Fires

Army and joint forces employ various types of air-to-surface capabilities, to include fixed-wing aircraft, rotary-wing aircraft, and unmanned aircraft systems (UASs). These systems provide lethal and nonlethal effects, standoff weapons, and target acquisition capabilities that can be employed to detect and create integrated effects against adversary targets.

Fixed-Wing Aircraft

Fixed-wing aircraft provide flexibility, range, speed, lethality, precision, and the ability to mass fires at a desired time and place. The capacity of aircraft to deliver precision guided munitions can limit collateral damage, as well as strike otherwise inaccessible targets. Aircraft also provide surveillance and combat assessment.

Rotary-Wing Aircraft

Rotary-wing aircraft employ a variety of weapons. They provide attack, reconnaissance, and terminal guidance for other weapon platforms. Army attack aviation conducts two basic types of attack missions: attacks against enemy forces in close, friendly contact with other Army maneuver forces and attacks against enemy forces out of friendly contact with other Army forces.

UAS

The long endurance of UAS necessary to support their intelligence, surveillance, and reconnaissance missions enables them to provide extended support to many types of missions. UAS can participate in supporting close air support (CAS), air interdiction, and other joint fires missions. Specific tasks may include target acquisition and marking, terminal guidance of ordnance, providing precision coordinates for Global Positioning System (GPS)-aided munitions, delivery of onboard precision-guided ordnance, battle damage assessment, signal intelligence, communication data relays, and retargeting such as shoot-look-shoot missions. UAS should be requested, tasked, routed, controlled, and deconflicted in a manner similar to methods used for fixed-winged and rotary-winged manned aircraft, with exceptions made for their unmanned nature (e.g., inability to see and avoid other air traffic). See JP 3-09, JP 3-09.3, and ATP 3-60.2 for additional information on joint missions and UAS integration.

Air Force Assets

Air Force assets, with their inherent speed, range, and precision attack capabilities, are combat multipliers for the ground commander. The destruction of decisive points, forces, and capabilities by striking enemy military targets such as fielded land forces, command and control (C2) nodes, vital logistics, or supporting infrastructure degrades the enemy system and contributes to an enemy incapable of effective resistance.

See following pages (pp. 6-14 to 6-15) for an overview and further discussion of Air Force assets for air-to-surface fires.

6-6 (Fires) I. Execute Fires Across the Domains

Army Surface-to-Surface Capabilities

Ref: ADP 3-19, Fires (Jul '19), pp. 2-1 to 2-2.

Rockets

The multiple launch rocket system (MLRS) supplements cannon artillery by delivering a large volume of fires in a very short period of time against high-payoff targets. MLRS is used for counterfire and deliberate attacks against enemy air defenses, light materiel, and personnel targets. The all-weather MLRS fires free-flight and guided rockets and missiles. Free-flight or guided rocket options include warheads with either unitary high-explosive or dual-purpose improved conventional munitions. The basic free-flight rocket munitions have a maximum range of 26 kilometers, while the extended-range rocket may engage targets to about 45 kilometers. The Guided Multiple Launch Rocket System provides commanders with increased accuracy and greater range (up to 70 kilometers), reducing the number of rockets required to create desired effects on a target. The M270A1 MLRS can carry 12 rockets and the M142 High Mobility Artillery Rocket System (referred to as HIMARS) can carry 6 rockets. However, their extremely high altitude of delivery (apex of missile trajectory, maximum ordinate) requires close coordination with air planners and liaisons to ensure aircraft are not in the vicinity during launches and descents.

Missiles

The Army Tactical Missile System (ATACMS) provides long-range, surface-to-surface fire support. ATACMSs are fired from an MLRS (two missiles) or high mobility artillery rocket system platform (one missile) and may consist of antipersonnel/antimaterial submunitions or a unitary high-explosive warhead. The ATACMS retains the responsiveness of rockets, though it possesses a much greater range (up to 300 kilometers). The ATACMS antipersonnel/antimaterial warhead is designed to engage soft targets and the unitary high-explosive warhead is designed to engage fixed infrastructure while minimizing collateral damage. The ATACMS's accuracy and all-weather capability, coupled with extended range and quick response time, make it a formidable system against dynamic targets. Due to the range and altitude of the ATACMS and Guided multiple launch rocket system, target engagements require detailed airspace coordination and integration.

For more discussion on MLRS/High Mobility Artillery Rocket System and ATACMS, refer to ATP 3-09.60.

Cannon Artillery

Cannon artillery is usually the most available fire support system within the AO, capable of performing counterfire, interdiction, and suppression of enemy air defenses. Cannon artillery provides near immediate response times, 24-hour availability, and 360-degree coverage. Cannon artillery offers both area fires and precision fires (such as the Excalibur and precision guidance kit).

For more on cannon artillery, refer to ATP 3-09.23, ATP 3-09.50, and ATP 3-09.70.

Mortars

Army maneuver formations have organic mortar platoons and sections. The primary role of mortars is to provide immediate, responsive, indirect fires in support of maneuver companies or battalions. The mobility of mortar systems makes them well suited for close support of maneuver. Mortars can also be used for final protective fire, obscuration, and illumination. U.S. mortar munitions include a 120-millimeter precision munition; some multinational mortar units also have precision-guided munitions of different calibers. The maneuver commander decides how and when mortars, as a key fire support asset, will be integrated into operations.

For more on mortars, refer to TC 3-22.90.

(Fires) I. Execute Fires Across the Domains 6-7

Air Force Assets (Air-to-Surface)

Ref: ADP 3-19, Fires (Jul '19), pp. 2-3 to 2-4.

Air Force assets, with their inherent speed, range, and precision attack capabilities, are combat multipliers for the ground commander. The destruction of decisive points, forces, and capabilities by striking enemy military targets such as fielded land forces, command and control (C2) nodes, vital logistics, or supporting infrastructure degrades the enemy system and contributes to an enemy incapable of effective resistance. Detailed integration of air and surface capabilities create synergistic effects that are greater than the sum of individual air and surface operations. Ground commanders are the ultimate authority for the use of all supporting fires in their respective operational area. Air operations are not associated with a particular type of aircraft. Each weapons system has unique characteristics that should be considered based on the nature of the threat, targets to be attacked, desired effects, and environmental conditions. Many of the assets used to interdict forces deep in the enemy rear area can also be used to support the close fight. Fighters, bombers, and remotely piloted aircraft/UAS are examples of joint air assets that can achieve desired effects for the supported commander. There are two distinct types of air operations for engaging enemy land forces that are typically coordinated with the ground scheme of maneuver to maximize effects on the enemy.

Counterland operations are defined as "airpower operations against enemy land force capabilities to create effects that achieve joint force commander (JFC) objectives." The aim of counterland operations is to dominate the surface environment using airpower. By dominating the surface environment, counterland operations can assist friendly land maneuver while denying the enemy the ability to resist. Although most frequently associated with support to friendly surface forces, counterland operations may also be conducted independent of friendly surface force objectives or in regions where no friendly land forces are present. For example, recent conflicts in the Balkans, Afghanistan, and Iraq illustrate situations where counterland operations have been used absent significant friendly land forces or with small numbers of special operations forces (SOF) providing target cueing. This independent attack of adversary land operations by airpower often provides the key to success when seizing the initiative, especially in the opening phase of an operation.

Counterland Operations

Air Interdiction (AI)

Close Air Support (CAS)

Refer to AFOPS2: The Air Force Operations & Planning SMARTbook, 2nd Ed. (Guide to Curtis E. LeMay Center & Joint Air Operations Doctrine). Topics and references of the 376-pg AFOPS2 include airpower fundamentals and principles (Vol 1), command and organizing (Vol 3); command and control (Annex 3-30/3-52), airpower (doctrine annexes), operations and planning (Annex 3-0), planning for joint air operations (JP 3-30/3-60), targeting (Annex 3-60), and combat support (Annex 4-0, 4-02, 3-10, and 3-34).

Air Interdiction

Air interdiction is defined as "air operations conducted to divert, disrupt, delay, or destroy the enemy's military surface capabilities before it can be brought to bear effectively against friendly forces, or to otherwise achieve objectives that are conducted at such distances from friendly forces that detailed integration of each air mission with the fire and movement of friendly forces is not required." (JP 3-03) Air forces employ such weapons as projectiles, missiles, unguided munitions, precision-guided munitions, EW systems, and sensors from airborne platforms in the air interdiction role. Submission of a preplanned air support request allows the request to process from the Army and enter into the joint air tasking cycle. Air interdiction missions can be executed in a number of tactics, techniques, and procedures, or methods to include strike, coordination and reconnaissance, and kill box operations.

Close Air Support (CAS)

CAS is defined as "air action by fixed- and rotary-wing aircraft against hostile targets that are in close proximity to friendly forces and that require detailed integration of each air mission with the fire and movement of those forces." (JP 3-09.3) Based on threats and the availability of other means of fire support, synchronizing CAS in time, space, and purpose with supported ground forces requires detailed and continuous integration. The supported commander establishes the target priority, effects, and timing of CAS fires. The terminal attack controller either a (joint terminal attack controller or forward air controller (Airborne) is qualified to talk-on CAS aircrew to the target and issue weapons release clearance to attacking aircraft. CAS provides commanders with flexible and responsive fire support. Using CAS, commanders can take full advantage of battlefield opportunities by massing firepower to maintain the momentum of an offensive action or reduce operational and tactical risks. The mobility and speed of aircraft provide commanders with a means to strike the enemy swiftly and unexpectedly. CAS distribution begins with the senior ground forces commander's inputs to the JFC's broad air apportionment decision. CAS is tasked to support ground forces based on the preplanned air support requests. During the joint air tasking cycle planners adhere to the apportionment decision by the JFC. The number of air support requests for CAS that are sourced is limited by the JFC air apportionment decision and the joint force air component commander's interpretation of that JFC decision (allocation), and anticipated number of air support requests (ASRs) for CAS provided from the ground component. The senior ground forces commander (corps, JFLCC) then makes a distribution decision for use of CAS. The CAS distribution decision is the ground commander's guidance for the employment of CAS assets among competing requirements and how each air support operations center will source immediate ASRs from CAS missions on the air tasking order. For example if 1AD is the main effort and has the preponderance of CAS dedicated by the senior ground forces commander (LCC) in the CAS distribution decision gives the air support operations center guidance on first call on resources. The distribution decision only affects those sorties assigned as CAS missions that are provided to support the Army.

The Army process preplanned and immediate ASRs through the Army air-ground system to identify air support requirements to the supporting air component. Pre-planned ASRs are processed per the battle rhythm in sufficient time to meet the planning stages of the joint air tasking cycle and sourced on the initial published air tasking order. Immediate ASRs arise after the air tasking order is published and must be sourced by assets already tasked on the air tasking order. Scheduled and/or on-call air missions are tasked to support preplanned ASRs. Scheduled missions are planned against targets on which air attacks are delivered at a specific time. On-call missions (X-airborne alert and G-ground alert) are planned against target types for which a need can be anticipated for a timeframe. On-call air missions are preferably tasked to support immediate ASRs to satisfy dynamic targeting requirements.

(Fires) I. Execute Fires Across the Domains 6-9

III. Surface-to-Air Fires

Ref: ADP 3-19, Fires (Jul '19), pp. 2-4 to 2-5.

Surface-to-air fires capabilities include active defense weapons that are employed in both area and point defenses. ADA delivers precision surface-to-air missiles to defend friendly forces, fixed and semi-fixed assets, population centers and key infrastructure against air and missile threats. ADA executes Army AMD operations in support of joint counterair efforts. The role of ADA is to deter and defeat the range of aerial threats in order to assure allies, ensure operational access, and defend critical assets and deployed forces in support of unified land operations (FM 3-01).

The air defense officer serves as the AMD expert for the maneuver commander at every echelon. Within BCTs, that service member is the senior air defense officer located in the Brigade ADAM Cell. The senior air defense officer in Divisions and Corps will be assigned to the AMD section within the G3. The senior air defense officer has the responsibility for AMD coordination and integration, which supports the critical asset list/defended asset list development process for the commander. See FM 3-01.

Ground units also take active and passive measures to defend against aerial threats when limited air defense assets are available. Active measures include small arms engagement of aerial threats. Passive measures include camouflage, concealment, or deception to prevent detection from aerial threats. Together, these measures are referred to as combined arms for air defense. For more information on combined arms for air defense, see FM 3-01 or ATP 3-01.8.

Surface-to-air weapons are normally used in a defensive role and offer optimal fires against aerial threats at different ranges and altitudes, especially upon entering a preplanned and fully established engagement zone. ADA forces operate these weapon systems based on directives and guidelines established by fire control orders, ROE, and weapons control status. ADA weapon systems and personnel are employed at strategic, operational, and tactical levels to defend the force against the full range of air and missile threats. ADA performs the AMD functions of planning, coordinating and executing surface-to-air fires for supported commanders. ADA forces are expeditionary in nature, forward stationed, and operate these systems worldwide during joint combined arms operations. ADA capabilities include both high-to-medium air defense (also known as HIMAD) and short-range air defense (SHORAD) systems.

High-to-Medium Air Defense (HIMAD)

High-to-medium air defense (also known as HIMAD) capabilities include Patriot, Terminal High Altitude Area Defense system, and AN/TPY-2 forward-based mode radar system batteries. Patriot is a multi-mission system that provides AMD of combat land forces and other critical assets. Patriot forces are capable of defending against ballistic missiles, cruise missiles, unmanned aircraft, tactical air-to-surface missiles, large-caliber rockets, and fixed-and rotary-wing aircraft. Terminal High Altitude Area Defense system is an upper tier system that provides the capability to engage and negate the short-, medium- and intermediate-range ballistic missile threats within and outside the atmosphere. The AN/TPY-2 is a high precision, long range, phased-array radar. In its forward-based mode of deployment, the AN/TPY-2 radar detects ballistic missiles early in their flight and provides precise tracking information. All of these systems are considered land-based contributors to the ballistic missile defense program. They are deployed to joint areas of operations to protect national and strategic interests, defend the force and other critical assets, and act as deterrent while simultaneously providing active AMD.

Short-Range Air Defense (SHORAD)

SHORAD capabilities include Avenger; Stinger; counter-rocket, artillery, and mortar; and Sentinel. Avenger is a mobile lightweight weapon system used to counter enemy reconnaissance, surveillance, and target acquisition efforts and low-level fixed-and rotary-wing threats. Stinger is an infrared homing, fire-and-forget missile. It is mounted in missile pods on the Avenger and is employed by dismounted Stinger teams. Counter-rocket, artillery, and mortar is a system that consists of sensor, interceptor, and C2 systems. It is a fast reacting, short-range system used to detect and destroy incoming rockets and artillery and mortar rounds in the air before they hit their ground targets, or simply to provide early warning. The Sentinel radar is a 360-degree phased array radar that provides persistent air surveillance and fire control quality data. It can acquire, track, and classify cruise missiles, UAS, and fixed- and rotary-wing aircraft. SHORAD forces are positioned with maneuver formations in the close area, where maneuver force commanders plan to conduct decisive operations. Air defense airspace management (known as ADAM) cell personnel in the brigade combat team, for instance, plan and coordinate the support of SHORAD or other ADA forces and relay pertinent AMD information and early warning of enemy air activity to subordinate maneuver formations. Short-range-air defense forces generally defend assets in the division and brigade areas, while Patriot and Terminal High Altitude Area Defense units maintain coverage of assets in the division, corps, and theater areas.

Surface-to-air fires defend the force and support complementary efforts of Army and joint surface-to¬surface, air-to-surface, and other fires and effects coordinated with cyberspace operations and electronic attack. Networked surface-to-air and air-to-surface capabilities assist commanders with the ability to surveil deep into enemy territory allowing friendly forces to see first, assume a safe posture, provide alert, and support attack operations. ADA systems also have a capability to provide launch point determinations to support attack operations (offensive counterair).

The defending force's surveillance and firepower must be capable of defending throughout the entire AO in all directions. Surface-to-air fires may be challenged by series of complex or multiple integrated attacks in geographic areas where an advanced military or competing threat resides. These attacks may vary depending on region. Complex integrated attacks may include a mix of capabilities such as coordinated air¬to-surface missiles, surface-to-surface weapons, unmanned aircraft systems, fixed-wing aircraft, and rotary-wing aircraft. Complex integrated attacks will likely be supported by enemy activities in other domains, such as jamming efforts and special operations forces' attacks in the land domain.

All of the surface-to-air systems use a common engagement sequence to defend against aerial threats. The sequence begins with the surveillance of the airspace by electronic or visual means, followed by the detection of an aerial object. The object is tracked and then subjected to identification procedures which may be electronic or manual (visual identification). Once the object has been identified as a hostile (enemy), an evaluation is conducted of its intended target, predicted impact point, and expected time of arrival. The best weapon, paired with an appropriate sensor, is assigned to engage the target and, after the engagement, an assessment is conducted of the need for reengagement.

ADA personnel provide planning expertise at all echelons. These tasks include integrating procedures and positioning surface-to-air assets in support of operations plans. AMD planning establishes optimum layered defenses and allows for quick and efficient responses against air attacks defending friendly forces and critical infrastructure over considerably large geographic areas. Surface-to-air fires correctly classify, discriminate, and identify threatening air and missile targets, then engage designated threats with the appropriate number and type of interceptors.

(Fires) I. Execute Fires Across the Domains 6-11

IV. Space Operations

Many lethal and nonlethal fires capabilities depend on space capabilities to support, integrate, and deliver fires. Army space capabilities are integrated throughout the fires warfighting function, providing robust and reliable planning, contributing to target development, and providing positioning, navigation, and timing (PNT), satellite communications, imagery, geolocation, weather, and terrain capabilities.

- GPS enables precision guided munitions, command and control systems, and near real-time situational awareness for lethal and nonlethal fires.

- Satellite communications enables real time communications between commanders and forces to enable immediate redirection of fires over extended distances to shape the operations.

- Weather satellites provide a variety of data points necessary for predicting effects of meteorological conditions on fires.

- Combined, PNT and satellite communications supports fires through the systems interfaces on the Advanced Field Artillery Tactical Data System.

In space operations, the fires warfighting function includes space control operations that create a desired effect on enemy space systems and across multiple domains. Space control plans and capabilities use a broad range of response options to provide continued, sustainable use of space. Space control contributes to space deterrence by employing a variety of measures to assure the use of space and attribute enemy attacks. These include terrestrial fires to defend space operations and assets. A capability for, or employment of, fires may deter threats and/or contain and de-escalate a crisis.

Offensive space control are offensive operations conducted for space negation (JP 3-14). Negation in space operations, are measures to deceive, disrupt, degrade, deny, or destroy space systems. (JP 3-14). Offensive space control actions targeting an enemy's space-related capabilities and forces could employ reversible or nonreversible means, and are considered a form of fires.

For more on offensive space control, refer to FM 3-14.

V. Special Operations

Special operations forces execute a diverse set of missions across warfighting functions to produce scalable lethal and nonlethal effects, either in support of a combatant commander's campaign plan or as part of a joint, Army, or other Service effort. Army special operations forces contribute to the fires warfighting function by providing unique contributions for understanding the OE, nominating and developing targets and recommending effects, and providing specific lethal and nonlethal capabilities such as psychological operations, civil affairs, or surgical strike capabilities to the supported commander.

A special operations task force may establish a joint fires element. A joint fires element is an optional staff element that provides recommendations to the operations directorate to accomplish fires planning and synchronization (JP 3-60). When the joint fires element is established, it becomes the fires coordination link between commands. The joint fires element is responsible for planning joint fires and executing the targeting process within the special operations task force. It is part of the current operations division and consists of organic intelligence, sustainment, plans, communications, aviation, Special Forces, civil affairs, Ranger, and psychological operations personnel; conventional force liaisons; representatives of attached units; and augmentees from the Services, multinational partner units, and government agencies that can achieve lethal and nonlethal effects, integrate into the targeting process, and advise on their organizations' capabilities.

Establishing liaisons between special operations and conventional force fires elements helps mitigate the tempo of armed conflict and the subsequent rapid information flow.

6-12 (Fires) I. Execute Fires Across the Domains

VI. Cyberspace Operations and Electronic Warfare

Ref: ADP 3-19, Fires (Jul '19), pp. 2-6 to 2-6.

Friendly, enemy, adversary, and host nation networks, communications systems, computers, cellular phone systems, social media websites, and technical infrastructures are all part of cyberspace. Cyberspace operations are the employment of cyberspace capabilities where the primary purpose is to achieve objectives in or through cyberspace (JP 3-0). The interrelated cyberspace missions are Department of Defense information network operations, defensive cyberspace operations, and offensive cyberspace operations.

Electronic attack involves the use of electromagnetic energy, directed energy, or anti-radiation weapons to attack personnel, facilities, or equipment with the intent of degrading, neutralizing, or destroying enemy combat capability and is considered a form of fires. Electronic attack includes:

- Actions taken to prevent or reduce an enemy's effective use of the electromagnetic spectrum.
- Employment of weapons that use either electromagnetic or directed energy as their primary destructive mechanism.
- Offensive and defensive activities, including countermeasures.

See pp. 2-16 to 2-17 for an overview and further discussion.

Cyberspace Electromagnetic Activities (CEMA)

Cyberspace electromagnetic activities is the process of planning, integrating, and synchronizing cyberspace and electronic warfare operations in support of unified land operations (ADP 3-0). Incorporating cyberspace electromagnetic activities (CEMA) throughout all phases of an operation is key to obtaining and maintaining freedom of maneuver in cyberspace and the EMS while denying it to enemies and adversaries. CEMA synchronizes capabilities across domains and warfighting functions and maximizes complementary effects in and through cyberspace and the EMS. Intelligence, signal, information operations, cyberspace, space, and fires operations are critical to planning, synchronizing, and executing cyberspace and EW operations.

CEMA adhere to the joint principles of operations, and of these principles, mass, unity of effort, surprise, and security are the most relevant. Commanders and staffs ensure compliance with relevant authorities and associated legal frameworks before conducting cyberspace and EW operations. When conducting cyberspace operations, the Army acts according to command lines of authority. For example, the Army provides forces for offensive cyberspace operations (OCO), but does not execute these operations except as part of the joint force and as approved by the JFC. OCO and defensive cyberspace response actions are subject to authorities that reside with the President and Secretary of Defense.

Refer to CYBER1: The Cyberspace Operations & Electronic Warfare SMARTbook (Multi-Domain Guide to Offensive/Defensive CEMA and CO). Topics and chapters include cyber intro (global threat, contemporary operating environment, information as a joint function), joint cyberspace operations (CO), cyberspace operations (OCO/DCO/DODIN), electronic warfare (EW) operations, cyber & EW (CEMA) planning, spectrum management operations (SMO/JEMSO), DoD information network (DODIN) operations, acronyms/abbreviations, and glossary.

VII. Information Operations (IO)

Ref: ADP 3-19, Fires (Jul '19), p. 2-8.

Information operations is the integrated employment, during military operations, of information-related capabilities in concert with other lines of operation to influence, disrupt, corrupt, or usurp the decision-making of adversaries and potential adversaries while protecting our own (JP 3-13).

Information operations, as an integration and synchronization staff function, plans and oversees the coordinated delivery of information-related capabilities to achieve cognitive effects against adversary and enemy decision-makers across the conflict continuum while simultaneously establishing the conditions that allow for more timely and better-informed friendly decision-making. Intrinsic information related capabilities include (FM 3-13):

- Military deception.
- Cyberspace electromagnetic activities (to include: cyberspace operations, EW, and spectrum management operations).
- Military information support operations.
- Special technical operations.
- Space Operations.
- Public Affairs.
- Combat camera.
- Civil Affairs.
- Operations security.
- Soldier and leader engagements, to include police engagement.

Commanders can also designate other enabling information related capabilities (both lethal and nonlethal) to control the flow of information to adversary/enemy decision-makers and protect friendly command and control means. These activities and capabilities include:

- Physical attack (to include lethal fires and maneuver).
- Presence, posture, and profile.
- Communication synchronization.
- Cybersecurity.
- Foreign disclosure.
- Physical security.
- Special access programs.
- Civil military operations.
- Intelligence.

See pp. 2-13 to 2-18 for further discussion of information operations.

Refer to INFO1: The Information Operations & Capabilities SMARTbook (Guide to Information Operations & the IRCs). INFO1 chapters and topics include information operations (IO defined and described), information in joint operations (joint IO), information-related capabilities (PA, CA, MILDEC, MISO, OPSEC, CO, EW, Space, STO), information planning (information environment analysis, IPB, MDMP, JPP), information preparation, information execution (IO working group, IO weighted efforts and enabling activities, intel support), fires & targeting, and information assessment.

II. Integrate Army, Multi-national & Joint Fires

Chap 6

Ref: ADP 3-19, Fires (Jul '19), chap. 3.

I. Fires in the Operations Process

The fires warfighting integrates with the other warfighting functions as part of the operations process. The operations process is the major command and control activities performed during operations: planning, preparing, executing, and continuously assessing (ADP 5-0). As part of the operations process, commanders and staffs plan to execute fires through subordinate planning processes such as fire support planning, airspace planning and management, electromagnetic spectrum management, multinational integration, and air and missile defense planning and integration. Leaders further synchronize fires through rehearsals.

A. Integrating Fires into Planning

Integration of fires begins during mission analysis, supported by continuously updated estimates, and incorporates post-execution assessment. Commanders at all levels are responsible for the effective integration of fires. The scheme of fires, part of the concept of the operation developed during the military decision-making process, specifies how the commander wants to shape the OE in support of his requirements and objectives.

Fire support planning is the continuous process of analyzing, allocating, integrating, synchronizing, and scheduling fires to describe how the effects of fires facilitate maneuver force actions (FM 3-09). This process facilitates the maneuver commander's ability to synchronize fire support with maneuver and employ fire support resources to achieve their objectives. Coordination of fire support begins with the commander's intent and concept of the operation and continues simultaneously with the development of the scheme of maneuver.

For more information on fire support planning, refer to FM 3-09.

Planning for the integration of fires flows from higher echelons to lower echelons. When building their plan, higher echelons should attempt to anticipate the needs of their subordinates for fires capabilities and request and allocate those capabilities on behalf of the subordinate HQ as much as possible. This allows the subordinate commander and staff to plan with known assets instead of building a plan with requested assets that may need to be altered significantly if those requests aren't approved. Commanders should also generally avoid assigning fires tasks to fires assets that are organic, assigned, or in direct support to subordinate units to allow the subordinate commander maximum flexibility in their own planning.

When integrating fires into their plan, each echelon must consider both the needs of their own commander, as well as their role in executing the plan of the echelons above them. Lower echelons will generally have more tactical information available to them and are therefore responsible for refining and executing their portion of the plan of the higher echelon to ensure that the intent of the higher commander is met. An example of this type of refinement is refinement of the location of a target from a general location to a specific location based on the subordinate commander's understanding of the terrain and use of obstacles. To integrate fires with the other elements of combat power, planners must build an environment that is permissive for the use of fires capabilities. In the physical domains, this is accomplished primarily through the use of control measures to delineate responsibilities. A control measure is a means of regulating forces or warfighting functions (ADP 6-0). This delineation

Fires
(ADP 3-19)

(Fires) II. Integrate Army, Multinational & Joint Fires 6-15

of responsibilities begins with the assignment of an AO. An area of operations is an operational area defined by a commander for land and maritime forces that should be large enough to accomplish their missions and protect their forces (JP 3-0).

Commanders responsible for an AO will typically be the primary creator of effects within that AO. Therefore, commanders must take care not to assign subordinate commanders an AO that is larger than their area of influence, which is a geographical area wherein a commander is directly capable of influencing operations by maneuver or fire support systems normally under the commander's command or control (JP 3¬0). An AO that is too large may create a gap within the AO that an enemy can operate from relatively free from the effects of friendly forces while limiting the ability of the higher HQ to create effects in that same terrain.

Once a commander has been assigned an AO, they use additional control measures to create an environment that enable the use of fires capabilities. These control measures may be permissive or restrictive in nature. They will primarily consist of maneuver control measures, FSCMs, ACMs, and air defense measures (ADMs).

For more on the use of FSCMs, refer to FM 3-09.

Effects created in the cyber domain, including the electromagnetic spectrum, as well as the space domain and information environment also require careful planning and management to avoid duplication of effort and the creation of unintended effects. The use of fires in these domains can easily create effects outside of a commander's AO. Therefore, they are typically constrained by using authorities. The authority to create effects in these domains will normally be held at a higher level, including the theater or national strategic level for some effects. Commanders must balance the authorities needed to create effects in these domains by subordinate commanders with the potential risks associated with creating unintentional effects.

B. Fires Preparation

Preparation consists of those activities performed by units and Soldiers to improve their ability to execute an operation (ADP 5-0). Preparation begins during planning with activities that are required to set conditions for the execution of operations such as information collection, movement of forces, terrain management, and sustainment preparation. These activities will continue after completion of the operation order brief to subordinate units with a confirmation brief, which is used to ensure subordinates the commander's intent, the mission, and the concept of operations. Preparation activities also include rehearsals, which allow confirmation of a shared understanding as well as synchronization of operations prior to execution. The four types of rehearsals are the backbrief, the combined arms rehearsal, the support rehearsal, and the battle drill or standard operating procedures rehearsal.

C. Fires Assessment

Assessment is the determination of the progress toward accomplishing a task, creating an effects, or achieving an objective (JP 3-0). Assessment takes place throughout the operations process, allowing the commander and staff to analyze collected information to make decisions that allow the unit to create required effects and achieve objectives. Assessment consists of three activities: monitoring, evaluating, and recommending or directing action for improvement. Monitoring is continuous observation of those conditions relevant to the current operation (ADP 5-0). It begins during planning as the staff gathers available information to conduct analysis. Evaluating is using criteria to judge progress toward desired conditions and determining why the current degree of progress exists (ADP 5-0).

6-16 (Fires) II. Integrate Army, Multinational & Joint Fires

D. Airspace Planning and Integration

Ref: ADP 3-19, Fires (Jul '19), pp. 3-2 to 3-3.

Airspace planning occurs throughout the operations process (planning, preparing, executing and assessing of operations) by consolidating the requirements of airspace users. Airspace control is a continuous activity of the operations process and an integral part of risk management. All warfighting functions and liaisons represented in a commander's staff are integral to the integration of airspace use and users. Units' fires cell and airspace element coordination is vital to effective air-ground integration.

Refer to FM 3-52, Airspace Control, for additional information.

The UAP is the integrated set of ACMs to support Army operations submitted to the airspace control authority for integration into a future ACO. The UAP is developed throughout the operations process by consolidating ACM contributions from participating warfighting functions.

Refer to ATP 3-52.1 for more on UAP development.

All commanders inherently have airspace management responsibilities to control their assigned airspace users and to coordinate the use of airspace. Commanders assigned an AO are responsible for performing airspace management of Army and supporting airspace users and to process a UAP to their higher HQ. The Army air-ground systems enables Army commanders and staffs to coordinate and integrate the actions of Army airspace users over the AO regardless of whether they have been assigned airspace control responsibility for a volume of airspace.

A commander who meets specific criteria may exercise airspace control when delegated an assigned volume of airspace by the airspace control authority. One common criteria requires the implementation of a joint air-ground integration center (JAGIC) supported with an Air Force air support operations center. Airspace planning with the airspace element is essential to responsive fires. The JAGIC is the execution node for fires and airspace control.

Refer to ATP 3-91.1 (The Joint Air-Ground Integration Center) for more information about JAGIC.

Close coordination is required to integrate airspace use with the employment of fires. Fire support agencies normally establish FSCMs. Integration and deconfliction of airspace and joint fires normally occurs during mission planning where FSCMs, ACMs, and other appropriate coordination measures are disseminated through command, airspace control, air and missile defense, and fire support channels. Real-time coordination, integration, and deconfliction of airspace and joint fires with airspace control elements and C2 nodes are essential in developing situations.

For more information on fire support coordination and FSCMs, refer to FM 3-09.

See pp. 3-29 to 3-32 for an overview and discussion of Army airspace command and control (A2C2) from FM 3-52.

(Fires) II. Integrate Army, Multinational & Joint Fires 6-17

II. Integrating Multinational Fires

Ref: ADP 3-19, Fires (Jul '19), pp. 3-3 to 3-4.

The U.S. conducts multinational operations as part of an alliance or coalition. U.S. commanders must integrate multinational fires capabilities in concert with all other elements of combat power from all contributing allies and partners. Multinational operations is a collective term to describe military actions conducted by forces of two or more nations, usually undertaken within the structure of a coalition or alliance (JP 3-16).

Alliance

An alliance is the relationship that results from a formal agreement between two or more nations for broad, long-term objectives that further the common interests of the members.

Coalition

A coalition is an arrangement between two or more nations for common action. Coalitions are typically ad hoc; formed by different nations, often with different objectives; usually for a single problem or issue, while addressing a narrow sector of common interest. Operations conducted with units from two or more coalition members are referred to as coalition operations.

Each ally or coalition partner can bring unique fires capabilities to any operation, and these capabilities come with important considerations, employment options, caveats, and challenges. Understanding multinational capabilities and all associated considerations allows commanders to employ multinational fires assets in concert with U.S. fires assets to create effects and achieve objectives.

Interoperability is the ability to act together coherently, effectively, and efficiently to achieve tactical, operational, and strategic objectives (JP 3-0). Multinational interoperability for fires must incorporate human, procedural, and technical means to create effects from within a multinational force.

Human interoperability addresses the fundamental interaction of people from multiple nations to achieve common objectives. Multinational organizations must develop a program to exchange liaison officers and rely on human interaction in addition to any digital liaison. Important human interoperability considerations include:

- Liaison and exchange officers at appropriate echelon.
- Integration of embedded officers into appropriate organizations such as the JAGIC (refer to ATP 3-91.1 for additional information on the JAGIC).
- Language skills among partners and allies and the use of interpreters.
- Authorities vested in exchanged personnel.

Procedural interoperability is the relevant coordination and synchronization of multinational assets to create the desired effects according to established and agreed upon protocol. Planners must understand when procedure differs between nations and within multinational organizations, and adhere to the correct protocol in employing multinational assets. Some procedural interoperability considerations include:

- Common understanding of accepted procedures (U.S., North Atlantic Treaty Organization, American, British, Canadian, Australian, and New Zealand Armies Program, or other agreements).
- Common understanding of the ROE.

6-18 (Fires) II. Integrate Army, Multinational & Joint Fires

- Common coordination measures to include maneuver control measures, FSCMs, and ACMs.
- Common risk management, collateral damage estimation, and battle damage assessment methodology.
- Centralized or decentralized command and control.
- Use of and entry into the joint targeting cycle.
- National approval processes and the role of host nation authorities.
- Integration of multinational capabilities into planning and execution.
- Counterfire procedures.
- Force protection (including countering UAS).
- Joint terminal attack controller and joint fires observer authorities for multinational CAS.

Technical interoperability includes the interface of systems used to link platforms to acquire targets and create effects while providing command and control across the multinational enterprise. Technical interoperability is usually the most difficult to achieve due to the varied approaches among nations to technical challenges, and the necessarily restrictive information sharing policies between allies and partners. Some important technical interoperability considerations are:

- Levels of interoperability among nations (deconflicted, compatible, or integrated).
- Common operating picture across the multinational force.
- Multinational sensor-to-shooter links (integrated fires network).
- Collaborative target development.
- Integrated sensor management.
- Integrated terrain and airspace management.
- Integrated AMD capabilities.
- Ammunition interoperability.

All of the considerations for multinational interoperability listed above require extensive discussion, agreement, and exercise to function and improve across any multinational force. Frequent multinational and bilateral exercises will produce lessons learned for all allied and partner nations to implement better understanding, and sustain and enhance multinational cooperation and interaction to develop and practice a common approach to creating effects for a multinational force.

For more on multinational operations, refer to JP 3-16 and FM 3-16.

Refer to TAA2: Military Engagement, Security Cooperation & Stability SMARTbook (Foreign Train, Advise, & Assist) for further discussion. Topics include the Range of Military Operations (JP 3-0), Security Cooperation & Security Assistance (Train, Advise, & Assist), Stability Operations (ADRP 3-07), Peace Operations (JP 3-07.3), Counterinsurgency Operations (JP & FM 3-24), Civil-Military Operations (JP 3-57), Multinational Operations (JP 3-16), Interorganizational Cooperation (JP 3-08), and more.

(Fires) II. Integrate Army, Multinational & Joint Fires 6-19

III. Army Targeting Process (D3A)

Ref: ADP 3-19, Fires (Jul '19), pp. 3-7 to 3-9.

The Army targeting process organizes the efforts of the commander and staff to accomplish key targeting requirements. This methodology is referred to as the D3A. D3A assists the commander and staff decide which targets must be acquired and engaged and to help develop options to engage those targets. Options may include lethal or nonlethal, organic or supporting assets at all levels, including maneuver, electronic attack, psychological operations, attack aircraft, surface-to-surface fires, air to surface fires, other information-related capabilities, or a combination of these options.

The D3A methodology is an integral part of the MDMP. As the MDMP is conducted, targeting becomes more focused based on the commander's guidance and intent. Certain targets may require special considerations or caution, because engaging them improperly could create unintended effects. Examples include targets that should be handled with sensitivity due to potential political and or diplomatic repercussions and targets located in areas with a high risks of collateral damage, to include weapons of mass destruction facilities. These measures are incorporated in the coordinating instructions and appropriate annexes of the operation plan or operation order.

A. Decide

Decide is the first function in targeting and occurs during the planning portion of the operations process. It is the most important function, requiring close interaction between the commander, intelligence, plans, operations, the fires cell, and staff judge advocate. It begins during the mission analysis portion of the MDMP and continues throughout the operation.

B. Detect

Detect is the second function in targeting and occurs initially during the prepare portion of the operations process, continuing throughout the operation. A key resource for fires planning and targeting is the intelligence generated through information collection to answer the targeting information requirements. Commanders express requirements for target detection and action as priority intelligence and information requirements. During large-scale combat operations, it might be challenging to prioritize the detection of targets and could require the opening of windows of opportunity for specific collection capabilities in support of fires. High-payoff targets must be integrated and support associated priority intelligence requirements. Their priority depends on the importance of the target to the friendly course of action and target acquisition requirements. Targets are prioritized through a quantitative and qualitative valuation methodology. An example of a valuation methodology is the target value analysis process that prioritizes targets based on the target's criticality, accessibility, recuperability, vulnerability, effect, and recognizability. Targeting working groups incorporate priority intelligence and information requirements that support acquisition of high-payoff targets into the overall information collection plan along with named areas of interest, target areas of interest, and engagement areas.

C. Deliver

Deliver is the third function in targeting and occurs primarily during the execution portion of the operations process. The main objective is to engage targets in accordance with the commander's guidance or engagement authority's direction. The selection of a weapon system or a combination of weapons systems leads to the tactical decision of time of engagement and then the technical solution for the selected weapon.

D. Assess

Assess is the fourth function of targeting and occurs throughout the operations process. The commander and staff assess the results of mission execution. The assessment process is continuous and directly tied to the commander's decisions throughout planning, preparation, and execution of operations.

6-20 (Fires) II. Integrate Army, Multinational & Joint Fires

Operations Process & Targeting Relationship

Fires are an integral part of the operations process—the major mission command activities performed during operations: planning, preparing, executing, and continuously assessing the operation (ADP 5-0). The commander drives the operations process.

Army targeting uses the functions decide, detect, deliver, and assess (D3A) as its methodology. Its functions complement the planning, preparing, executing, and assessing stages of the operations process. Army targeting addresses two targeting categories—deliberate and dynamic.

Operations Process	D3A	Targeting Task
Continuous Assessment / Planning	Decide	• Perform target value analysis to develop fire support, high-value targets, and critical asset list. • Provide fires running estimates and information/influence to the commander's targeting guidance and desired effects.
		• Designate potential high-payoff targets. • Deconflict and coordinate potential high-payoff targets. • Develop high-payoff target list/defended asset list. • Establish target selection standards and identification matrix (air and missile defense). • Develop attack guidance matrix, fire support, and cyber/electromagnetic activities tasks. • Develop associated measures of performance and measures of effectiveness.
		• Refine high-payoff target list. • Refine target selection standards. • Refine attack guidance matrix and surface-to-air-missile tactical order. • Refine fire support tasks. • Refine associated measures of performance and measures of effectiveness. • Develop the target synchronization matrix. • Draft airspace control means requests.
		• Finalize the high-payoff target list. • Finalize target selection standards. • Finalize the attack guidance matrix. • Finalize the targeting synchronization matrix. • Finalize fire support tasks. • Finalize associated measures of performance and measures of effectiveness. • Submit information requirements to staff and subordinate units.
Preparation	Detect	• Collect information (surveillance, reconnaissance). • Report and disseminate information. • Update information requirements as they are answered. • Focus sensors, locate, identify, maintain track, and determine time available. • Update the high-payoff target list, attack guidance matrix, targeting synchronization matrix, identification matrix (air and missile defense) and surface-to-air-missile tactical order as necessary. • Update fire support tasks. • Update associated measures of performance and measures of effectiveness. • Target validated, deconfliction and target area clearance resolved, target execution/engagement approval.
Execution	Deliver	• Order engagement. • Execute fires in accordance with the attack guidance matrix, the targeting synchronization matrix, identification matrix (air and missile defense), and surface-to-air-missile tactical order. • Monitor/manage engagement.
Assess	Assess	• Assess task accomplishment (as determined by measures of performance). • Assess effects (as determined by measures of effectiveness). • Reporting results. • Reattack/reengagement recommendations.

Legend: D3A – decide, detect, deliver, and assess

Ref: Adapted from ADRP 3-09, Fires (Aug '12), table 3-2, p. 3-2 (not provided in ADP 3-19).

Refer to BSS6: The Battle Staff SMARTbook, 6th Ed. (Plan, Prepare, Execute, & Assess Military Operations) for fires and targeting (D3A - decide, detect, deliver and assess) as it relates to the operations process. In-depth topics include high-payoff target list, intelligence collection plan, target selection standards, attack guidance matrix, attack of targets, tactical and technical decisions, and combat assessments.

(Fires) II. Integrate Army, Multinational & Joint Fires 6-21

IV. Joint Targeting

Ref: ADP 3-19, Fires (Jul '19), pp. 3-9 to 3-11.

Joint targeting is dependent in part on joint planning through publication of the campaign or contingency plan, operation order, or fragmentary order. Plans and orders provide the context for targeting. Geographic combatant commands maintain a database for targets within their areas of responsibility that relate to their campaign plans and contingency plans. Detailed foundational intelligence products to include dynamic threat assessments, joint intelligence preparation of the operational environment, and country assessments facilitate detailed targeting, beginning with target systems analysis. Many products used to support a contingency or military operation are developed, maintained, and continuously updated as foundational information for specific targets. A combatant command can normally provide a subordinate JFC with a list of targets, and perhaps target folders, applicable to a plan for a joint operations area within their area of responsibility.

The joint targeting cycle is a six-phase iterative process:

- **Phase 1—Commander's objectives, Targeting guidance, and intent**. The JFC develops and issues targeting guidance. This guidance includes targeting priorities, time-sensitive targets criteria and procedures, component critical targets, target acquisition and identification criteria, authorized.

- **Phase 2—Target development and prioritization**. Target development is the systematic examination of potential target systems and their components, individual targets, and even elements of targets to determine the necessary type and duration of the action that must be exerted on each target to create an effect that is consistent with the commander's specific objectives.

- **Phase 3—Capabilities analysis**. This phase of the joint targeting cycle involves evaluating all available capabilities against targets' critical target elements to determine the appropriate options available to the component commander for target engagement and developing the best possible solution under given circumstances.

- **Phase 4—Commander's decision and force assignment**. The force assignment process at the component level integrates previous phases of joint targeting and fuses capabilities analysis with available forces, sensors, and weapons systems.

- **Phase 5—Mission planning and force execution**. Upon receipt of component tasking orders, detailed unit-level planning must be performed for the execution of operations. The joint targeting process supports this planning by providing component planners with direct access to detailed information on the targets, supported by the nominating component's analytical reasoning that linked the target with the desired effect (phase 2).

- **Phase 6—Combat assessment**. The combat assessment phase is a continuous process that assesses the effectiveness of the activities that occurred during the first five phases of the joint targeting cycle.

Integrating Army Targeting with Joint Targeting

LCCs contribute to the joint targeting cycle by assisting the JFC in formulating guidance, integrating land component fires with other joint fires to support JFC operations, conducting target development, synchronizing and coordinating the use of collection assets, engaging targets, and providing feedback as part of the assessment process. These functions remain constant regardless of how the joint force is organized (functional or Service components). Coordination and communication between the components, theater analyst, and multinational partners is critical executing fire plans and engaging targets of opportunity.

The LCC HQ is responsible for integrating the D3A targeting processes into the joint targeting cycle. Additional target development steps are required when nominating a target into the joint targeting cycle.

The LCC HQ consolidates subordinate tactical level targeting nominations (developed through D3A) for inclusion into the joint targeting cycle while bridging the target development gaps required for phases II and III of the joint targeting cycle (see JP 3-60 and also see figure 3-1 for more information). A critical intelligence gap between D3A and joint targeting cycle is the capability to conduct intermediate, and advanced target development on tactical target nominations in accordance with CJCSI 3370.01C. Target nominations must meet the JFC's target validation criteria through the joint targeting decision board. The joint integrated prioritize target list is used by all components to task assets available to best create the desired effects against targets.

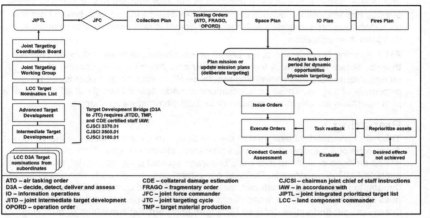

Ref: ADP 3-19, Fires (Jul '19), fig. 3-1. Example LCC D3A Target nominated to joint targeting cycle.

In order to meet the requirement for intermediate target development within phase II of the joint targeting cycle the LCC staff must certify intelligence staff in joint intermediate target development (JITD) guidance as defined by the current updated Chairman of the Joint Chiefs of Staff Instruction on target development.

In order to meet the requirement for advanced target development within phase III of the joint targeting cycle the LCC staff must certify intelligence staff and establish accredited target material production work centers according to guidance defined by the current updated CJCSIs on target development and collateral damage methodology.

Target development in a joint environment generally requires the following things:

- Component analysts trained and certified in joint intermediate target development, target material production, collateral damage estimation, weaponeering, and battle damage assessment. Component analysts require training and require access to intelligence and targeting databases and repositories (many found only on Top Secret architecture). This includes training on the creation and maintenance of electronic target folders which are databased in the Modernized Integrated Database (referred to as the MIDB). The two interfaces for the Modernized Integrated Database are the National production Workshop and the Joint Targeting Toolbox.
- Training on all three steps in target development:
  - Target system analysis
  - Entity level target development
  - Target list management

For more on Joint Target Development, reference CJCSI 3370.01 Target Development Standards and JP 3-60 Joint Targeting.

(Fires) II. Integrate Army, Multinational & Joint Fires 6-23

V. Air and Missile Defense Planning/Integration

Surface-to-air planning and integration considers the activities and capabilities of Army, joint, and multinational AMD elements. AMD operations are often joint efforts to which all Services contribute and which are integrated at the theater level to accomplish the JFC's counterair related missions.

Planning

AMD planning begins at the theater level and addresses the various aspects of AMD capabilities and airspace requirements. ADA staff personnel from the Army air and missile defense command participate in the development of war plans to shape specific regions well in advance of conflict; they ensure the surface-based counterair capabilities are integrated with the other domains and performed continuously throughout an operation.

AMD planning considerations are based on capabilities of projected air and missile threats, sensor coverage by various AMD air and ground assets, sharing of air picture information, and networking requirements to support and coordinate the engagement of targets. These considerations are addressed throughout all phases of a joint operation as operations expand or tactical circumstances change.

Preparing

The ADA commanders assigned to defend designated critical assets at the various Army echelons translate the defense plan into defense designs. They plan integrated defense designs to maximize coverage against projected aerial threats and to execute AMD engagements. Defense considerations include early warning of aerial threats, to trigger passive defense measures by affected units; situational awareness of the airspace in the areas of operations, with respect to both friendly and enemy usage; and active defense (engagements) against surveilling or attacking aerial platforms. Coordination is effected with the supported commander of the fixed/semi-fixed asset or maneuver formation to ensure reciprocal understanding of the availability and use of supporting capabilities and the supported commander's intent and plans. Collaboration between commanders is also effected in positioning ADA assets in the design of defenses.

Executing

AMD command and control actions are executed by Army and joint commanders. Command of ADA units is exercised by Army commanders. Control of AMD fires is exercised in accordance with the JFC's directives or by delegated authorities.

Assessing

The assessment process continuously collects and evaluates all available information on friendly and enemy forces to support decisions made by the commander. Assessments are embedded in all planning, preparing, and executing activities to ensure timely and appropriate actions consistent with current or evolving situations. ADA commanders and staffs evaluate plans and operations, modifying them as necessary; new priorities designated for defense or different levels of protection specified for assets, for example, may dictate the movement of ADA resources. Branches and sequels to evolving plans are also considered for contingencies. Defense designs are adjusted to account for tactics being or anticipated to be employed by the enemy, the availability of ADA forces, and the flow of friendly operations. Control of AMD firing units during periods of high intensity enemy air activities may necessitate the decentralization of engagement authority for all air and missile threats to battery commanders. Assessments of engagement results, a fundamental step in the ADA engagement sequence, confirm the kill of the target or need for reengagement.

6-24 (Fires) II. Integrate Army, Multinational & Joint Fires

Chap 7: The Sustainment Warfighting Function

Ref: ADP 4-0, Sustainment (Jul '19), chap. 1, ADP 3-0, Operations (Jul '19), pp. 5-5 to 5-6 and FM 3-0, Operations (Oct. '22), p. 2-3.

The sustainment warfighting function is the related tasks and system that provide support and services to ensure freedom of action, extended operational reach, and prolong endurance (ADP 3-0). Sustainment employs capabilities from all domains and enables operations through each domain. Sustainment determines the limits of depth and endurance during operations. Sustainment demands joint and strategic integration, and it should be meticulously coordinated across echelons to ensure continuity of operations and that resources reach the point of employment.

Sustainment employs an integrated network of information systems linking sustainment to operations. As a result, commanders at all levels see an operational environment, anticipate requirements in time and space, understand what is needed, track and deliver what is requested, and make timely decisions to ensure responsive sustainment. Because the situation is always changing, sustainment requires leaders capable of improvisation. Because sustainment operations are often vulnerable to enemy attacks, sustainment survivability depends on active and passive measures and maneuver forces for protection.

Successful sustainment enables freedom of action by increasing the number of options available to the commander. Sustainment is essential for retaining and exploiting the initiative. The sustainment warfighting function consists of four elements: logistics, financial management, personnel services and health service support:

A. Logistics

Logistics is planning and executing the movement and support of forces. It includes those aspects of military operations that deal with: design and development; acquisition, storage, movement, distribution, maintenance, and disposition of materiel; acquisition or construction, maintenance, operation, and disposition of facilities; and acquisition or furnishing of services. The explosive ordnance disposal tasks are discussed under the protection warfighting function. Army logistics elements are:

- Maintenance
- Transportation
- Supply
- Field Services
- Distribution
- Operational contract support
- General engineering

Refer to SMFLS5: The Sustainment & Multifunctional Logistics SMARTbook (Guide to Operational & Tactical Level Sustainment). SMFLS5 topics include the sustainment warfighting function; sustainment operations), sustainment execution (logistics, financial management, personnel services, & health services support); sustainment planning; brigade support; division, corps & field army sustainment; theater support; joint logistics; and deployment & redeployment.

(Sustainment) Warfighting Function 7-1

II. Sustainment Overview

Ref: ADP 4-0, Sustainment (Jul '19).

ADP 4-0, Sustainment, is the Army's doctrine for sustainment in support of operations. The endurance of Army forces is primarily a function of their sustainment and is essential to retaining and exploiting the initiative. Sustainment provides the support necessary to maintain operations until mission accomplishment. The relationship between sustainment and operation is depicted in introductory figure-1 the facing page.

Fundamentals of Sustainment

For the Army, sustainment is the provision of logistics, financial management, personnel services, and health service support necessary to maintain operations until successful mission completion. Sustainment is accomplished through the coordination, integration, and synchronization of resources from the strategic level through the tactical level in conjunction with our joint and multinational partners.

Sustainment operations enable force readiness. Sustainment operations maintain Army forces by equipping it with materiel, funding it with required resources, staffing it with trained Soldiers and leaders, and by providing it with the force health protection needed.

Army sustainment is based on an integrated process (people, systems, materiel, health service support, and other support) inextricably linking sustainment to operations. The concept focuses on building an operational ready Army, delivering it to the CCDR as part of the joint force, and sustaining its combat power across the depth of the operational area and with unrelenting endurance.

Principles of Sustainment

The principles of sustainment shown below are essential to maintaining combat power, enabling strategic and operational reach, and providing Army forces with endurance. While these principles are independent, they are also interrelated and must be synchronized in time, space, and purpose. The principles of sustainment and the principles of logistics are the same.

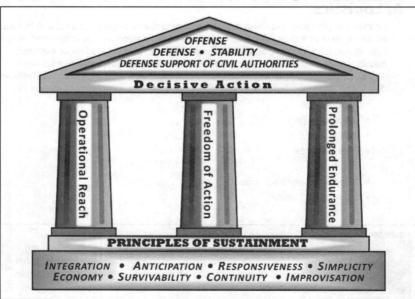

Ref: ADP 4-0 (Jul '19), fig. 1-1. Principles of sustainment

Sustainment Underlying Logic

Sustainment Elements

Logistics
- Maintenance
- Transportation
- Supply
- Field Services
- Distribution
- Operational Contract Support
- General Engineering Support

Personnel Services
- Human Resources Support
- Legal Support
- Religious Support
- Music Support
- Mortuary Affairs

Financial Management
- Finance Operations
- Resource Management

Health Service Support
- Casualty Care
  - Organic Medical Support
  - Area Medical Support
  - Hospitalization
  - Dental Treatment
  - Behavioral Health
  - Clinical Laboratory Services
  - CBRN Patient Treatment
- Medical Logistics
- Medical Evacuation

Sustainment in Joint Operations

Sustainment is the provision of logistics and personnel services to maintain operations until mission accomplishment and redeployment of the forces.

Sustainment in joint operations provides the joint task force flexibility, endurance, and the ability to extend operational reach. (JP 4-0)

Unified Action
The synchronization, coordination, and/or integration of governmental and non-governmental entities with military operations to achieve unity of effort. (JP-1)

Sustainment of Unified Action
Joint Interdependence: The purposeful reliance by Service forces on another Service's capabilities. (JP 1)

Unified Land Operations
Simultaneous execution of offense, defense, stability, and defense support of civil authorities across multiple domains to shape the operational environment, prevent conflict, prevail in large-scale ground combat, and consolidate gains as part of unified action (ADP 3-0).

Strategic Base leverages National capabilities to generate Theater Capabilities.

Sustainment Capabilities

Strategic to Tactical Linked by
- MED COM (DS)
- TSC
- MED BDE (SPT)
- ESC
- AFSB
- HRSC
- FMC
- SUST BDE

Army Joint Interdependence Capabilities
- Setting the Theater
- Common User Logistics
- Army Support to Other Services
- Reception, Staging, Onward movement, and Integration

DECISIVE ACTION
Offensive – Defensive – Stability – Defense Support to Civil Authorities

Enabling CCDR & ARFOR to Conduct
- Freedom of Action
- Operational Reach
- Prolonged Endurance

Setting the Theater

US Army Strategic Roles — Win

| Shape the Operational Environment | Prevent Conflict | Prevail in Large-Scale Combat Operations | Consolidate Gains |

LEGEND:
- ADP = Army doctrinal publication
- AFSB = Army field service brigade
- ARFOR = Army forces
- BDE = brigade
- CCDR = component commander
- COM = command
- (DS) = deployment support
- ESC = expeditionary sustainment command
- FMC = financial management company
- FMSC = financial management support center
- HRSC = human resource support center
- JP = joint publication
- MED = medical
- SPT = support
- SUST = sustainment
- TSC = theater sustainment command

Refer to SMFLS5: The Sustainment & Multifunctional Logistics SMARTbook (Guide to Operational & Tactical Level Sustainment). SMFLS5 topics include the sustainment warfighting function; sustainment operations), sustainment execution; sustainment planning; brigade support; division, corps & field army sustainment; theater support; joint logistics; and deployment & redeployment.

(Sustainment) Warfighting Function 7-3

B. Financial Management

Financial management leverages fiscal policy and economic power across the range of military operations. Financial management encompasses finance operations and resource management.

C. Personnel Services

Personnel services are those sustainment functions related to Soldiers' welfare, readiness, and quality of life. Personnel services complement logistics by planning for and coordinating efforts that provide and sustain personnel. Personnel services include—

- Human resources support
- Legal support
- Religious support
- Band support

D. Health Service Support

Army Health System (AHS) support includes both health service support and force health protection that are critical capabilities embedded within formations across all warfighting functions. The force health protection mission falls under the protection warfighting function and will not be covered in this publication (see ADP 3-37).

Health service support encompasses all support and services performed, provided, and arranged by the Army Medical Department to promote, improve, conserve, or restore the behavioral and physical well-being of Army personnel and as directed, unified action partners (UAPs). Health service support includes the following—

- Casualty care, which encompasses a number of medical functions, including:
    - Medical treatment (organic and area medical support).
    - Hospitalization.
    - Dental care (treatment aspects).
    - Behavioral health/neuropsychiatric treatment.
    - Clinical laboratory services.
    - Treatment of chemical, biological, radiological, and nuclear patients).
- Medical evacuation (including medical regulating).
- Medical logistics (including blood management).

7-4 (Sustainment) Warfighting Function

I. Sustainment of Unified Land Operations

Ref: ADP 4-0, Sustainment (Jul '19), chap. 3.

Operational Context

Any operational environment consists of many interrelated variables and sub variables, as well as the relationships among those variables and sub variables. How the many entities and conditions behave and interact with each other within an operational environment is difficult to discern and always results in differing circumstances. Different actor or audience types do not interpret a single message in the same way. Therefore, no two operational environments are the same (ADP 3-0).

Unified Land Operations

Unified land operations require the integration of U.S. military operations with that of multinational partners and our UAPs. The goal of unified land operations is to establish conditions that achieve the joint force commander's end state by applying landpower as part of unified action (ADP 3-0).

The sustainment warfighting function is nested within all four of the Army' strategic roles (shape operational environments, prevent conflict, prevail in large-scale ground combat operations, and consolidate gains). FM 4-0 describes in detail how the sustainment warfighting function supports the strategic roles during operations described in FM 3-0. The sustainment warfighting function is also essential for conducting unified land operations and providing resources for generating and maintaining combat power. Sustainment provides maneuver commanders with operational reach, freedom of action, and operational endurance needed to maintain the initiative in conducting unified land operations.

Sustainment Support of the Army Strategic Roles

Sustainment supports the Army strategic role of shaping operational environments by setting the theater and supporting military engagements. Sustainment activities during shaping operational environments include establishing logistics partnerships, enhancing interoperability, establishing or refining HNS agreements, and gaining access to potential critical infrastructure nodes. Sustainment supports the Army strategic role of preventing conflict by tailoring forces to the type of operation, geographic location, permissiveness of the environment, threat, and a host of other planning considerations determined during continued analysis of the operational environment and mission variables. In addition, refinement of plans and logistics estimates to support expected operations occur during prevent conflict. Sustainment supports the Army strategic role of prevailing in large-scale ground combat operations by providing the freedom of action, prolonged endurance, and extended operational reach required to conduct sustained defensive and offensive operations. During the Army strategic role of consolidate gains, sustainment provides support to combat operations while establishing security, restoring combat power, and preparing for continued operations to destroy remaining enemy forces.

Refer to SMFLS5:The Sustainment & Multifunctional Logistics SMARTbook (Guide to Operational & Tactical Level Sustainment). SMFLS5 topics include the sustainment warfighting function; sustainment operations), sustainment execution (logistics, financial management, personnel services, & health services support); sustainment planning; brigade support; division, corps & field army sustainment; theater support; joint logistics; and deployment & redeployment.

I. Operational Reach

Operational reach is the distance and duration across which a unit can successfully employ military capabilities (JP 3-0). The limit of a unit's operational reach is its culminating point. Operational reach enables commanders to determine where to engage the enemy by giving them the ability to strike enemy decisive points to achieve decisive force at the appropriate time and place. Extending operational reach is critical to achieving the necessary freedom of action enabling the commander to seize and retain the initiative. Extended operational reach allows the commander to present the adversary with multiple dilemmas, confounding his decision making, and challenging their ability to act effectively.

Sustainment enables operational reach. Extending operational reach is a paramount concern for commanders. Commanders and staff increase operational reach through deliberate, focused operational design, and the allocating appropriate sustainment resources. This requires strategic sustainment capabilities such as materiel, supplies, health services, and other support and global distribution systems to deploy, maintain, and conduct operations over great distances for extended periods of time. Army forces can increase the joint force's ability to extend operational reach by securing and operating bases in the AOR, the use of contracted and local procurements, and the use of aerial delivery.

A. Army Pre-Positioned Stocks (APS)

Pre-positioning of stocks in potential theaters provides the capability to rapidly supply and resupply forces until air and sea lines of communication are established. Army pre-positioned stocks are located at or near the point of planned use or at other designated locations. This reduces the initial amount of strategic lift required for power projection, to sustain the war fight until the line of communication with CONUS is established, and industrial base surge capacity is achieved (ATP 3-35.1).

B. Theater Opening

Theater opening is the ability to establish and operate ports of debarkation (air, sea, and rail), to establish a distribution system and sustainment bases, and to facilitate port throughput for the reception, staging, onward movement and integration of forces within a theater of operations. Preparing for theater opening operations requires unity of effort among the various commands and a seamless strategic¬to-tactical interface. It is a complex joint process involving the GCC and strategic and joint partners such as USTRANSCOM and DLA. Theater opening functions set the conditions for effective support and lay the groundwork for subsequent expansion of the theater distribution system.

See following pages (pp. 7-8 to 7-9) for further discussion.

C. Theater Closing

Theater closing is the process of redeploying Army forces and equipment from a theater, the drawdown and removal or disposition of Army non-unit equipment and materiel, and the transition of materiel and facilities back to host nation or civil authorities. Theater closing begins with the termination of joint operations.

See pp. 7-10 to 7-11 for further discussion.

7-6 (Sustainment) I. Sustainment of Unified Land Operations

Basing

Ref: ADP 4-0, Sustainment (Jul '19), pp. 3-8 to 3-9.

A base camp is an evolving military facility that supports military operations of a deployed unit and provides the necessary support and services for sustained operations (ATP 3-37.10). Basing directly enables and extends operational reach, and involves the provision of sustainable facilities and protected locations from which units can conduct operations. Army forces typically rely on a mix of bases and/or base camps to deploy and employ combat power to operational depth. Options for basing range from permanent basing in CONUS to permanent or contingency (non-permanent) basing OCONUS. A base camp is an evolving military facility that supports military operations of a deployed unit and provides the necessary support and services for sustained operations.

Bases or base camps may have a specific purpose (such as serving as an intermediate staging base, logistics base, or forward operating base) or they may be multifunctional. A base or base camp has a defined perimeter and established access controls, and should take advantage of natural and man-made features.

Bases or base camps may be joint or single service and will routinely support both U.S. and multinational forces, as well as interagency partners, operating anywhere along the range of military operations. Commanders often designate a single commander as the base or base camp commander that is responsible for protection, terrain management, and day-to-day operations of the base or base camp. This allows other units to focus on their primary function. Units located within the base or base camp are under the tactical control of the base or base camp commander for base security and defense.

Within large echelon support areas, controlling commanders may designate base clusters for mutual protection and mission command. Within a support area, a designated unit such as a BCT or maneuver enhancement brigade provides area security, terrain management, movement control, mobility support, clearance of fires, and required tactical combat forces. Operational area security operations focus on the protected force, base, base camp, route, or area. This allows sustainment units to focus on their primary function. Sustainment commanders and planners must constantly coordinate with supported operational staffs to synchronize sustainment operations to include all activities of the base camp life cycle and the basing strategy.

Refer to ATP 3-37.10 for more information on base camps

Intermediate Staging Bases (ISB)

An intermediate staging base is a tailorable, temporary location used for staging forces, sustainment and/or extraction into and out of an operational area (JP 3-35). While not a requirement in all situations, the intermediate staging base may provide a secure, high-throughput facility when circumstances warrant. The commander may use an intermediate staging base as a temporary staging area en route to a joint operation, as a long-term secure forward support base, and/or secure staging areas for redeploying units, and noncombatant evacuation operations. An intermediate staging base is task organized to perform staging, support, and distribution functions as specified or implied by the CCDR and the theater Army operations order.

Forward Operating Bases (FOB)

Forward operating bases extend and maintain the operational reach by providing secure locations from which to conduct and sustain operations. They not only enable extending operations in time and space; they also contribute to the overall endurance of the force. Forward operating bases allow forward deployed forces to reduce operational risk, maintain momentum, and avoid culmination. Forward operating bases are generally located adjacent to a distribution hub. This facilitates movement into and out of the operational area while providing a secure location through which to distribute personnel, equipment, and supplies.

(Sustainment) I. Sustainment of Unified Land Operations 7-7

Theater Opening

Ref: ADP 4-0, Sustainment (Jul '19), pp. 3-6 to 3-9.

Theater opening is the ability to establish and operate ports of debarkation (air, sea, and rail), to establish a distribution system and sustainment bases, and to facilitate port throughput for the reception, staging, onward movement and integration of forces within a theater of operations. Preparing for theater opening operations requires unity of effort among the various commands and a seamless strategic¬to-tactical interface. It is a complex joint process involving the GCC and strategic and joint partners such as USTRANSCOM and DLA. Theater opening functions set the conditions for effective support and lay the groundwork for subsequent expansion of the theater distribution system.

When given the mission to conduct theater opening, a sustainment brigade, and a mix of functional battalions and multi-functional CSSBs are assigned based on mission requirements. The sustainment brigade will participate in assessing and acquiring available HN infrastructure capabilities and contracted support and coordinating with military engineers for general engineering support.

For additional information refer to ATP 4-93 and ATP 4-0.1.

Port Opening

Port opening is a subordinate function of theater opening. Port opening is the ability to establish, initially operate and facilitate throughput for ports of debarkation to support unified land operations.

The port opening process is complete when the port of debarkation and supporting infrastructure is established to meet the desired operating capacity for that node. Supporting infrastructure can include the transportation needed to support port clearance of cargo and personnel, holding areas for all classes of supply, and the proper in-transit visibility systems established to facilitate force tracking and end-to-end distribution.

Port opening and port operations are critical components for preparing and conducting theater opening. Commanders and staffs coordinate with the HN to ensure seaports and aerial ports possess sufficient capabilities to support arriving vessels and aircraft. USTRANSCOM is the port manager for deploying U.S. forces (ATP 4-0.1).

Joint Task Force Port Opening (JTF-PO)

The joint task force-port opening is a joint capability designed to rapidly deploy and initially operate aerial and seaports of debarkation, establish a distribution node, and facilitate port throughput within a theater of operations (JP 4-0). It is a standing task force that is a jointly trained, ready set of forces constituted as a joint task force at the time of need. The Army contribution to the Joint Task Force-Port Opening is the rapid port opening element which deploys within hours to establish air and seaports of debarkation in contingency response operations. The rapid port opening element also provides in-transit visibility and cargo clearance.

The joint task force-port opening facilitates joint reception, staging, onward movement, and integration and theater distribution by providing an effective interface with the theater joint deployment and distribution operations center and the sustainment brigade for initial aerial port of debarkation (APOD) operations. It is designed to deploy and operate for up to 60 days. As follow-on theater logistics capabilities arrive, the joint task force-port opening will begin the process of transferring mission responsibilities to arriving sustainment brigade forces or contracted capabilities to ensure the seamless continuation of airfield and distribution operations.

Seaports

USTRANSCOM's SDDC is the single port manager for all common user seaports of debarkation (SPODs) and as the single port manager it develops policy and advises the GCC on port management, recommends ports to meet operational demands, and is primarily

---

7-8 (Sustainment) I. Sustainment of Unified Land Operations

responsible for the planning, organizing, and directing the operations at the seaport. The TSC and its subordinate sustainment brigades, terminal battalions and seaport operating companies perform the port operator functions at SPODs. These functions can include port preparations and improvement, cargo discharge and upload operations, harbor craft services, port clearance and cargo documentation activities. If the operational environment allows, SDDC may have the ability to contract locally for port operator support eliminating or decreasing the requirement for the TSC and its subordinate units.

The single port manager may have OPCON of a port support activity that is an ad hoc organization consisting of military and/or contracted personnel with specific skills to add in port operations. The TSC and SDDC will coordinate the port support activity requirements. It assists in moving unit equipment from the piers to the staging/marshaling/loading areas, assisting the aviation support element with movement of helicopters in preparation for flight from the port, providing limited maintenance support for equipment being offloaded from vessels, limited medical support, logistics support, and security for port operations.

Ideally, the SPOD will include berths capable of discharging large medium speed roll-on/roll-off ships. The SPOD can be a fixed facility capable of discharging a variety of vessels, an austere port requiring ships to be equipped with the capability to conduct their own offloading, or beaches requiring the conducting of joint logistics over-the-shore operations. Whatever the type of SPOD, it should be capable of accommodating an armored BCT.

When vessels arrive at the SPOD, the TSC and/or SDDC is responsible for discharging the unit equipment, staging the equipment, maintaining control and in-transit visibility, and releasing it to the unit. This includes minimum standards that are critical for the physical security/processing of DOD sensitive conventional arms, ammunition, and explosives, including non-nuclear missiles and rockets.

The theater gateway personnel accounting team and supporting HR company and platoons will normally operate at the SPOD as well as movement control teams to facilitate port clearance of personnel and equipment. The movement control team that has responsibility for the SPOD, coordinates personnel accounting with the supporting CSSB or sustainment brigade for executing life support functions for personnel who are transiting into or out of the theater.

Aerial Ports

Airfields supporting strategic air movements for deployment, redeployment, and sustainment are designated aerial ports. Aerial ports are further designated as either an aerial port of embarkation for departing forces and sustainment, or as an APOD for arriving forces and sustainment. Reception at the APOD is coordinated by the senior logistics commander and executed by an Air Force contingency response group/element and an arrival/departure airfield control group. It is an ad hoc organization established to control and support the arrival and departure of personnel, equipment, and sustainment cargo at airfields and must be a lead element when opening an APOD. Elements of a movement control team and an inland cargo transfer company typically operate the arrival/departure airfield control group however; any unit can perform the mission with properly trained personnel and the appropriate equipment. USTRANSCOM's Air Mobility Command is the single port manager for all common user APODs

Basing *(See p. 7-7.)*

A base camp is an evolving military facility that supports military operations of a deployed unit and provides the necessary support and services for sustained operations (ATP 3-37.10). Basing directly enables and extends operational reach, and involves the provision of sustainable facilities and protected locations from which units can conduct operations. Army forces typically rely on a mix of bases and/or base camps to deploy and employ combat power to operational depth. Options for basing range from permanent basing in CONUS to permanent or contingency (non-permanent) basing OCONUS. A base camp is an evolving military facility that supports military operations of a deployed unit and provides the necessary support and services for sustained operations.

(Sustainment) I. Sustainment of Unified Land Operations 7-9

Theater Closing

Ref: ADP 4-0, Sustainment (Jul '19), pp. 3-15 to 3-17.

Theater closing is the process of redeploying Army forces and equipment from a theater, the drawdown and removal or disposition of Army non-unit equipment and materiel, and the transition of materiel and facilities back to host nation or civil authorities. Theater closing begins with the termination of joint operations.

Terminating Joint Operations

Terminating joint operations is an aspect of the CCDR's functional or theater strategy that links to achievement of national strategic objectives (JP 5-0). Based on the President's strategic objectives that compose a desired national strategic end state, the supported CCDR can develop and propose termination criteria. The termination criteria describe the standards to be met before conclusion of a joint operation. These criteria help define the desired military end state, which normally represents a period in time or set of conditions beyond which the President does not require the military instrument of national power as the primary means to achieve remaining national objectives. Termination criteria should account for a wide variety of operational tasks that the joint force may need to accomplish, to include disengagement, force protection (including force health protection support to conduct retrograde cargo inspections and pest management operations), transition to post-conflict operations, reconstitution, and redeployment. While there may be numerous terminating tasks the Army must achieve, the discussion below is deliberately broad and not all-inclusive. The discussion focuses on redeployment, drawdown of non-unit materiel, and transitioning of materiel, facilities and capabilities to HN or civil authorities.

Redeployment

Redeployment involves the return of personnel, equipment, and materiel to home and/or demobilization stations and is considered as an operational movement critical in reestablishing force readiness (ATP 3-35). Under the sustainable readiness model deployment and redeployment of forces in support of extended operations is a cyclic process. However, for terminating joint operations, Army forces may be completely redeployed from the joint operational area. Many of the procedures used in the initial deployment of forces to theater apply during redeployment. Unlike cyclic deployment where units fall in on positioned unit equipment and sets, termination redeployment efforts require movement of unit sets to aerial or seaports of embarkation for shipment to home station or other designated locations. After completion of military operations, redeploying forces move to designated assembly areas or directly to redeployment assembly areas. The same elements that operate and manage the theater distribution system during deployment and sustainment will usually perform support roles during redeployment.

Two critical aspects of equipment and materiel redeployment are property book accountability and asset visibility. Furthermore, the identification of how much equipment is on the ground, location of the equipment, type of equipment, condition of the equipment, and reporting procedures will allow for timely planning, as it will impact mode of transportation, resources, timeline, personnel, storage capabilities, and the like. Moreover, the accurate reporting of equipment on property books, locations, and condition will influence strategic-level decision making in terms of funding, field or sustainment reset, and disposal of equipment.

Drawdown

Planning for drawdown of non-unit equipment and materiel should occur early in the operational and strategic planning process. Drawdown planning entails more than returning equipment to CONUS. At the strategic level, the requirement for specific types of equipment may necessitate the redistribution of equipment to another AOR.

Even though equipment drawdown is an important mission in the redeployment operation, it may not be the Army's or the GCC main priority; thus, prioritization of equipment redistribution/disposition must be established early on to maximize distribution capacity

7-10 (Sustainment) I. Sustainment of Unified Land Operations

and velocity. A challenge is visibility of strategic-level materiel requirements synthesized into the already established priority timeline. Overcoming this challenge requires strategic-level collaboration between partners including Service Headquarters, GCCs, USAMC, DLA, and USTRANSCOM to effectively, efficiently and strategically reset both Joint and Army forces. Through this partnership, the nation's resources are preserved for other security needs.

As planners begin the process of reducing forces in a theater of operations, they must develop a balance between operational capability and sustainment capability. There is a natural tendency to eliminate the sustainment and enabler forces first because they do not provide an inherent capability to engage with the population or enemy. However, as the sustainment and enabling force are withdrawn, there is a direct impact on the operational forces in the form of reduced operational reach and requirements for assumption of additional missions.

To provide unity of effort and ensure operational freedom of action through rapid return, repair, redistribution, and combat power regeneration for the Army, a USAMC Responsible Reset Task Force provides a comprehensive solution for drawdown. Reset is a coordinated effort to methodically plan and execute the timely, repair, redistribution, and/or disposal of non-unit equipment, non-consumable and materiel identified as excess to theater requirements, to home station, sources of repair, or storage or disposal facilities. Through the phased redeployment of forces, the Responsible Reset Task Force mission will reset the Army in the shortest time possible.

The TSC and ESC work closely with DLA in the close out of materiel in the theater. The DLA support team serves as the single point of contact to the TSC and ESC. The DLA support teams are tasked to provide support to the theater closure plans, and are focused on providing support to all echelons based on the priorities of effort. During theater closure, DLA provides support in the form of adjusting the flow of supply classes I, II, III (B) (P), IV, VIII and IX to ensure support to the warfighter. DLA could, if requested, provide a theater consolidation and shipping point for departing forces. Additionally, DLA assists Army forward operating base closure by providing expeditionary disposal remediation teams to provide expert advice and oversight to U.S. Forces on the preparation for and the closure of Army units.

Closing Operational Contracts

The supporting contracting organization will be required to terminate and close out existing contracts and orders and includes the redeployment of contractor personnel. Ratifications and claims must be processed to completion. Contracting for life support services and retrograde support may continue until the last Soldiers and U.S. personnel depart, but standards of support should be reduced as much as possible prior to final contract closeout. In some operations, the supporting contracting organization may be required to assist in the transition of contracted support (the contracts themselves are not transferable) to the Department of State, a multi-national partner, or to the HN. This transition of contract support may include limited continuation of existing contracts in support of high priority Department of State operations. Because of the nature of contract support transition and/or closeout during termination operations, contingency contracting officers will often be some of the last Soldiers to leave the area of operations.

Port Closing

USTRANSCOM, through SDDC is responsible for providing and managing strategic common-user sealift and terminal services in support GCC's drawdown or termination requirements. As the single port manager for USTRANSCOM, it is SDDC's responsibility to integrate and synchronize strategic and theater re-deployment execution and distribution operations within each CCDR's area of responsibility. Ensures drawdown and/or termination requirements are met through the use of both military and commercial transportation assets based on the supported commander business rules and Joint Deployment and Distribution Enterprise best business practices.

(Sustainment) I. Sustainment of Unified Land Operations 7-11

II. Freedom of Action

Preparation for the sustainment of operations consists of activities performed by units to improve their ability to execute an operation. Preparation includes but is not limited to plan refinement, rehearsals, information collection, coordination, inspections, and movements. For sustainment to be effective, several actions and activities are performed across the levels of war to properly prepare forces for operations.

A. Sustainment Preparation

Sustainment preparation of the operational environment is the analysis to determine infrastructure, physical environment, and resources in the operational environment that will optimize or adversely impact friendly forces means for supporting and sustaining the commander's operations plan. The sustainment preparation of the operational environment assists planning staffs to refine the sustainment estimate and concept of support. It identifies friendly resources (HNS, contractible, or accessible assets) or environmental factors (endemic diseases, climate) that impact sustainment.

See facing page.

B. Sustainment Execution

Execution is putting a plan into action by applying combat power to accomplish the mission (ADP 5-0). It focuses on actions to seize, retain, and exploit the initiative.

Sustainment determines the depth and duration of Army operations. It is essential to retaining and exploiting the initiative and it provides the support necessary to maintain operations until mission accomplishment. Failure to provide sustainment could cause a pause or culmination of an operation resulting in the loss of the initiative. It is essential that sustainment planners and operation planners work closely to synchronize all of the warfighting functions, in particular sustainment, to allow commanders the maximum freedom of action.

Sustainment plays a key role in enabling the simultaneous offensive, defensive, and stability or defense support of civil authorities tasks that occur as part of unified land operations. For example, general engineering support provides construction support to protect key assets such as personnel, infrastructure, and bases. Horizontal and vertical construction enables assured mobility of transportation networks and survivability operations to alter or improve cover and concealment to ensure freedom of action, extend operational reach, and endurance of the force. Legal personnel supporting rule of law activities may find themselves working closely with HN judicial, law enforcement, and corrections systems personnel.

III. Endurance

Endurance refers to the ability to employ combat power anywhere for protracted periods. Endurance stems from the ability to create, protect, and sustain a force, regardless of the distance from its base and the austerity of the environment. Endurance involves anticipating requirements and continuity of integrated networks of interdependent sustainment organizations. Prolonged endurance is enabled by an effective distribution system and the ability to track sustainment from strategic to tactical level.

Reconstitution operations are extraordinary actions that commanders plan and implement to restore attrited units' combat effectiveness commensurate with the mission requirements and available resources. Reconstitution restores combat power to the levels necessary to maintain endurance and continue operations.

See pp. 7-24 to 7-25 for discussion of reconstitution operatons.

Distribution is the primary means to prolong endurance. Distribution is the operational process of synchronizing all elements of the logistic system to deliver the "right things" to the "right place" at the "right time" to support the geographic combatant commander (JP 4-0).

See following page (p. 7-14) for discussion of distribution.

7-12 (Sustainment) I. Sustainment of Unified Land Operations

Sustainment Preparation

Ref: ADP 4-0, Sustainment (Jul '19), pp. 3-12 to 3-13.

Preparation for the sustainment of operations consists of activities performed by units to improve their ability to execute an operation. Preparation includes but is not limited to plan refinement, rehearsals, information collection, coordination, inspections, and movements. For sustainment to be effective, several actions and activities are performed across the levels of war to properly prepare forces for operations.

Negotiations and Agreements

Negotiating and establishing agreements with HN resources is an important task of mission command for sustainment operations. Through negotiation and agreements, Army forces are able to reduce the military sustainment footprint and/or all military sustainment resources to focus on higher priority operations that may not be conducive to civilian support functions. Negotiation of agreements enables access to HNS resources identified in the requirements determination phase of planning. This negotiation process may facilitate force tailoring by identifying available resources (such as infrastructure, transportation, warehousing, and other requirements) which if not available would require deploying additional sustainment assets to support.

Host nation support agreements may include pre-positioning of supplies and equipment, OCONUS training programs, and humanitarian and civil assistance programs. These agreements are designed to enhance the development and cooperative solidarity of the HN and provide infrastructure compensation should deployment of forces to the target country be required. The pre-arrangement of these agreements reduces planning times in relation to contingency plans and operations.

Sustainment Preparation of the Operational Environment

Sustainment preparation of the operational environment is the analysis to determine infrastructure, physical environment, and resources in the operational environment that will optimize or adversely impact friendly forces means for supporting and sustaining the commander's operations plan. The sustainment preparation of the operational environment assists planning staffs to refine the sustainment estimate and concept of support. It identifies friendly resources (HNS, contractible, or accessible assets) or environmental factors (endemic diseases, climate) that impact sustainment. Some of the factors considered (not all-inclusive) are as follows:

- **Geography.** Information on climate, terrain, and endemic diseases in the AO to determine when and what types of equipment are needed. For example, water information determines the need for such things as early deployment of well-drilling assets and water production and distribution units.

- **Supplies and Services.** Information on the availability of supplies and services readily available in the AO. Supplies (such as subsistence items, bulk petroleum, and barrier materials) are the most common. Common services consist of bath and laundry, sanitation services, and water purification.

- **Facilities.** Information on the availability of warehousing, cold-storage facilities, production and manufacturing plants, reservoirs, administrative facilities, hospitals, sanitation capabilities, and hotels.

- **Transportation.** Information on road and rail networks, inland waterways, airfields, truck availability, bridges, ports, cargo handlers, petroleum pipelines, materials handling equipment, traffic flow, choke points, and control problems.

- **Maintenance.** Availability of host nation maintenance capabilities.

- **General Skills.** Information on the general skills such as translators and skilled and unskilled laborers.

(Sustainment) I. Sustainment of Unified Land Operations 7-13

Distribution

Ref: ADP 4-0, Sustainment (Jul '19), pp. 3-13 to 3-14.

Distribution is the primary means to prolong endurance. Distribution is the operational process of synchronizing all elements of the logistic system to deliver the "right things" to the "right place" at the "right time" to support the geographic combatant commander (JP 4-0).

The distribution system consists of a complex of facilities, installations, methods, and procedures designed to receive, store, maintain, distribute, manage, and control the flow of military materiel between point of receipt into the military system and point of issue to using activities and units.

The joint segment of the distribution system is referred to as global distribution. Global distribution is defined as the process that coordinates and synchronizes fulfillment of joint force requirements from point of origin to point of employment (JP 4-09). It provides national resources (personnel and materiel) to support the execution of joint operations.

The Army segment of the distribution system is referred to as theater distribution. Theater distribution is the flow of personnel, equipment, and materiel within theater to meet the GCC's missions (JP 4-09). The theater segment extends from the ports of debarkation or source of supply (in theater) to the points of need.

Distribution management synchronizes and optimizes transportation, its networks, and materiel management with the warfighting functions to move personnel and materiel from origins to the point of need in accordance with the supported commander's priorities. Distribution management includes the management of transportation and movement control, warehousing, inventory control, order administration, site and location analysis, packaging, data processing, accountability for equipment (materiel management), people, and communications.

Refer to ATP 4-0.1, Army Theater Distribution, for additional information on distribution and distribution management.

A support team from the Medical Logistics Management Center accomplishes the distribution management of medical materiel. The Medical Logistics Management Center support team collocates with the DMC of the TSC and ESC to provide the MEDCOM (DS) with visibility and control of all class VIII.

In-Transit Visibility

In-transit visibility is the ability to track the identity, status, and location of DOD units, and non-unit cargo (excluding bulk petroleum, oils, and lubricants) and passengers; patients and personal property from origin to consignee, or destination across the range of military operations (JP 4-01.2). This includes force tracking and visibility of convoys, containers/pallets, transportation assets, other cargo, and distribution resources within the activities of a distribution node.

Retrograde of Materiel

Another aspect of distribution is retrograde of materiel. Retrograde of materiel is an Army logistics function of returning materiel from the owning or using unit back through the distribution system to the source of supply, directed ship-to location, and/or point of disposal (ATP 4-0.1). Retrograde includes turn-in/classification, preparation, packing, transporting, and shipping. Contractor equipment is considered when planning retrograde. To ensure these functions are properly executed, commanders must enforce supply accountability and discipline and utilize the proper packing materials. Retrograde of materiel can take place as part of theater distribution operations and as part of redeployment operations. Retrograde of materiel must be continuous and not be allowed to build up at supply points/nodes.

Early retrograde planning is essential and necessary to preclude the loss of materiel assets, minimize environmental impact, and maximize use of transportation capabilities. Planners must consider environmental issues when retrograding hazardous material.

7-14 (Sustainment) I. Sustainment of Unified Land Operations

Chap 7

II. Sustain Large-Scale Combat

Ref: FM 4-0, Sustainment Operations (Jul '19), chap. 5 and FM 3-0, Operations (Oct. '22), pp. 6-18 to 6-19.

I. Overview

Large-scale combat operations require greater sustainment than other types of operations. Their high tempo and lethality significantly increase maintenance requirements and expenditure of supplies, ammunition, and equipment. Large-scale combat incurs the risk of mass casualties, which increase requirements for health service support, mortuary affairs, and large-scale personnel and equipment replacements. Large-scale combat operations demand a sustainment system to move and distribute a tremendous volume of supplies, personnel, and equipment.

See p. 1-87 for an overview of large-scale combat operations from FM 3-0 (2022).

Army sustainment is a key enabler of the joint force on land. Army forces provide sustainment to other elements in the joint force according to the direction of the JFC. The JFC has the overall responsibility for sustainment throughout a theater, but the JFC headquarters executes many of its sustainment responsibilities through the TSC. When directed, Army sustainment capabilities provide the bulk of Army support to other services through executive agency, common-user logistics, lead Service, and other common sustainment resources.

Capabilities from other domains enable sustainment of Army forces. Air sustainment capabilities provide responsive sustainment for high priority requirements. Maritime-enabled sustainment supports large-scale requirements. Space- and cyberspace-enabled networks facilitate rapid communication of sustainment requirements and precise distribution.

Successful sustainment operations strike a balance between protecting sustainment capabilities and providing responsive support close to the forward line of troops. A well-planned and executed logistics operation permits flexibility, endurance, and application of combat power. Plans must anticipate and mitigate the risk posed by enemy forces detecting and attacking friendly sustainment capabilities. Sustainment formations pursue operations security, survivability, and protection with the same level of commitment as all other forces. While most rear and support operations are economy of force endeavors when allocating combat power in divisions and corps, the continuity and survivability of those operations are vital to deep and close operations.

Dispersion of assets and redundancy help protect sustainment formations. Dispersing sustainment formations makes it less likely that enemy long-range fires can destroy large quantities of material. Dispersion also creates flexibility, as several nodes can execute the sustainment concept without a single point of failure. However, dispersed sustainment operations complicate C2 and can be less efficient than a massed and centralized approach. Commanders balance the risk between dispersion and efficiency to minimize exposure to enemy fires while maintaining the ability to enable the supported formation's tempo, endurance, and operational reach.

Commanders must plan for the possibility of heavy losses to personnel, supplies, and equipment. Even with continuous and effective sustainment support, units may rapidly become combat ineffective due to enemy action. Commanders at all levels must be prepared to conduct reconstitution efforts to return ineffective units to a level of effectiveness that allows the reconstituted unit to perform its future mission. Reconstitution is an operation that commanders plan and implement to restore units to a desired level of combat effectiveness commensurate with mission requirements and available resources (ATP 3-94.4).

Sustainment (ADP 4-0)

(Sustainment) II. Large-Scale Combat 7-15

II. Sustainment Synchronization

Logistics, financial management, personnel services, and health service support require coordination and synchronization at every stage of the planning process. This synchronization is crucial in large-scale combat operations with its inherent distributed nature. Only by integrating and synchronizing sustainment functions can the sustainment system achieve required effects at the speed, the volume, velocity, and lethality of large-scale combat operations.

Sustainment commanders and staffs present synchronized courses of action commensurate with sustainment capabilities to allow as much freedom of action as possible. Limitations such as insufficient infrastructure or non-availability of key classes of supply have a bearing on the commander's ability to execute the mission and are accounted for in the planning process. Sustainment leaders also coordinate, synchronize, and integrate the sustainment plan with joint and other unified action partners to ensure continuous linkage with strategic-level providers. A successful sustainment plan will extend operational reach, prevent culmination or loss of the initiative, manage transitions, exploit possible opportunities, and mitigate risk.

Identifying and accepting prudent risk is a principle of mission command. Throughout the operations process, commanders and staffs use risk management to identify and mitigate risks associated with all hazards that have the potential to injure or kill friendly and civilian personnel, damage or destroy equipment, or otherwise impact mission effectiveness. For sustainment commanders and staffs, identifying and mitigating risk must always include not only risk to finite and limited sustainment capabilities, but also how those capabilities are employed to enable freedom of action and extended operational reach.

Sustainment synchronization remains the focus as sustainment commanders plan for and coordinate support through such continuing activities as rhythm of military operations events, information collection, liaison, meetings, protection efforts, and reporting. For the purposes of sustaining large-scale combat operations, two of these—liaison and reporting—require special emphasis.

Liaison refers to contact or intercommunication maintained between elements of military forces or other agencies to ensure mutual understanding and unity of purpose and action. Most commonly used for establishing and maintaining close communications, liaison continuously enables direct, physical communications between commands.

Sustainment commanders and staffs have the continuous requirement to coordinate with higher, lower, adjacent, supporting, and supported units and civilian organizations. The sustainment liaisons participate in boards, bureaus, cells, centers, and working groups, especially in the case of the TSC with the ASCC, the ESC with the corps, the DSB with the division, the BSB with the BCT, and ASB with the CAB. While the use of liaisons taxes organic staff manpower in sustainment organizations, their presence and active participation is essential to sustaining large-scale combat operations and mitigating the effects of dispersion, threat disruption of communications, and accelerated tempo.

Both maneuver and sustainment commanders rely on logistics and personnel status reports to identify support requirements and capabilities to enable large-scale combat operations. Sustainment staffs use data from sustainment estimation tools, higher headquarters orders, and documents such as country studies to develop running estimates. A running estimate is the continuous assessment facts, assumptions, constraints and limitations concerning the current situation and operational environment used to determine if the current operation is proceeding according to the commander's intent and if planned future operations are supportable. Using sustainment information systems, commodity managers include information in running estimates such as quantity on-hand, quantity consumed, expected quantity on-hand, expected consumption to anticipate requirements and assist in synchronization. Each staff element and CP functional cell maintains a running estimate focused on how its specific areas of expertise are postured to support future operations.

7-16 (Sustainment) II. Large-Scale Combat

III. Threats to Sustainment Units

Ref: FM 4-0, Sustainment (Jul '19), pp. 5-5 to 5-6.

A threat is any combination of actors, entities, or forces that have the capability and intent to harm U.S. forces, U.S. national interest, or the homeland (ADP 3-0). Threats may include individuals, groups of individuals, paramilitary or military forces, nation-states, or national alliances. In general, a threat can be categorized as an enemy or an adversary.

FM 4-0 is focused on sustaining large-scale combat operations against peer threats. A peer threat is an adversary or enemy with capabilities and capacity to oppose U.S. forces across multiple domains world-wide or in a specific region where it enjoys a position of relative advantage. Peer threats present credible challenges to sustainment forces through the use of information warfare, isolation, systems warfare, preclusion and sanctuary. Other considerations such as contested LOC and anti-access/area de-nial techniques coupled with challenges across multiple domains (air, land, maritime, space and cyberspace) challenge sustainment support to operations. Sustainment commanders must consider the OE and all the factors that affect their ability to sustain operations during planning.

Peer threats have the ability to influence and direct irregular forces, criminal elements, and hostile populations. Peer threats have the ability to impose disruptive effects in cyberspace that will challenge Army sustainment during pre-deployment, deployment, employment, and redeployment. These disruptive effects may occur at unit home-stations, ports of embarkation, while in transit to the theater, and upon arrival at ports of debarkation as well as within the theater.

Peer threats employ their capabilities across multiple domains to attack U.S. vulnera-bilities, including sustainment facilities, networks and formations. Peer threats use their capabilities to create lethal and nonlethal effects throughout an OE. During combat operations, threats seek to inflict significant damage across multiple domains in a short period of time. Peer threats seek to delay U.S. forces long enough to achieve their goals and end hostilities before U.S. forces reach culmination. One effective way to delay U.S. forces, which generally operate on very long LOCs, is to disrupt sustain-ment operations and nodes.

Threat forces may employ tactics that force the U.S. Army into conducting large-scale combat operations in urban areas. Currently more than 50 percent of the world's population lives in urban areas, and this is likely to increase to 70 percent by 2050, making large-scale combat operations in cities likely. Large-scale combat operations in urban terrain is complex and resource intensive. In most urban operations, the terrain, the dense population, military forces, and unified action partners will further complicate sustainment operations.

For additional information on operations in urban areas, refer to ATP 3-06.

Threat use of subterranean spaces and structures (any space or structure located below ground) as a means to covertly maintain the initiative against a more powerful military opponent may occur during large-scale combat operations. Such spaces and structures can be used for command and control, defensive networks, operations, stor-age, production, or protection. Continued improvements in the construction of subter-ranean environments have increased their usefulness and their proliferation.

For additional information, refer to ATP 3-21.51.

See p. 7-21 for discussion of risks during large-scale combat.

(Sustainment) II. Large-Scale Combat 7-17

IV. Large-Scale Defensive Operations

Ref: FM 4-0, Sustainment (Jul '19), chap. 6.

As a component of large-scale combat operations, the defense is a combination of highly complex tasks that place tremendous and continuous demands on Army sustainment organizations. Situational awareness, mission analysis, and detailed planning are keys to successful support operations. Commanders take advantage of the time available during a defense to build combat power. However, the time available is likely to be unknown since the enemy typically has the initiative. As a result, sustainment organizations and the functions executed play a critical role in supporting the defense and the success of subsequent future operations.

See pp. 1-110 to 1-117 for discussion of defensive operations from FM 3-0.

Sustaining Defensive Operations

- Plan class IV for transitions from offense to defense
- Expect Increases in class V
- Plan for pre-positioning of supplies
- Plan retrograde support
- Increased demand for class VIII
- Mass casualties
- Large scale personnel replacements in a short period

Defending commanders combine the three types of defensive tasks to fit the situation. All three types of defense use mobile and static elements. In mobile defenses, static positions help control the depth and breadth of the enemy penetration and retain ground from which to launch counterattacks. In area defenses, commanders closely integrate patrols, security forces, and reserve forces to cover gaps among defensive positions. Commanders reinforce positions as necessary and counterattack, as directed. In retrograde operations, some units conduct area or mobile defenses or security operations to protect other units that execute carefully controlled maneuver or movement rearward. These units use static elements to fix, disrupt, turn, or block the attackers. Mobile elements are used to counterattack and destroy the enemy.

Commanders execute defensive tasks for various reasons, such as to retain decisive terrain or deny a vital area to the enemy, weaken or fix the enemy as a prelude to offensive actions, or increase the enemy's vulnerability by forcing the enemy to concentrate subordinate forces. The ultimate purpose of the defense is to create conditions for a counteroffensive whereby it allows Army forces to regain the initiative.

The transition to offensive operations will normally be as a planned operation but may also be a hasty operation conducted to capitalize on tactical opportunities or an identified enemy weakness. Because of this, sustainment commanders and leaders must stay cognizant of the status of the operation. Leaders must also use all available time to execute required sustainment functions knowing the mission may change quickly and frequently.

Sustainment Fundamentals

Enemy commanders look for opportunities to counter corps and division defensive tasks. The enemy will seek to employ special purpose forces, irregular forces, electronic warfare, long-range artillery, rockets, missiles, information capabilities, and cyberspace electromagnetic activities to disrupt sustainment activities. The enemy, to assist in targeting of sustainment units and locations, may exploit use of electronic signals, such as cell phones and geotagged photos. Sustainment commanders must be aware of these unintended threats and focus on those efforts that would help set the conditions necessary to regain the initiative during defensive operations.

7-18 (Sustainment) II. Large-Scale Combat

All sustainment functions are planned and executed to support defensive operations and build combat power to prepare for future offensive operations. The exact type and extent of support operations and the organizations executing them will vary by echelon based on the support requirement. Even though defensive operations may be the main effort, simultaneous offensive operations with their support requirements are also likely to be ongoing.

Sustainment commanders and staffs plan for increased requirements in class IV, V, and IX items to support the defensive effort and build class III (bulk and package) stocks to prepare to transition to offensive operations. Sustainment planners anticipate where the greatest need might occur during operations and consider pre-positioning sustainment stocks far forward to reduce response times for critical support. Planners also consider alternative methods for delivering sustainment in emergencies. Sustainment of defensive tasks requires a coordinated planning effort designed to maximize synchronization, integration, and continuity of support at all echelons. Commanders and staffs at every echelon must anticipate operational requirements, be responsive in requisitioning and distributing resources, and be prepared to improvise tactics and techniques for execution that ensures responsiveness even in unexpected situations.

Planning Considerations

Sustainment planning is both a continuous and a cyclical activity of the operations process. For sustainment planning, the most important factors are requirements, capabilities, and shortfalls. As outlined in the paragraphs below, planning considerations assist planners in identifying specific support or operational requirements based upon available information. Many planning considerations affect the ability to execute large-scale defensive operations. These considerations must be recognized, analyzed in the time available, and prioritized based on the commander's intent. Planning considerations must encompass all warfighting functions to ensure the plan is integrated across all functions and domains. A planning consideration may have various levels of effects that drive support requirements across all warfighting functions.

Just as significant as the commodity requirements for supporting a large-scale defense are the many implications for tactics, techniques, and procedures employed by sustainment forces. Sustainers anticipate how terrain, defensive obstacles, fire support coordination measures, and movement restrictions will affect sustainment operations. These factors are considered in all distribution management and movement control plans. Planners expect to weight sustaining operations support for spoiling attacks, counter attacks, and follow-on offensive operations. This may require sustainers to weight the main defensive effort by cross-leveling sustainment assets. In some cases, sustainers pre-position classes I, IIIB/P, IV, V, VIII, and IX stocks, as well as water centrally and well forward, but always balance forward positioning of sustainment assets with the need for rapid mobility. While supporting covering, guard, screening forces, counter and spoiling attack forces, sustainers plan for support elements to operate outside the unit boundaries and beyond the forward line of own troops. Sustainers also take into account operational control measures to include passage of lines with maneuver forces in perimeter defense. Finally, sustainment leaders identify sustainment forces that will support the defense reserve force in all types of defense tasks. Commanders determine what risk is acceptable in attaching sustainment units to that reserve force.

Large-scale defensive operations also place a burden on medical resources due to the magnitude and lethality of forces involved. Medical units anticipate large numbers of casualties in a short period of time due to the capabilities of modern conventional weapons and the possible employment of weapons of mass destruction. These mass casualty situations can exceed the capabilities of organic and DS medical assets. To mitigate this risk, planners should anticipate the possibility for mass casualty situations and coordinate with area support medical units to help absorb the acute rise in battlefield injuries. Unit commanders must plan for and ensure the availability of casualty evacuation assets to augment the available ambulances in the event of a mass-casualty situation.

(Sustainment) II. Large-Scale Combat 7-19

V. Large-Scale Offensive Operations

Ref: FM 4-0, Sustainment (Jul '19), chap. 7.

Sustainment commanders and their staffs prepare to support each offensive tasks. Sustainment determines the depth, duration, and endurance of Army operations, and plays a key role in enabling decisive action. Failure to provide adequate sustainment during offensive operations can result in a tactical pause, culmination of offensive operations, and prevent consolidation of gains. Operational and sustainment planners at each echelon of command work closely to synchronize sustainment support to allow commanders the freedom of action to maneuver and provide extended operational reach for the offense.

See pp. 1-118 to 1-126 for discussion of offensive operations from FM 3-0.

Sustaining Offensive Operations

- Continually update running estimates
- Support offense and consolidate gains simultaneously
- Understand enemy threat and challenges
- Increased class III (bulk and package) and class IX requirements
- Increased casualties and personnel replacements over extended battlefield

Sustainment Fundamentals

Offensive tasks involve an intense operational tempo, requiring sustainers to continually update their running estimates to anticipate friction points on the battlefield. Sustainers need to be able to accurately envision the offensive operation in time and space to accurately forecast operational requirements. Continuous coordination between planners at the various echelons is required for mission success.

If offensive momentum is not maintained, the enemy may recover from the shock of the first assault, gain the initiative, and mount a successful counterattack. Maintaining an understanding of offensive operations and future operations allows sustainment planners to simultaneously transition between offensive operations and the consolidation of gains. What starts out as a movement to contact could rapidly turn into a lengthy pursuit of enemy forces requiring extended operational reach to capitalize on opportunities. This requires robust planning and consideration for all possible outcomes.

Offensive operations require situational understanding of the enemy threat. Sustainment commanders should not assume unobstructed LOCs and should anticipate challenges across multiple domains. These commanders prepare for challenges of degraded sustainment systems, interdicted LOCs, and challenges from an enemy that has equal or overmatch capabilities. Sustainment commanders and planners prepare to push forward critical supplies in an OE where degraded systems and communications exist.

If the force is to maintain the initiative and combat power necessary for the successful performance of offensive tasks, the continued forward movement of units and sustainment support is critical. Maintaining the initiative in the close area often results in significant numbers of bypassed enemy forces and remnants of defeated units as friendly forces maneuver deep into enemy areas by avoiding enemy units in well prepared positions. The fluidity and rapid tempo of operations pose challenges when planning for the area security of support and consolidation areas.

Enemy commanders look for opportunities to counter or at least hinder the performance of corps and division offensive tasks. Enemy commanders attempt to strike deeply into friendly support and consolidation areas using multiple combinations of lethal and nonlethal effects from multiple domains. The enemy will seek to employ special purpose forces, irregular forces, electronic warfare, long-range artillery, rockets, missiles,

7-20 (Sustainment) II. Large-Scale Combat

information capabilities, and cyberspace electromagnetic activities to disrupt sustainment activities. Sustainment commanders remain aware of conventional enemy units and other elements bypassed during the advance of friendly forces and the threat presented by their presence in support and consolidation areas.

Sustainment units synchronize with maneuver units to ensure security of support and consolidation areas. Corps and division headquarters must plan to keep CPs operating, sustainment capabilities functional, respective LOCs open, and supply stocks at an acceptable level. The conduct of noncontiguous operations increases the difficulty of these tasks, as does the lack of friendly host nation security forces.

Sustainment of offensive tasks is a high-intensity operation. Sustainment commanders and staffs plan for increased requirements in class III (B), IX items, and personnel replacements to sustain the pace and tempo of operations. Plan and rehearse command and control, forward positioning, orders issuance, personnel accounting, logistical support, processing, and transportation of replacements, and most critically maneuver unit rapid integration of replacements. Sustainment planners anticipate where the greatest need might occur during offensive operations. Planners consider positioning sustainment units in close proximity to operations to reduce response times for critical support. Planners also consider alternative methods for delivering sustainment in emergencies. Extended LOCs require analysis of how to best emplace forward sustainment elements to support the commander. It is important to clearly lay out key actions for rehearsing offensive operations for example casualty evacuation routes, logistics release points, support area displacement times and locations, detainee collection points and holding areas, and fuel and ammunition resupply points to foresee potential problems and means to mitigate them.

Risks during Large-Scale Combat

Risk, uncertainty and chance are inherent in all military operations. Sustainment professionals must seek to understand, balance and take risks rather than avoid risks to ensure sustainment of the operational force. Sustainment commanders must assess and mitigate risk continuously throughout large-scale combat operations. The following is a sample list of risk considerations during large-scale offensive operations:

- Are sustainment forces properly dispersed and camouflaged? Are movements in and out of sustainment areas coordinated to avoid drawing attention to the area?

- Does the force have a sufficient number of mobile fueling vehicles to maintain offensive momentum? At what point will a loss of tankers cause mission failure?

- Are sufficient quantities of the correct class V available for rapid replenishment? Are munition dumps established forward and their contents dispersed?

- Are sustainment systems hardened against cyber-attack? How do you validate requirements received through electronic systems? Does the threat have the capability to change information verses directed denial of service attacks?

- Do medical units have sufficient class VIII to address mass casualty events?

- Are sufficient recovery vehicles available and placed to support rapid transportation of disabled vehicles to maintenance collection points?

- Does the enemy have plans to leave stay behind forces to interdict sustainment lines of supply? Do friendly forces have sufficient EOD assets available and positioned to remove enemy ordnance or IEDs emplaced on the MSRs?

- Are reinforcements available by skill/grade, and accessible in sufficient quantity to replace losses and maintain units at strength? Which units are the resourcing priority at what points during the operation?

(Sustainment) II. Large-Scale Combat 7-21

Current sustainment systems possess vulnerabilities and connectivity requirements that may make them susceptible to disruption and deliberate targeting by threat forces, both lethally and non-lethally. To mitigate this vulnerability and maintain an accurate readiness COP, organizations develop the rhythm of military operations, data cut-off times, as-of times, and reporting times. Commanders and staffs also balance the timeliness and potential latency of reporting with the amount of time needed to analyze data when evaluating unit readiness and combat capability.

Sustainment enterprise resource planning systems and associated decision support tools help provide near real-time status with minimal staff effort required to gather and display information from multiple databases. Integrating this information with command and control systems is crucial to give the sustainment leaders and supported commanders and staffs the identical current COP. The value of integrated sustainment information systems and command and control systems is that everyone on the network can see and use the same reported information to plan and control operations.

Sustainment Rehearsals

Sustainment rehearsals are critical to synchronization and the success and accomplishment of the mission. Conducting sustainment rehearsals immediately after the combined arms rehearsal ensures understanding and synchronization of the unit's maneuver and sustainment plan as it traverses the battlefield. It is critical that the combined arms team and all elements of sustainment are represented and participate in sustainment rehearsals to ensure all sustainment commodities understand how these integrate with other elements of sustainment to accomplish the mission. The sustainment rehearsal helps synchronize the sustainment warfighting functions with the other warfighting functions to create a common understanding of the plan.

VI. Support Area

The support area is a smaller, subordinate AO inside the commander's overall AO. The support area is normally, but not always, positioned within and surrounded by the consolidation area. It is where most of an echelon's sustaining operations occur. The geographic size of a support area is based on mission and operational variables and is difficult to quantify. These variables include the number of units assigned to the support area, the existing threat, and the amount of terrain that can be influenced by the unit assigned support area responsibility. As an example for a division support area, if it is assumed to be a brigade-sized area, it will be approximately 10 square kilometers. This number is for general planning consideration and to give readers an idea of the geographic scope of a division support area and the impact it has on command and control and protection. It should be understood that division support area size may vary widely. The corps support area will be significantly larger.

See related discussion on p. 1-58 of support areas.

Within the joint security area, strategic enablers such as USTRANSCOM, USAMC, DLA and each of their individual subordinate components link strategic support activities with theater support activities. Examples of these activities include synchronizing strategic and operational distribution of equipment, supplies and personnel; managing materiel and establishing contracts, establishing theater fuel farms and managing excess property turn-in. USASOC coordinates operational support requirements while monitoring SOF activities within the theater. The TSC, ESC and its attached sustainment brigade conduct RSOI for units arriving in theater and support the movement of those units forward to corps and division areas. MEDCOM (DS) provides command and control of all EAB medical units providing direct or GS to the corps and division areas. Other sustainment forces in the joint security area support activities including -classes I and III (Bulk) distribution, APOD and SPOD operations, personnel services, financial management activities, and other support tasks.

7-22 (Sustainment) II. Large-Scale Combat

Within a division support area, a MEB is normally designated to control the area. If a MEB is not available, a BCT must be designated to control the area. For the corps support area, a MEB is normally designated to control the area. If a MEB is not available, a corps MP BDE with augmentation can control the corps support area. Support area control responsibilities include area security, terrain management, information collection, integration, and synchronization, civil affairs operations, movement control, mobility support, and clearance of fires, personnel recovery, airspace control, and minimum-essential stability tasks. This allows sustainment units to focus on their primary functions.

The corps headquarters is likely to position assets in the division support area to facilitate division operations and enable freedom of action. The division headquarters orchestrates the sustainment and protection tasks essential to ensuring freedom of action in the division close and deep areas. Planning in the support area largely influences current and future operations in the deep, close, and consolidation areas. The support area is not a single large base. It is a base cluster comprised of multiple bases, each established by units assigned to the support area. The MEB is responsible for terrain management to include placement and integrated protection of the bases.

Depending on the situation, including the threat, size of the support area, and number of units within the support and consolidation areas, division and corps commanders may employ a support area command post (SACP) in the support area to assist in controlling operations. The SACP enables division and corps commanders to exercise command and control over disparate functionally focused elements operating within the support and consolidation areas that may exceed the effective span of control of the MEB or division and corps main CPs.

At the corps level, the SACP is not a separate section in the corps' table of organization and equipment. The corps commanders form the SACP in the support area from the equipment and personnel from the main and tactical CPs when required. Normally, the deputy corps commander leads the corps SACP.

For the division, the SACP is organic to the division headquarters. The SACP consists of 14 personnel and has a command group, a movement and maneuver section, and a sustainment section. The DSB has a limited role in establishing the SACP. The DSB must however coordinate with the SACP for support prioritization. The primary role of the SACP is to provide command authority and general officer oversight of division support area operations and sustainment, medical and other division support activities. The SACP performs tasks and functions, as defined by the commander, based on operational & mission variables. The division SACP in the support area normally co-locates with the MEB, which provides the CP with signal connectivity, life support, security and workspace. Functions of the CP include planning and directing sustainment, terrain management, movement control, and area security. When augmented by the MEB staff, the CP may also plan and control combined arms operations with units under division or corps control, manage airspace, and employ fires.

Sustainment forces prepare for the various threats in the support area. Threats in the division support area are categorized by the three levels of defense required to counter them. Any or all threat levels may exist simultaneously in the division support area. All threats pose potential risks to sustainment and other support operations. Emphasis on base defense and security measures may depend on the anticipated threat level. A Level I threat is a small enemy force that can be defeated by those units normally operating in the echelon support area or by the perimeter defenses established by friendly bases and base clusters. A Level I threat for a typical base consists of a squad-sized unit or smaller groups of enemy soldiers, agents, or terrorists. Typical objectives for a Level I threat include supplying themselves from friendly supply stocks, disrupting friendly command and control nodes and logistics facilities, and interdicting friendly LOCs.

(Sustainment) II. Large-Scale Combat 7-23

VII. Reconstitution Operations

Ref: FM 4-0, Sustainment (Jul '19), pp. 5-35 to 5-37.

The scale, complexity, and increased destructive power of large-scale combat operations assumes an all-encompassing multi-domain fight, potentially resulting in the greater loss of personnel, weapon platforms, supplies, and equipment of our warfighting formations. Restoring combat power to the levels necessary, within a limited window of time to continue the fight, is the objective of normal, day-to-day sustainment actions, specifically personnel replacement operations, rebuild and maintain units at strength. Under exceptional conditions, with severely degraded units, constrained time, and limited or no personnel replacements, commanders may make the decision to execute reconstitution. Reconstitution operations are extraordinary actions that commanders take to restore degraded units to an acceptable level of combat effectiveness as determined by the commander, commensurate with mission requirements and available time and resources. Reconstitution must be planned and resourced during operations to shape and prevent. The commander directing the reconstitution mission uses assets under their control, along with those provided by higher echelons.

Reconstitution is not a sustainment operation, although sustainment plays an integral part. Reconstitution is an operational event enabled by all warfighting functions. Activities such as personnel and equipment restoration, health service support to restore combat power, and collective training, are involved to ensure combat readiness. The reconstitution effort should be thoroughly planned and understood by all involved to ensure success. Any maneuver, maneuver support, or sustainment unit may require reconstitution. Therefore, planners at all levels of command must have a contingency plan prepared to execute reconstitution operations, when necessary.

Two Types of Reconstitution

Reconstitution operations consists of two elements: reorganization and regeneration.

Reorganization involves the cross leveling of forces and resources following combat operations to increase combat effectiveness of an attrited unit.

Regeneration involves the large-scale replacement of personnel, weapon platforms, equipment, and supplies; as well as the reestablishment and/or replacement of the chain of command, and the conduct of mission essential training. Regeneration operations take place at EAB, using corps, theater and strategic capabilities and resources to enable restoration of combat power.

For additional information, refer to FM 4-0, appendix C.

Regeneration, being the more complex of the reconstitution elements, returns degraded Army units into large-scale combat operations. Planning for reconstitution operations is necessary prior to beginning of large-scale combat operations. A critical aspect of regeneration is that of decision authority to direct units to conduct regeneration operations. Regeneration will require the removal of attrited units from the battlefield to an area that allows regeneration activities out of contact. Commanders must balance the resources required to move attrited units out of combat versus the overall mission, time constraints, and resources required.

Operational readiness thresholds that trigger both entry and exit criteria for regeneration are critical in determining the scope of operations. Operational readiness thresholds are measures used to determine a units capability to perform the missions or functions for which it is organized or designed. Thresholds to enter or exit regeneration must be determined by an operational. At the theater level, the operational plans should address combat power degradation thresholds to aid the commander in determining regeneration operations. Commanders must also consider further degradation of a unit during retrograde operations.

7-24 (Sustainment) II. Large-Scale Combat

Regeneration Task Force (RTF)

During large-scale combat operations, regeneration operations are resource intensive and METT-TC dependent. Designation of a specialized regeneration task force (RTF) allows for the rapid (days to weeks) execution of large-scale replacement operations. The RTF provides the synchronization of command and control, security, and sustainment support activities to a specified location, either within the corps consolidation area, joint security area, or outside of the theater, to rebuild a degraded unit's combat power.

The RTF predetermines maintenance collection points suitable for Heavy Equipment Transporter System on and/or off-loading during hours of darkness with limited time available. Equipment towed or carried from the battle area goes to a central location that facilitates sorting of equipment. The RTF determines special requirements for vehicles that require special road clearances prior to beginning regeneration.

Sustainment commanders must consider time constraints required to support regeneration versus sustainment of large-scale combat operations. Utilization of sustainment assets to support regeneration operations may affect the execution of distribution operations of classes I, III, and V supplies. The RTF may need to request echelon above brigade transportation assets to support regeneration.

Other sustainment planning considerations are as follows—

- Regeneration of units is an exceptionally intensive form of reconstitution requiring the direction of maneuver commanders two levels up, and substantial sustainment support.
- Attrited units are not able to support any sustainment requirements.
- Planners and the RTF should request equipment from theater reserve stocks.
- Movement control support should be requested by the RTF to assist with planning the attrited unit's move.
- Planners should standardize information required by the RTF from an attrited unit.
- Reconstitution plans should reflect procedures for managing replacements of Soldiers in attrited units.
- Reconstitution plans should include combat stress counselors and psychologists in the RTF.
- Planners should plan for the recovery of an attrited unit's disabled vehicles and crews from forward areas to the regeneration site.
- Planners should prepare for attrited units arriving to the regeneration site with a form of chain of command. However, regardless of how the attrited unit is organized, the RTF regenerates all the entities within the attrited unit.
- Planners should ensure the RTF should has a 24-hour operating capability.
- The regeneration site should be as close as possible to the attrited unit location to minimize amount of time and difficulty reaching the site.
- The site should be far enough away to prevent easy enemy interdiction and provide the safest feasible location for personnel executing regeneration operations.
- The site should contain enough space to conduct regeneration operations.
- Planners should be prepared to coordinate with the unit's parent organization. For example, the RTF may be required to coordinate a division property book office to establish correct hand receipts.
- Planners and the RTF should plan for signal support when conducting reconstitution in austere locations.
- Planners and the RTF should request decontamination support and prepare to handle decontaminated personnel and equipment.

(Sustainment) II. Large-Scale Combat 7-25

Units must also employ cover and concealment to prevent observation and detection of sustainment equipment and bases by reconnaissance elements, reconnaissance aircraft, drones, or attack aircraft. Cover and concealment includes signature management and emissions control. Dispersion aids in concealment and limits destruction and losses in the event of an attack. Units at all echelons must conduct CBRN defense preparation.

At higher echelons military police units from the MEB enhance protection capabilities by performing area security within the support areas. These units perform response-force operations to defeat Level II threats against bases and base clusters located in that support area. A Level II threat is an enemy force or activities that can be defeated by a base or base cluster's defensive capabilities when augmented by a response force (ATP 3-91). These units maintain contact with Level III threats in the division support area until the tactical combat force under the MEB's control can respond. Level III threat is an enemy force or activities beyond the defensive capability of both the base and base cluster and any local reserve or response force.

Corps and division commanders designate close, deep, support, and consolidation areas to describe the physical arrangement of forces in time, space, and focus. Commanders should always designate a close area and a support area, and designate a deep area and consolidation area as required. The consolidation area does not necessarily need to surround, nor contain, the support area base clusters, but typically, it does. The consolidation area requires a purposefully task organized, combined arms unit to conduct area security and stability tasks and employ and clear fires.

The consolidation area is the portion of the commander's area of operations that is designated to facilitate the security and stability tasks necessary for freedom of action in the close area and to support the continuous consolidation of gains (ADP 3-0). Corps and division commanders may establish a consolidation area, particularly in the offense as the friendly force gains territory, to exploit tactical success while enabling freedom of action for forces operating in the other areas. When designated, a consolidation area refers to an AO assigned to an organization that extends from its higher headquarters boundary to the boundary of forces in close operations where forces have established a level of control and large-scale combat operations have ceased.

Army forces consolidate gains as part of a combat operation to enable combat power for continued action against remaining enemy forces in support of a host nation and its civilian population, or as part of the pacification of a hostile state. These gains may include the establishment of population security temporarily by using the military as a transitional force, the relocation of displaced civilians, re-establishment of law and order, and restoration of key infrastructure. Concurrently, corps and divisions must be able to accomplish these activities while sustaining, repositioning, and reorganizing subordinate units to continue operations in the close and deep area.

Consolidation of gains activities may encompass a lengthy period of post-conflict operations prior to redeployment. Consolidation of gains may occur even if large-scale combat operations are occurring in other parts of an AO to exploit tactical success. Anticipation and early planning for activities after large-scale combat operations ease the transition process.

Commanders address the decontamination, disposal, and destruction of war materiel. Commanders must also address the removal and destruction of unexploded ordnance and the responsibility for demining operations. (The consolidation of friendly and available enemy mine field reports is critical to this mission.) In addition, if support areas are surrounded by a consolidation area, the higher echelon headquarters must clearly articulate the roles and responsibilities for controlling supply routes and clearance procedures. Additionally, the theater Army is prepared to provide medical support, emergency restoration of utilities, support to social needs of the indigenous population, and other humanitarian activities.

7-26 (Sustainment) II. Large-Scale Combat

Chap 8

Protection Warfighting Function

Ref: ADP 3-37, Protection (Jul '19), ADP 3-0, Operations (Jul '19), p. 5-7 and FM 3-0, Operations (Oct. '22), p. 2-3.

Protection safeguards friendly forces, civilians, and infrastructure and is inherent to command. The protection warfighting function enables the commander to maintain the force's integrity and combat power through the integration of protection capabilities throughout operational preparation, operations to shape, operations to prevent, large-scale ground combat operations, and operations to consolidate gains.

I. The Protection Warfighting Function

The protection warfighting function is the related tasks, systems, and methods that prevent or mitigate detection, threat effects, and hazards to preserve combat power and enable freedom of action. Protection encompasses everything that makes Army forces hard to detect and destroy. Protection requires commanders and staffs to understand threats and hazards throughout the operational environment, prioritize their requirements, and commit capabilities and resources according to their priorities. Commanders balance their protection efforts with the need for tempo and resourcing the main effort. They may assume risk in operations or areas that may be vulnerable, but that are considered low enemy priorities for targeting or attack. Commanders account for threats from space, cyberspace, and outside their assigned area of operations (AO) as they develop protection measures. Protection results from many factors, including operations security, dispersion, deception, survivability measures, and the way forces conduct operations.

Planning, preparing, executing, and assessing protection is a continuous and enduring activity. Defending networks, data, and systems; implementing operations security; and conducting security operations contribute to information advantages by protecting friendly information. Prioritization of protection capabilities is situationally dependent and resource informed.

The protection warfighting function includes the following tasks:

- Conduct survivability operations.
- Provide force health protection.
- Conduct chemical, biological, radiological, and nuclear operations.
- Provide explosive ordnance disposal support.
- Coordinate air and missile defense support.
- Conduct personnel recovery.
- Conduct detention operations.
- Conduct risk management.
- Implement physical security procedures.
- Apply antiterrorism measures.
- Conduct police operations.
- Conduct population and resource control.
- Conduct area security.
- Perform cyberspace security and defense.
- Conduct electromagnetic protection.
- Implement operations security.

See pp. 8-4 to 8-5 for an overview and discussion of primary protection tasks.

(Protection) Warfighting Function 8-1

Protection (ADP 3-37)

II. Protection (Overview)

Ref: ADP 3-37, Protection (Jul '19), pp. 1 to 2.

Protection is the preservation of the effectiveness and survivability of mission-related military and nonmilitary personnel, equipment, facilities, information, and infrastructure deployed or located within or outside the boundaries of a given operational area (JP 3-0). Protection serves as an Army warfighting function. A shared understanding of the joint protection function (see JP 3-0) enables Army leaders to integrate the Army's protection warfighter function with unified action partners. Army leaders must anticipate that joint support will be limited in larger-scale ground combat operations and must protect the force utilizing a combination of measures. The joint protection function focuses on preserving the joint force fighting potential in four primary ways:

- Active defensive measures to protect friendly forces, civilians, and infrastructure.
- Passive defensive measures to make friendly forces, systems, and facilities difficult to locate, strike, and destroy when active measures are limited or unavailable.
- The application of technology and procedures to reduce the risk of fratricide.
- Emergency management and response to reduce the loss of personnel and capabilities due to accidents, health threats, and natural disasters

Protection is not linear – planning, preparing, executing, and assessing protection is continuous and enduring. The protection warfighting function tasks are incorporated into the operations process in a comprehensive, layered, and redundant approach to achieve enduring force protection. Protection preserves capability, momentum, and tempo which are important contributors to operational reach. Synchronizing, integrating, and organizing protection capabilities and resources throughout the operations process preserves combat power and mitigates the effects of threats and hazards to enable freedom of action.

Role of Protection

Army forces gain, sustain, and exploit control over land to deny its use to an enemy. They do this with combined arms formations, possessing the mobility, firepower, and protection to defeat an enemy and establish control of areas, resources, and populations. Military activities and operations are inherently hazardous. Commanders and leaders conducting unified land operations must accept prudent risks every day based on the significance of the mission, the demand of the operation, and opportunity. In warfare, this reality defines the sacred trust that must exist between leaders and Soldiers regarding mission accomplishment and force protection. Force protection is preventive measures taken to mitigate hostile actions against Department of Defense personnel (to include family members), resources, facilities, and critical information (JP 3-0). A commander's inherent duty to protect the force should not lead to risk aversion or inhibit the freedom of action necessary for maintaining initiative and momentum or achieving decisive results during operations. Leaders must balance these competing responsibilities and make risk decisions based on experience, ethical and analytical reasoning, knowledge of the unit, and the situation.

Commanders and staffs synchronize, integrate, and organize capabilities and resources to preserve combat power and identify and prevent or mitigate the effects of threats and hazards. Protection integrates all protection capabilities to safeguard the force, personnel (combatants and noncombatants), systems, and physical assets of the United States and its mission partners. In addition to the primary protection task, commanders and staffs must coordinate, synchronize, and integrate additional protection capabilities and resources of unified action partners.

The goal of protection integration is to balance protection with the freedom of action throughout the duration of military operations. This is achieved by integrating reinforcing or complementary protection capabilities to mitigate or assume risk for identified and prioritized vulnerabilities.

8-2 (Protection) Warfighting Function

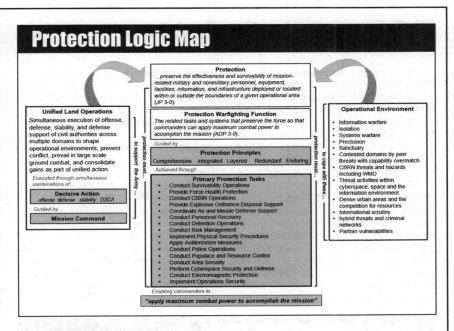

Ref: ADP 3-37 (Jul '19), fig. 1-1. Protection logic map.

Protection in Support of Army Operations

The Army accomplishes its mission by supporting the joint force in four strategic roles: shape OEs, prevent conflict, conduct large-scale ground combat, and consolidate gains. The strategic roles clarify the enduring reasons for which the United States (U.S.) Army is organized, trained, and equipped. The Army conducts operations across multiple domains and the information environment. All Army operations are multi-domain operations, and all battles are multi-domain battles. Multi-domain operations include airborne and air assault operations, air and missile defense, fire support, aviation, cyberspace electromagnetic activities, information operations, space operations, military deception, and information collection. Large-scale ground combat operations such as these entail significant operational risk, synchronization and capabilities convergence, and high operational tempo. Protection is a key consideration for operating in multiple domains.

Protection emphasizes the importance of planning and expanding protection priorities, to include protecting mission partners, civilian populations, equipment, resources, infrastructure, and cultural landmarks across the range of military operations. The synchronization, integration, and organization of protection capabilities and resources to preserve combat power from the effects of threats and hazards are essential. When properly integrated and synchronized, the tasks and systems that relate to protection effectively protect the force, preserve combat power, and increase the probability of mission success.

Protection is achieved through commanders and their units by changing tempo, taking evasive action, or maneuvering to gain positional advantage in relation to a threat. Formations often derive protection by exploiting terrain and weather conditions, or by using the cover of darkness to mask movement. The use of key physical terrain features supports protection measures and complements the positioning of forces during planning. The ability to protect and preserve the force and secure the AO is vital in seizing, retaining, and exploiting the initiative to shape the OE, prevent conflict, consolidate gains, and win wars as a part of unified action.

III. Primary Protection Tasks

Ref: ADP 3-37, Protection (Jul '19), table 1-1, pp. 1-5 to 1-6.

The Army accomplishes its mission by supporting the joint force in four strategic roles: shape OEs, prevent conflict, conduct large-scale ground combat, and consolidate gains. The strategic roles clarify the enduring reasons for which the United States (U.S.) Army is organized, trained, and equipped. The Army conducts operations across multiple domains and the information environment. All Army operations are multi-domain operations, and all battles are multi-domain battles. Multi-domain operations include airborne and air assault operations, air and missile defense, fire support, aviation, cyberspace electromagnetic activities, information operations, space operations, military deception, and information collection. Large-scale ground combat operations such as these entail significant operational risk, synchronization and capabilities convergence, and high operational tempo. Protection is a key consideration for operating in multiple domains.

Protection emphasizes the importance of planning and expanding protection priorities, to include protecting mission partners, civilian populations, equipment, resources, infrastructure, and cultural landmarks across the range of military operations. The synchronization, integration, and organization of protection capabilities and resources to preserve combat power from the effects of threats and hazards are essential. When properly integrated and synchronized, the tasks and systems that relate to protection effectively protect the force, preserve combat power, and increase the probability of mission success.

Table 1-1 below outlines a complete list of primary protection tasks:

Primary Protection Tasks

Primary Protection Task	Tasks
Conduct Survivability Operations (ATP 3-37.34, ATP 3-34.20, ATP 3-90.4)	• Employ camouflage, cover, and concealment • Establish fighting positions • Harden facilities • Construct protective works for explosive hazards • Clear routes and areas of explosive hazards
Provide Force Health Protection (ATP 4-02.8)	• Provide preventive medicine • Provide veterinary service • Provide combat and operational stress control • Provide preventive dentistry • Conduct area medical laboratory support
Conduct CBRN Operations (FM 3-11, ATP 3-11.32, ATP 3-11.36)	• Conduct CBRN information collection • Provide hazard awareness and understanding • Conduct CBRN defense • Respond to CBRN events
Provide EOD Support (ATP 4-32.1, ATP 4-32.3)	• Identify and collect information on explosive ordnance and hazards • Render-safe and dispose of explosive ordnance and hazards • Post blast investigations
Coordinate Air and Missile Defense Support (FM 3-01, FM 3-27)	• Plan ballistic missile defense • Conduct ballistic missile defense • Manage system configuration • Perform asset management • Provide global missile defense capabilities
Conduct Personnel Recovery (FM 3-50)	• Apply the fundamentals of Army personnel recovery • Support the Commander and Staff Organization and Operations • Coordinate Knowledge Management • Manage the Unit Personnel Recovery Program

8-4 (Protection) Warfighting Function

Primary Protection Tasks (Continued)

Primary Protection Task	Tasks
Conduct Detention Operations (FM 3-63)	• Conduct detainee operations • Confine U.S. military prisoners • Conduct host nation corrections training and support
Conduct Risk Management (ATP 5-19)	• Identify hazards • Assess hazards • Develop controls • Implement controls • Supervise and evaluate
Implement Physical Security Procedures (ATP 3-39.32)	• Deter • Detect • Assess • Delay • Respond
Apply Antiterrorism Measures (ATP 3-37.2)	• Establish an antiterrorism program • Collect, analyze, and disseminate threat information • Assess and reduce critical vulnerabilities • Increase antiterrorism awareness • Maintain defenses • Establish civil-military partnerships • Conduct terrorist threat/incident response planning • Conduct exercises, and evaluate/assess the plan
Conduct Police Operations (ATP 3-39.10)	• Perform law enforcement • Conduct criminal investigations • Conduct traffic management and enforcement • Employ forensics capabilities • Conduct police engagement • Conduct customs operations • Support host nation police development • Conduct civil law enforcement support • Conduct border control and boundary security support
Conduct Populace and Resources Control (ATP 3-39.30, ATP 3-57.10, ATP 3-07.6)	• Support dislocated civilian operations • Support noncombatant evacuation operations • Protect and maintain critical infrastructure • Enforce resource control measures
Conduct Area Security (JP 3-10, ADP 3-90)	• Conduct area and base security operations • Conduct critical installation and facilities security • Provide protective services for selected individuals • Conduct response force operations • Secure supply routes and convoys • Conduct support area operations • Establish local security
Perform Cyberspace Security and Defense (JP 3-12, FM 3-12, FM 6-02, ATP 6-02.71)	• Perform cybersecurity activities • Conduct defensive cyberspace operations-internal defensive measures
Conduct Electromagnetic Protection JP 6-01, FM 3-12, ATP 6-02.70)	• Conduct electronic protection actions • Conduct defensive electronic attack • Conduct electromagnetic spectrum management
Implement Operations Security (JP 3-13.3, ATP 3-13.3)	• Conduct operations security • Analyze threat • Implement measures and counter measures

(Protection) Warfighting Function 8-5

IV. Survivability (One of the Five Dynamics of Combat Power)

Ref: FM 3-0, Operations (Oct. '22), pp. 2-3 to 2-6, ADP 3-37, Protection (Jul '19), 2-1 to 2-2 and ADP 3-90, Offense and Defense (Jul '19), p. 4-16.

Combat power is the total means of destructive and disruptive force that a military unit/ formation can apply against an enemy at a given time (JP 3-0). It is the ability to fight. The complementary and reinforcing effects that result from synchronized operations yield a powerful blow that overwhelms enemy forces and creates friendly momentum. Army forces deliver that blow through a combination of five dynamics: leadership, firepower, information, mobility and survivability. See pp. 2-4 to 2-6 for further discussion.

Survivability

Survivability is a quality or capability of military forces which permits them to avoid or withstand hostile actions or environmental conditions while retaining the ability to fulfill their primary mission (ATP 3-37.34). It represents the degree to which a formation is hard to kill. Survivability is relative to a unit's capabilities and the type of enemy effects it must withstand, its ability to avoid detection, and how well it can deceive enemy forces. Survivability is also a function of how a formation conducts itself during operations. For example, an infantry BCT's survivability against indirect fire is contingent on it not being detected, being dispersed, digging in, and adding overhead cover when stationary. An armor BCT's survivability is a function of logistics, security, and avoiding situations that constrain its mobility or freedom of action.

Leaders assess survivability as the ability of a friendly force to withstand enemy effects while remaining mission capable. Armor protection, mobility, tactical skill, avoiding predictability, and situational awareness contribute to survivability. Enforcement of operations security techniques and avoiding detection while initiating direct fire contact on favorable terms also increases survivability. Situational awareness regarding the nine forms of contact and minimizing friendly signatures contributes to survivability.

To increase survivability, units employ air defense systems, reconnaissance and security operations, modify tempo, take evasive action, maneuver to gain positional advantages, decrease electromagnetic signatures, and disperse forces. Dispersed formations improve survivability by complicating targeting and making it more difficult for enemy forces to identify lucrative targets. Tactical units integrate procedures for the use of camouflage, cover, concealment, and conducting electromagnetic protection—including noise and light discipline. During large-scale combat operations, survivability measures may include radio silence, communication through couriers, or alternate forms of communication. Space-based missile warning systems provide early warning of adversary artillery and missile attacks, allowing friendly forces to seek cover. Application of chemical, biological, radiological, and nuclear (CBRN) defense measures increase survivability in CBRN environments.

Conduct Survivability Operations

Survivability is a quality or capability of military forces which permits them to avoid or withstand hostile actions or environmental conditions while retaining the ability to fulfill their primary mission (ATP 3-37.34). Personnel and physical assets have inherent survivability qualities or capabilities that can be enhanced through various means and methods. These qualities are especially important where elements that are targeted by threats and other protection capabilities are in limited supply. Survivability and survivability operations are not interchangeable. Survivability refers to a quality or capability, while survivability operations are a specific group of tasks that enhance survivability.

8-6 (Protection) Warfighting Function

Protection (ADP 3-37)

Units conduct survivability within the limits of their capabilities. When existing terrain features offer insufficient cover and concealment, altering the physical environment to provide or improve cover and concealment enhances survivability. Similarly, using natural or artificial materials such as camouflage may confuse or mislead the enemy or adversary. Together, these are called survivability operations—those protection activities that alter the physical environment by providing or improving camouflage, cover, and concealment. While such activities often have the added benefit of providing shelter from the elements, survivability focuses on providing camouflage, cover, and concealment. Movement, such as rapid dispersal, is used with cover and concealment to enhance protection.

Survivability operations enhance the ability to avoid or withstand hostile actions by altering the physical environment. They accomplish this by providing or improving camouflage, cover, and concealment via the following four tasks:

- Employing camouflage, cover, concealment, and movement.
- Constructing fighting positions.
- Hardening facilities.
- Constructing protective positions.

Constructing survivability positions against threats from indirect and direct fire may also require actions to protect forces and equipment from other explosive hazards. Enemy forces employ moving or mobile improvised explosive devices (vehicle borne, personnel-borne, airborne, and waterborne) against stationary targets in a unit's area of operation. Units harden structures against the effects of mobile improvised explosive devices by creating standoff distance between a vehicle-borne improvised explosive device attack against a high occupancy structure such as a living area or headquarters. Captured enemy ammunition and bulk explosives located in the area of operations and under friendly force control may require protective barriers constructed around them or they require manpower to assist in proper disposal using demolitions. These actions to construct barriers, walls, shields, or berms enhance a unit's protection.

Considerations in the Defense

An attacking enemy force usually has the initiative. A defending commander must take a wide range of actions to reduce the risk of losses, including developing a survivability plan. Survivability in the defense prioritizes hardening command posts, artillery positions, air and missile defenses, and other critical equipment and supply areas. It also includes preparing individual, crew-served, and combat vehicle fighting positions.

To avoid detection and destruction by enemy forces, units move frequently and quickly establish survivability positions. To provide flexibility, units may need primary, alternate, and supplementary positions. Units enhance their survivability using concealment, military deception, decoy or dummy positions, dispersion, and field fortifications. Commanders increase security during defensive preparations because an enemy force will attack lightly defended areas whenever possible.

Survivability tasks include using engineer equipment to help in constructing trenches, command post shelters, and artillery, firing, radar, and combat vehicle fighting positions. Commanders use dispersion to limit the damage done by enemy attacks. Enemy forces should never be able to put a unit out of action with just a single attack. Dispersed troops and vehicles force attacking forces to concentrate on a single small target that may be missed. The wider the dispersion of unit personnel and equipment is, the greater potential for limiting damage it has. Commanders protect supply stocks against blast, shrapnel, incendiaries, and CBRN contamination using dispersion and constructing survivability positions. Units also use cover to limit the amount of damage and casualties that they can receive because of an enemy attack.

Refer to ATP 3-37.34 for information on survivability and ATP 3-34.20 for information on countering explosive hazards.

(Protection) Warfighting Function 8-7

V. Protection Integration in the Operations Process

Ref: ADP 3-37, Protection (Jul '19), fig. 1-2, p. 1-4.

Protection is integrated throughout the operations process to provide a synchronization of efforts and an integration of capabilities. The protection warfighting function tasks are incorporated into the process in a layered and redundant approach to complement and reinforce actions to achieve force protection.

Protection within the Operations Process

Plan
- Establish protection working group
- Conduct initial assessments
- Establish protection priorities
- Organize protection tasks
- Develop a PPL
- Develop scheme of protection
- Refine running estimate
- Synchronize protection within combat power
- Integrate protection throughout the operations process

Prepare
- Revise and refine the plan
- Emplace systems to detect threats to the PPL
- Direct OPSEC measures
- Prepare and improve survivability positions
- Liaison and coordinate with adjacent units
- Train with defended assets
- Implement vulnerability reduction measures
- Rehearse

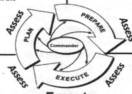

Execute
- Ensure that the protection focus supports the decisive operation
- Review and recommend changes to protection related CCIRs
- Review changes to graphic control measures and boundaries for the increased risk of fratricide
- Evaluate the effectiveness of tracking for constraints on personnel recovery
- Monitor the employment of security forces for gaps in protection or unintended patterns
- Evaluate the effectiveness of liaison personnel for protection actions
- Evaluate movement coordination and control to protect critical routes
- Monitor adjacent unit coordination procedures for terrain management vulnerabilities
- Monitor readiness rates of response forces involved in fixed-site protection
- Monitor force health protection

Ref: ADP 3-37 (Jul '19), fig. 1-2. Integration of protection throughout the operations process.

A. Plan *(See pp. 8-9 to 8-18.)* Planning is the first step toward effective protection. Commanders consider the most likely threats and hazards and decide which personnel, physical assets, and information to protect.

B. Prepare *(See pp. 8-19 to 8-22.)* During the preparation phase, protection focuses on deterring and preventing the enemy or adversary from actions that would affect combat power and the freedom of action. The implementation of protection tasks with ongoing preparation activities assists in the prevention of negative effects.

C. Execute *(See pp. 8-23 to 8-24.)* The continuous and enduring character of protection makes the continuity of protection tasks and systems essential during execution. Commanders implement control measures and allocate resources that are sufficient to ensure protection continuity and restoration.

D. Assess *(See pp. 8-25 to 8-26.)* Assessing protection is an essential, continuous activity that occurs throughout the operations process. While a failure in protection is typically easy to detect, the successful application of protection may be difficult to assess and quantify.

Chap 8
I. Protection Planning

Ref: ADP 3-37, Protection (Jul '19), chap. 2. See also p. 7-6.

Planning is the first step toward effective protection. Commanders consider the most likely threats and hazards and then decide which personnel, physical assets, and information to protect. They establish protection priorities for each phase or critical event of an operation. An effective scheme of protection and risk decisions is developed based on the information that flows from mission analysis. Mission analysis provides commanders with a better understanding of the situation and problem. Commanders and staffs identify what the command must accomplish, when and where it must be done and, most importantly, why it must be carried out—the purpose for the operation. This understanding of the situation and problem allows commanders to identify and analyze threats and hazards and develop a scheme of protection.

I. Initial Assessments

Initial protection planning requires various assessments to establish protection priorities. Assessments include threats, hazards, vulnerability, and criticality. These assessments are used to determine which assets can be protected given no constraints and which assets can be protected with available resources. There are seldom sufficient resources to simultaneously provide all assets the same level of protection. For this reason, commanders make decisions on acceptable risks and provide guidance to the staff so that they can employ protection capabilities based on the protection priorities.

Initial Assessments (Protection)

 A Threat and Hazard Assessment *(p. 8-10)*

 B Criticality Assessment *(p. 8-10)*

 C Vulnerability Assessement *(p. 8-14)*

Protection planning is a continuous process that must include an understanding of the threats and hazards that may impact operations from the deep area back to the strategic support area. Protection capabilities are aligned to protect critical assets and mitigate effects from threats and hazards. The protection cell and protection working group must prioritize the protection of critical assets during operations to shape, operations to prevent, large-scale ground combat, and during the consolidation of gains that best supports the commander's end state.

During shaping operations, the focus is on cooperation, prevention, and deterrence. Within planning, the protection staff conducts continuous assessments to understand the environment, cooperate with and support partners to build a network of

(Protection) I. Planning 8-9

protection, and safeguard the force through active and passive measures and training and exercises.

During operations to prevent, the main purpose is to deter adversary actions. The staff estimates the protection assets necessary for future operations and the increased threat according to the commander's priorities, forces availability, and the adversary's perceived intent. Units prioritize protection capabilities and align them to defend critical assets. It is imperative to conduct information operations to deny adversaries the ability to obtain information.

Planning prioritization considerations for large-scale ground combat operations includes efforts to conserve and increase combat power; protect nodes that are critical to force projection and sustainment; counter enemy fires and maneuver by making personnel, systems, and units difficult to locate, strike, and destroy; gain air, space, and electromagnetic spectrum superiority; use defensive information operations; and protect airports, seaports, lines of communication networks, and base camps.

Success in consolidating gains is obtained through setting the conditions for a stable environment. Staffs should weigh the prioritization of protection capabilities required to support large-scale and stability operations with the simultaneous protection of consolidation areas. During operations to consolidate gains, prioritization considerations are focused on security tasks to stabilize the area and protect the force, bases, routes, areas, and critical infrastructure.

An important aspect of protection planning for corps and divisions involves the support and consolidation areas. If conditions in the support area degrade, it is detrimental to the success of operations. A degraded support area also inhibits the ability to shape the deep area for the brigade combat teams involved in close operations. Therefore, the protection of support areas requires planning considerations equal to those in the close areas. When the support area is located inside a division consolidation area, the unit that is assigned responsibility for the consolidation area provides significant protection for support area units.

A. Threat and Hazard Assessment

Personnel from all staff sections and warfighting functions help conduct threat and hazard analysis. This analysis comprises a thorough, in-depth compilation and examination of information and intelligence that address potential threats and hazards in the area of operations. The integrating processes (intelligence preparation of the battlefield, targeting, and risk management) provide an avenue to obtain the threats and hazards that are reviewed and refined. Threat and hazard assessments are continuously reviewed and updated as the operational environment changes.

See following pages for an overview and further discussion of threat and hazard assessments. (Threats and hazards are defined on facing page.)

B. Criticality Assessment

A criticality assessment identifies key assets that are required to accomplish a mission. It addresses the impact of a temporary or permanent loss of key assets or the unit ability to conduct a mission. A criticality assessment should also include high-population facilities (recreational centers, theaters, sports venues) that may not be mission-essential. It examines the costs of recovery and reconstitution, including time, expense, capability, and infrastructure support. The staff gauges how quickly a lost capability can be replaced before providing an accurate status to the commander. The general sequence for a criticality assessment is—

Step 1. List the key assets and capabilities.

Step 2. Determine if critical functions or combat power can be substantially duplicated with other elements of the command or an external resource.

8-10 (Protection) I. Planning

Threats and Hazards

Ref: ADP 3-37, Protection (Jul '19), pp. 1-10 to 1-11.

The protection warfighting function preserves the combat power potential and survivability of the force by providing protection from threats and hazards.

- **Hostile actions.** Threats from hostile actions include any capability that forces or criminal elements have to inflict damage upon personnel, physical assets, or information. These threats may include improvised explosive devices, suicide bombings, network attacks, mortars, asset theft, air attacks, or CBRN weapons.
- **Nonhostile activities.** Nonhostile activities include hazards associated with Soldier duties within their occupational specialty, Soldier activity while off duty, and unintentional actions that cause harm. Examples include on- and off-duty accidents, OPSEC violations, network compromises, equipment malfunctions, or accidental CBRN incidents.
- **Environmental conditions.** Environmental hazards associated with the surrounding environment could potentially degrade readiness or mission accomplishment. Weather, natural disasters, and diseases are common examples.

Threats

The various actors in any area of operations can qualify as a threat, enemy, adversary, neutral, or friendly. Land operations often prove complex because actors intermix, often with no easy means to distinguish one from another.

- A **threat** is any combination of actors, entities, or forces that have the capability and intent to harm United States forces, United States national interests, or the homeland (ADP 3-0). Threats may include individuals, groups of individuals (organized or not organized), paramilitary or military forces, nation-states, or national alliances.
- A **peer threat** is an adversary or enemy with the capabilities and capacity to oppose U.S. forces across multiple domains worldwide or in a specific region where they have a position of relative advantage. Peer threats possess roughly equal combat power in geographical proximity to a conflict area with U.S. forces.
- An **enemy** is a party identified as hostile against which the use of force is authorized (ADP 3-0). An enemy is also called a **combatant** and is treated as such under the law of war.
- An **adversary** is a party acknowledged as potentially hostile to a friendly party and against which the use of force may be envisaged (JP 3-0)
- An **insider threat** is a person with placement and access who intentionally causes loss or degradation of resources or capabilities or compromises the ability of an organization to accomplish its mission through espionage, providing support to international terrorism, or the unauthorized release or disclosure of information about the plans and intentions of United States military forces (AR 381-12).
- A **neutral** is a party identified as neither supporting nor opposing friendly or enemy forces (ADRP 3-0)
- A **friendly** is a contact positively identified as friendly (JP 3-01)
- A **hybrid threat** is the diverse and dynamic combination of regular forces, irregular forces, terrorist forces, and/or criminal elements unified to achieve mutually benefitting effects (ADP 3-0).

Hazards

A hazard is a condition with the potential to cause injury, illness, or death of personnel; damage to or loss of equipment or property; or mission degradation (JP 3-33). Hazards are usually predictable and preventable and can be reduced through effective risk management efforts.

(Protection) I. Planning 8-11

Threat and Hazard Assessment

Ref: ADP 3-37, Protection (Jul '19), pp. 3-2 to 3-4.

Personnel from all staff sections and warfighting functions help conduct threat and hazard analysis. This analysis comprises a thorough, in-depth compilation and examination of information and intelligence that address potential threats and hazards in the area of operations. The integrating processes (intelligence preparation of the battlefield, targeting, and risk management) provide an avenue to obtain the threats and hazards that are reviewed and refined. Threat and hazard assessments are continuously reviewed and updated as the operational environment changes.

Considerations for the threat and hazard assessment include—

- Enemy and adversary threats
  - Operational capabilities
  - Intentions
  - Activities
- Foreign intelligence and security service threats (*refer to ATP 2-22.2-1*)
- Criminal activities
- Civil disturbances
- Medical and safety hazards
- CBRN weapons and toxic industrial material
- Other relevant aspects of the operational environment
- Incident reporting and feedback points of contact

The threat and hazard assessment results in a comprehensive list of threats and hazards and determines the likelihood or probability of occurrence of each threat or hazard. Table 3-1 shows examples of potential threats and hazards in an area of operations. In the context of assessing risk, the higher the probability or likelihood of a threat or hazard occurring, the higher the risk of asset loss.

Potential Threats and Hazards

Area of Concern	Potential Threats and Hazards
Survivability	• Hostile actions (lethal and nonlethal capabilities) • Weather effects or environmental conditions
Force health protection	• Endemic and epidemic diseases • Environmental factors • Diseases from animal bites, poisonous plants, animals, or insects • Risks associated with the health, sanitation, or behavior of local populace
CBRN	• CBRN weapons • Toxic industrial materials
EOD	• Explosive ordnance and hazards (friendly and enemy) • Adversary attacks on personnel, vehicles, or infrastructure
Air and missile defense	• Artillery • Mortars • Rockets • Ballistic and cruise missiles • Fixed- and rotary-wing aircraft • Unmanned aircraft systems

8-12 (Protection) I. Planning

Potential Threats and Hazards (Continued)

Area of Concern	Potential Threats and Hazards
Personnel recovery	• Events that separate or isolate individuals or small groups of friendly forces from the main force • Weather effects or environmental conditions
Detention operations	• U.S. military prisoners • Detainees
Risk management	• Hazards associated with enemy or adversary activity • Accident potential • Weather or environmental conditions • Equipment
Physical security	• Adversary attacks on personnel, vehicles, or infrastructure • Insider threats • Weather or environmental conditions • Criminals
AT	• Improvised explosive devices • Suicide/mail bombs • Snipers • Standoff weapons • Active shooters • Insider threats
Police operations	• Criminals • Active shooters • Insider threats • Bombs
Populace and resources control	• Criminals • Dislocated civilians • Insider threats
Area Security	• Adversary attacks on personnel, vehicles, or infrastructure • Explosive ordnance and hazards (friendly and enemy) • Criminals • Dislocated civilians • Unmanned aircraft systems
Cyberspace Security and Defense	• Network attacks, exploitations, intrusions or effects of malware • Known system vulnerabilities • Insider threats
Electromagnetic Protection	• Electromagnetic energy, directed energy, and antiradiation weapons • Electromagnetic interference, jamming • Electromagnetic deception
Operations Security	• Enemy collection plans • Friendly vulnerabilities • Indicators

Ref: ADP 3-37 (Jul '19), table 3-1. Potential threats and hazards.

(Protection) I. Planning **8-13**

Step 3. Determine the time required to substantially duplicate key assets and capabilities in the event of temporary or permanent loss.

Step 4. Set priorities for the response to threats toward personnel, physical assets, and information.

The protection cell staff and working group continuously update criticality assessments during the operations process. As the staff develops or modifies a friendly course of action (COA), information collection efforts confirm or deny information requirements. As the mission or threat changes, initial criticality assessments may also change, increasing or decreasing the subsequent force vulnerability. The protection cell members monitor and evaluate these changes and begin coordination among the staff to implement modifications to the protection concept or recommend new protection priorities. Priority intelligence requirements, running estimates, measures of effectiveness (MOEs), and measures of performance (MOPs) are continually updated and adjusted to reflect the current and anticipated risks associated with the OE.

C. Vulnerability Assessment

A vulnerability assessment is an evaluation (assessment) to determine the magnitude of a threat or hazards effect on an installation, personnel, a unit, an exercise, a port, a ship, a residence, a facility, or other site. It identifies the areas of improvement required to withstand, mitigate, or deter acts of violence or terrorism or attacks against threats. The staff addresses who or what is vulnerable and how it is vulnerable against threats. The vulnerability assessment identifies physical characteristics or procedures that render critical assets, areas, infrastructures, or special events vulnerable to known or potential threats and hazards. The general sequence is—

Step 1. List assets and capabilities and the threats against them

Step 2. Determine the common criteria for assessing vulnerabilities

Step 3. Evaluate the vulnerability of assets and capabilities

Vulnerability evaluation criteria may include the degree to which an asset may be disrupted, quantity available (if replacement is required due to loss), dispersion (geographic proximity), and key physical characteristics.

DOD has created several decision support tools to perform criticality assessments in support of the vulnerability assessment process, including—

MSHARPP (Mission, Symbolism, History, Accessibility, Recognizability, Population, and Proximity)

MSHARPP is a targeting analysis tool that is geared toward assessing personnel vulnerabilities, but it also has application in conducting a broader analysis. The purpose of the MSHARPP matrix is to analyze likely terrorist targets and to assess their vulnerabilities from the inside out.

CARVER (criticality, accessibility, recuperability, vulnerability, effect, and recognizability)

The CARVER matrix is a valuable tool in determining criticality and vulnerability. For criticality purposes, CARVER helps assessment teams and commanders (and the assets that they are responsible for) to determine assets that are more critical to the success of the mission. This also helps determine which resources should be allocated to protect critical assets (personnel, infrastructure, and information). The CARVER targeting matrix assesses a potential target from a terrorist perspective to identify what the enemy might perceive as a good (soft or valuable) target.

Refer to ATP 3-37.2 for more information on MSHARPP and CARVER.

8-14 (Protection) I. Planning

II. Scheme of Protection Development

Ref: ADP 3-37, Protection (Jul '19), pp. 3-9 to 3-10.

The scheme of protection describes how protection tasks support the commander's intent and concept of operations, and it uses the commander's guidance to establish the priorities of support to units for each phase of the operation. A commander's initial protection guidance may include protection priorities, civil considerations, protection task considerations, potential protection decisive points, high-risk considerations, and prudent risk.

The protection cell (supported by the protection working group) develops the scheme of protection after receiving guidance and considering the principles of protection in relation to mission variables, the incorporation of efforts, and the protection required. The scheme of protection is based on the mission variables, thus it includes protection priorities by area, unit, activity, or resource. It addresses how protection is applied and derived during all phases of an operation. For example, the security for routes, bases/base camps, and critical infrastructure is accomplished by applying protection assets in dedicated, fixed, or local security roles; or it may be derived from economy-of-force protection measures, such as area security techniques. It also identifies areas and conditions where forces may become fixed or static and unable to derive protection from their ability to maneuver. These conditions, areas, or situations are anticipated; and the associated risks are mitigated by describing and planning for the use of response forces.

The protection cell considers the following items, at a minimum, as it develops the scheme of protection:

- Protection priorities.
- Work priorities for survivability assets.
- Air and missile defense positioning guidance.
- Specific terrain and weather factors.
- Information focus and limitations for security efforts.
- Areas or events where risk is acceptable.
- Protected targets and areas.
- Civilians and noncombatants in the AO.
- Vehicle and equipment safety or security constraints.
- Personnel recovery actions and control measures.
- FPCON status.
- Force health protection measures.
- Mission-oriented protective posture guidance.
- Environmental guidance.
- Scheme of information operations.
- Explosive ordnance and hazard guidance.
- Ordnance order of battle.
- OPSEC risk tolerance.
- Fratricide avoidance measures.
- Rules of engagement, standing rules for the use of force, and rules of interaction.
- Escalation of force and nonlethal weapons guidance.
- Operational scheme of maneuver.
- Military deception.
- Obscuration.
- Radiation exposure status.
- Contractors in the AO.

(Protection) I. Planning 8-15

III. Protection Prioritization List

Ref: ADP 3-37, Protection (Jul '19), pp. 3-6 to 3-9.

Protection prioritization lists are organized through the proper alignment of critical assets. The commander's priorities and intent and the impacts on mission planning determine critical assets. A critical asset is a specific entity that is of such extraordinary importance that its incapacitation or destruction would have a very serious, debilitating effect on the ability of a nation to continue to function effectively (JP 3-07.2). Critical assets can be people, property, equipment, activities, operations, information, facilities, or materials. For example, important communications facilities and utilities, analyzed through criticality assessments, provide information to prioritize resources while reducing the potential application of resources on lower-priority assets. Stationary weapons systems might be identified as critical to the execution of military operations and, therefore, receive additional protection. The lack of a replacement may cause a critical asset to become a top priority for protection.

Protection Priorities

Criticality, vulnerability, and recoverability are some of the most significant considerations in determining protection priorities *(previous page)* that become the subject of commander guidance and the focus of area security operations. The scheme of protection is based on the mission variables and should include protection priorities by area, unit, activity, or resource.

Although all military assets are important and all resources have value, the capabilities they represent are not equal in their contribution to decisive operations or overall mission accomplishment. Determining and directing protection priorities may involve the most important decisions that commanders make and their staffs support. There are seldom sufficient resources to simultaneously provide the same level of protection to all assets.

Most prioritization methodologies assist in differentiating what is important from what is urgent. In protection planning, the challenge is to differentiate between critical assets and important assets and to further determine what protection is possible with available protection capabilities.

The protection cell and working group use information derived from the commander's guidance, the intelligence preparation of the battlefield, targeting, risk management, warning orders, the **critical asset list (CAL)** and **defended asset list (DAL)**, and the mission analysis to identify critical assets. Critical assets at each command echelon must be determined and prioritized.

The protection prioritization list is a key protection product developed during initial assessments. The protection cell and working group must use criticality, threat vulnerability, and threat probability to prioritize identified critical assets. Once the protection working group determines which assets are critical for mission success, it recommends protection priorities and establishes a protection prioritization list. It is continuously assessed and revised throughout each phase or major activity of an operation:

8-16 (Protection) I. Planning

Criticality

Criticality is the degree to which an asset is essential to accomplish the mission. It is determined by assessing the impact that damage to, or destruction of, the asset will have on the success of the operation. Damage to an asset may prevent, delay, or have no impact on the success of the plan.

- **Catastrophic**. Complete mission failure or the inability to accomplish the mission, death or total disability, the loss of major or mission-critical systems or equipment, major property or facility damage, mission-critical security failure, or unacceptable collateral damage.
- **Critical**. Severely degraded mission capability or unit readiness; total disability, partial disability, or temporary disability; extensive damage to equipment or systems; significant damage environment; security failure; or significant collateral damage.
- **Marginal**. Degraded mission capability or unit readiness; minor damage to equipment or systems, property, or the environment; lost days due to injury or illness; or minor damage to property or the environment.
- **Negligible**. Little or no adverse impact on mission capability, first aid or minor medical treatment, slight equipment or systems damage (remaining fully functional or serviceable), or little or no property or environmental damage.

Threat Vulnerability

Threat vulnerability measures the ability for a threat to damage the target (asset) using available systems (people and material). An asset's vulnerability is greater if a lower-level threat (Level I) can create damage or destruction that would result in mission failure or severely degrade its mission capability. If an asset can withstand a Level I or Level II threat, its vulnerability ability is less and may not require additional protection assets, depending on the asset's criticality. The following mitigating factors must be considered when assessing the vulnerability of a target: survivability (the ability of the critical asset to avoid or withstand hostile actions by using camouflage, cover [hardening], concealment, and deception), the ability to adequately defend against threats and hazards, mobility and dispersion, and recoverability (which measures the time required for the asset to be restored, considering the availability of resources, parts, expertise, manpower, and redundancies).

- **Level I threat**. Agents, saboteurs, sympathizers, terrorists, civil disturbances.
- **Level II threat**. Small tactical units. Irregular forces may include significant standoff weapons threats.
- **Level III threat**. Large tactical force operations, including airborne, heliborne, amphibious, infiltration, and major air operations.

Threat Probability

Threat probability assesses the probability that an asset will be targeted for surveillance or attack by a credible/capable threat. Determinations of the intent and capability of the threat are key in assessing the probability of attack.

- **Frequent**. Occurs very often; known to happen regularly. Examples are surveillance, criminal activities, cyberspace attacks, indirect fire, and small-arms fire.
- **Likely**. Occurs several times; a common occurrence. Examples are explosive booby traps/improvised explosive devices, ambushes, and bombings.
- **Occasional**. Occurs sporadically, but is not uncommon. Examples are air-to-surface attacks or insider threats, which may result in injury or death.
- **Seldom**. Remotely possible; could occur at some time. Examples are the release of CBRN hazards or the employment of WMD.
- **Unlikely**. Presumably, the action will not occur, but it is not impossible. Examples are the detonation of containerized ammunition or the use of a dirty bomb.

(Protection) I. Planning 8-17

IV. Protection Cell and Working Group

Commands utilize a protection cell and protection working group to integrate and synchronize protection tasks and systems for each phase of an operation or major activity.

A. Protection Cell

The protection cell membership does not require representatives from every functional element of protection. However, dedicated members should coordinate with other personnel and special staff elements as required, Primary members of the protection cell typically include the chief of protection, an air and missile defense officer, a personnel recovery officer, a provost marshal, a CBRN officer, an EOD officer, an engineer officer, and an AT officer.

B. Protection Working Group

The protection cell forms the core membership of the protection working group, which includes other agencies, as required. Protection cell and protection working group members differ in that additional staff officers are brought into the working group. These additional officers meet operational requirements for threat assessments, vulnerability assessments, and protection priority recommendations. The protection working group calls upon existing resources from the staff.

Protection working group meetings have the same purpose, regardless of the echelon. Protection functions at different echelons of command differ mostly in the size of the area of operations and the number of available protection capabilities. The protection working group—

- Determines likely threats and hazards from updated enemy tactics, the environment, and accidents.
- Determines vulnerabilities as assessed by the vulnerability assessment team.
- Establishes and recommends protection priorities, such as the CAL.
- Provides recommendations for the CAL and DAL.
- Reviews and coordinates unit protection measures.
- Recommends FPCONs and random AT measures.
- Determines required resources and makes recommendations for funding and equipment fielding.
- Provides input and recommendations on protection-related training.
- Makes recommendations to commanders on protection issues that require a decision.
- Performs tasks required for a force protection working group and a threat protection working group according to Department of Defense Instruction (DODI) 2000.16.
- Accesses assets and infrastructure that are designated as critical by higher headquarters.

Commanders augment the team with other unit specialties and unified action partners, depending on the operational environment and the unit mission. The chief of protection determines the working group agenda, meeting frequency, composition, input, and expected output

8-18 (Protection) I. Planning

Chap 8

II. Protection in Preparation

Ref: ADP 3-37, Protection (Jul '19), chap. 4.

The force is often most vulnerable to an enemy or adversary surprise attack during preparation. Preparation, operations to shape, and operations to prevent create conditions that improve friendly force opportunities for success. Preparation requires commander, staff, unit, and Soldier actions to ensure that the force is trained, equipped, and ready to execute operations. Preparation in support of protection is not a linear activity—protection preparation is a continuous and enduring activity. Preparation activities help commanders, staffs, and Soldiers to understand a situation and their roles in upcoming operations. Protection preparation requirements occur throughout operations to shape, operations to prevent, large-scale ground combat operations, and operations to consolidate gains. They focus on deterring and preventing the enemy or adversaries from taking actions that would affect combat power during future operations. The execution of protection tasks with ongoing preparation activities helps prevent negative effects. Commanders ensure the integration of protection warfighting function tasks to safeguard friendly forces, civilians, and infrastructure while forces prepare for operations. Active defense measures help deny the initiative to the enemy or adversary, while the execution of passive defense measures prepares the force against threat and hazard effects and accelerates the mitigation of those effects.

I. Protection Working Group

Preparation includes increased application and emphasis on protection measures. During preparation, operations to shape, and operations to prevent, the protection working group—

- Provides recommendations to refine the scheme of protection
- Makes changes to the protection prioritization list based on the commander's priorities and changes during the phase of an operation
- Recommends systems to detect threats to the critical assets
- Proposes the refinement of OPSEC measures
- Monitors quick-reaction force or tactical and troop movement
- Provides recommendations for improving survivability
- Liaisons and coordinates with adjacent and protected units
- Determines protection indicators and warnings for information collection
- Confirms backbriefs
- Analyzes and proposes vulnerability reduction measures
- Provides recommended revisions to tactical standard operating procedures
- Conducts personnel recovery rehearsals

During preparation, operations to shape, and operations to prevent, the protection working group ensures that the controls and risk reduction measures developed during planning have been implemented and are reflected in plans, standard operating procedures, and running estimates, even as the threat assessment is continuously updated. New threats and hazards are identified or anticipated based on newly assessed threat capabilities or changes in environmental conditions as compared with known friendly vulnerabilities and weaknesses. Commanders conduct after action re-

(Protection) II. Preparation 8-19

Protection (ADP 3-37)

II. Protection Considerations (Preparation)

Ref: ADP 3-37, Protection (Jul '19), pp 4-1 to 4-4.

As the staff monitors and evaluates the performance or effectiveness of a friendly COA, ground-and space-based information collection operations are used to collect information that may confirm or deny forecasted threat COAs. As the threat changes, the risk to the force changes. Some changes may require a different protection posture or the implementation or cessation of specific protection measures, activities, or restraints. The protection cell analyzes changes or variances that may require modifications to protection priorities and obtains guidance when necessary. Threat assessment is a dynamic and continually changing process. Protection planners stay alert for changing OE indicators and warnings that would signal new or fluctuating threats and hazards.

Detailed intelligence is used to develop threat assessments, and changes in the situation often dictate adjustments or changes to the plan when they exceed variance thresholds established during planning. During preparation, operations to shape, and operations to prevent, the staff continues to monitor and evaluate the overall situation because variable threat assessment information may generate new priority intelligence requirements, while changes in asset criticality could lead to new friendly force information requirements. Updated information requirements could be required based on changes to asset vulnerability and criticality when combined with the threat assessment.

Commanders exercising mission command direct and lead throughout the operations process. Commanders' actions during preparation, operations to shape and operations to prevent, may include—

- Reconciling the threat assessment with professional military judgment and experience.
- Providing guidance on risk tolerance and making risk decisions.
- Emphasizing protection tasks during rehearsals.
- Minimizing unnecessary interference with subunits to allow maximum preparatory time.
- Circulating throughout the environment to observe precombat inspections.
- Directing control measures to reduce risks associated with preparatory movement.
- Expediting the procurement and availability of resources needed for protection implementation.
- Requesting higher headquarters support to reinforce logistical preparations and replenishment.

Depending on the situation and the threat, some protection tasks may be conducted for short or long durations, covering the course of several missions or an entire operation. The staff coordinates the commander's protection priorities with vulnerability mitigation measures and clearly communicates them to—

- Higher headquarters and subordinate and adjacent units.
- Civilian agencies and personnel that are part of the force or those that may be impacted by the task or control.

Subordinate leaders also conduct integration processes and provide supervision to ensure that Soldiers understand their responsibilities and the significance of protection measures and tasks. This is normally accomplished through training, rehearsals, task organization, and resource allocation. Rehearsals, especially those using opposing force personnel, can provide a measure of protection plan effectiveness

8-20 (Protection) II. Preparation

Preparation Activities

Commanders, units, and Soldiers conduct preparation activities (as described in ADRP 5-0) to help ensure that the force is protected and prepared for execution. Protection is incorporated throughout all preparation activities, to include:

- Continue to coordinate & conduct liaison
- Conduct rehearsals
- Initiate information collection
- Conduct plans-to-operations transitions
- Initiate security operations
- Refine the plan
- Initiate troop movement
- Integrate new Soldiers and units.
- Initiate sustainment preparations
- Complete task organization
- Initiate network preparations
- Train
- Manage terrain
- Perform pre-operation checks and inspections
- Prepare terrain
- Continue to build partnerships and teams
- Conduct confirmation briefs
- Consider effects of protection activities in the information environment

Continue to Coordinate and Conduct Liaison

Coordination and liaison help ensure that leaders who are internal or external to the headquarters understand the unit role in upcoming operations and ensure that they are prepared to perform that role.

Initiate Information Collection

Throughout the operations process, commanders take every opportunity to improve their situational understanding. This requires aggressive and continuous information collection. Commanders and staffs continuously plan, task, and employ collection assets and forces to collect timely and accurate information that helps satisfy the commander's critical information requirements and other information requirements. For example, the protection working group uses staff analysis and coordination with higher headquarters to determine which critical assets or locations are likely to be attractive targets and require surveillance.

Initiate Security Operations

Commanders and staffs continuously plan and coordinate security operations throughout the conduct of operations. Security operations are those operations undertaken to—

- Provide an early and accurate warning of enemy or adversary operations
- Provide the force with the time and maneuver space necessary to react to the enemy or adversary
- Develop the situation so that commanders can effectively use the protected force

Security operations reflect increasing levels of combat power that can be applied to protect an asset or force from a directed threat, and they are typically conducted by operating forces designed to gain and exploit the initiative. The primary purpose of a **screen** operation is to provide early warning, thereby preventing surprise. **Guard and cover** operations involve combined arms units in combat, fighting to gain time with differing levels of capability and autonomy for independent action. **Operational area security** focuses on the protected force, installation, route, or area. **Local security** protection ranges from echelon headquarters to reserves and sustainment forces.

Manage and Prepare Terrain

Terrain management is the process of allocating terrain by establishing areas of operation, designating assembly areas, and specifying locations for units and activities to deconflict activities that might interfere with each other.

Integrate Information Operations

Information operations is the integrated employment, during military operations, of information-related capabilities in concert with other lines of operation to influence, disrupt, corrupt, or usurp the decision-making of adversaries and potential adversaries while protecting our own (JP 3-13).

(Protection) II. Preparation 8-21

views and war-game to identify changes to the threat. The protection working group lead maintains a list of prioritized threats, adverse conditions, and hazard causes. The challenge is to find the root cause or nature of a threat or hazard so that the most effective protection solution can be implemented and disseminated.

Subworking groups feed information to the protection working group and incorporate elements from other warfighting functions. Commanders augment the working groups with other unit specialties and unified action partners, depending on the OE and the unit mission. The lead for each working group determines the agenda, meeting frequency, composition, input, and expected output. Ultimately, the output from the working groups helps refine protection priorities, protection running estimates, assessments, EEFI, and the scheme of protection.

A. Antiterrorism Working Group

The AT working group is led by the AT officer and includes members from the protection working group, subordinate commands, host nation agencies, and other unified action partners. It—

- Develops and refines AT plans
- Oversees the implementation of the AT program
- Addresses emergent and emergency AT program issues

B. Counter Improvised Explosive Device Working Group

The counter improvised explosive device working group is led by the EOD officer and includes members from the protection working group, subordinate commands, host nation agencies, and other unified action partners. It—

- Disseminates improvised explosive device information (including best practices), improvised explosive device trend analysis, and improvised explosive device defeat equipment and training issues
- Determines operational tactics to analyze and defeat the area of operations improvised explosive device networks
- Recommends the protection working group improvised explosive device defeat initiatives relating to equipment, intelligence, and operations.
- Identifies improvised explosive device defeat requirements and issues throughout the unit, including separate and subordinate units

C. Chemical, Biological, Radiological, and Nuclear (CBRN) Working Group

The CBRN working group is led by the CBRN officer and includes members from the protection working group, subordinate commands, host nation agencies, and other unified action partners. It—

- Disseminates CBRN operations information, including trend analysis, defense best practices and mitigating measures, operations, the status of equipment and training issues, CBRN logistics, and consequence management and remediation efforts
- Refines the CBRN threat, hazard, and vulnerability assessments

8-22 (Protection) II. Preparation

Chap 8

III. Protection in Execution

Ref: ADP 3-37, Protection (Jul '19), chap. 5.

The execution of protection is continuous and must occur throughout operations to shape, operations to prevent, large-scale ground combat operations, and operations to consolidate gains, with a focus on deterring and preventing the enemy, adversaries, or hazards from actions that effect the force. Commanders implement control measures and allocate resources that are sufficient to ensure protection continuity and restoration. Employed mitigation measures that have been planned and prepared for allow the force to quickly respond and recover from the threat or hazard effects, ensuring a force that remains effective and continues the mission. Control measures may include restraint after careful and disciplined balancing decisions regarding the need for security and protection in the conduct of military operations.

I. Execution

Commanders who exercise mission command decide, direct, lead, access, and provide leadership to organizations and Soldiers during execution. As operations develop and progress, the commander interprets information that flows from systems for indicators and warnings that signal the need for the execution or adjustment of decisions. Commanders may direct and redirect the way that combat power is applied or preserved, and they may adjust the tempo of operations through synchronization. The continuous and enduring character of protection makes the continuity of protection capabilities essential during execution. Commanders implement control measures and allocate resources that are sufficient to ensure protection continuity and restoration.

The staff monitors the conduct of operations during execution, looking for variances from the scheme of maneuver and protection. When variances exceed a threshold value, adjustments are made to prevent a developing vulnerability or to mitigate the effects of the unforecasted threat or hazard. The status of protection assets is tracked and evaluated on the effectiveness of the protection systems as they are employed. Commanders maintain protection by applying comprehensive protection capabilities, from main and supporting efforts to decisive and shaping operations. Protection can be derived as a by-product or a complementary result of some combat operations (such as security operations), or it can be deliberately applied as commanders integrate and synchronize tasks that comprise the protection warfighting function.

The protection cell and working group monitor and evaluate several critical ongoing functions associated with execution for operational actions or changes that impact protection cell proponents, which include—

- Changes to threat and hazard assessments
- Changes in force vulnerabilities
- Changes to unit capabilities
- Relevancy of facts
- Validity of assumptions
- Reasons that new conditions affect the operation
- Running estimates
- Protection tasks
- System failures
- Resource allocations
- Increased risks

(Protection) III. Execution 8-23

Protection (ADP 3-37)

II. Protection in Support of Decisive Action

Ref: ADP 3-37, Protection (Jul '19), pp. 5-2 to 5-14.

In large-scale ground combat operations against a peer threat, commanders conduct decisive action to seize, retain, and exploit the initiative. Decisive action is the continuous, simultaneous execution of offensive, defensive, and stability operations or defense support of civil authority tasks (ADP 3-0). Operations conducted outside the United States and its territories simultaneously combine three elements—offense, defense, and stability. Within the United States and its territories, decisive action combines the elements of defense support of civil authorities and offense and defense to support homeland defense, when required.

Decisive action begins with the commander's intent and concept of operations. As a single, unifying idea, decisive action provides direction for the entire operation. Based on a specific idea of how to accomplish the mission, commanders and staffs refine the concept of operations during planning and determine the proper allocation of resources and tasks. Leaders must have a situational understanding in simultaneous operations due to the diversity of threats, the proximity to civilians, and the impact of information during operations. The changing nature of operations may require a surge of certain capabilities, such as protection, to effectively link decisive operations to shaping or stabilizing activities in the AO. In other operations, the threat may be less discernible, unlikely to mass, and immune to the center of gravity analysis, which requires a constant and continuous protection effort or presence.

Commanders must accept risk when exploiting time-sensitive opportunities by acting before adversaries discover vulnerabilities, take evasive or defensive action, and implement countermeasures. Commanders and leaders can continue to act on operational and individual initiative if they make better risk decisions faster than the enemy or adversary, ultimately breaking enemy or adversary will and morale through relentless pressure. Commanders can leverage information collection capabilities, such as geospatial intelligence products or processes, to minimize fratricide and increase the probability of mission accomplishment.

Accurate assessment is essential for effective decision making and the apportionment of combat power to protection tasks. Commanders fulfill protection requirements by applying comprehensive protection capabilities from main and supporting efforts to decisive and shaping operations. Protection can be derived inherently from combat operations (such as security operations), or it can be deliberately applied as commanders integrate and synchronize tasks and systems that comprise of the protection warfighting function in order to apply maximum combat power.

8-24 (Protection) III. Execution

Chap 8

IV. Protection Assessment

Ref: ADP 3-37, Protection (Jul '19), chap. 6.

Protection assessment is an essential activity that continuously occurs throughout the operations process. While a failure in protection is typically easy to detect, the successful application of protection may be difficult to assess and quantify.

I. Continuous Assessment

Assessment is the determination of the progress toward accomplishing a task, creating a condition, or achieving an objective (JP 3-0). Commanders typically base assessments on their situational understanding, which is generally a composite of several informational sources and intuition. Assessments help commanders determine progress toward attaining the desired end state, achieving objectives, and performing tasks. It also involves continuously monitoring and evaluating the OE to determine what changes might affect the conduct of operations.

Refer to The Battle Staff SMARTbook for discussion of the operations process to include assessment (and MOEs/MOPs).

II. Lessons-Learned Integration

The manner in which organizations and Soldiers learn from mistakes is key in protecting the force. Although the evaluation process occurs throughout the operations process, it also occurs as part of the after action review and assessment following the mission. Leaders at all levels ensure that Soldiers and equipment are combat-ready. Leaders demonstrate their responsibility to the sound stewardship practices and risk management principles required to ensure minimal loss of resources and military assets due to hostile, nonhostile, and environmental threats and hazards. Key lessons learned are immediately applied and shared with other commands. Commanders develop systems to ensure the rapid dissemination of approved lessons learned and TTP proven to save lives and protect equipment and information. The protection working group at each command echelon evaluates the integration of lessons learned and constantly coordinates protection lessons with other staff elements within and between the levels of command. Postoperational evaluations typically—

- Identify threats that were not identified as part of the initial assessment or identify new threats that evolved during the operation or activity.
- Assess the effectiveness of supporting operational goals and objectives. For example, determine if the controls positively or negatively impacted training or mission accomplishment and determine if they supported existing doctrine and TTP.
- Assess the implementation, execution, and communication of controls.
- Assess the accuracy of residual risk and the effectiveness of controls in eliminating hazards and controlling risks.
- Ensure coordination throughout the integration processes.
  - Was the process integrated throughout all phases of the operation?
  - Were risk decisions accurate?
  - Were risk decisions made at the appropriate level?
  - Did any unnecessary risks or benefits outweigh the cost in terms of expense, training benefit, or time?
  - Was the process cyclic and continuous throughout the operation?

(Protection) VI. Assessment 8-25

Protection (ADP 3-37)

III. Protection Considerations (Assessment)

Ref: ADP 3-37, Protection (Jul '19), pp. 6-1 to 6-2.

Assessment During Planning

The staff conducts analyses to assess threats, hazards, criticality, vulnerability, and capability to assist commanders in determining protection priorities, task organization decisions, and protection task integration.

Members of the protection cell evaluate COAs during the MDMP against the evaluation criteria derived from the protection warfighting function to determine if each COA is feasible, acceptable, and suitable in relation to its ability to protect or preserve the force.

Assessment During Preparation Activities

Assessment occurs during preparation, operations to shape, and operations to prevent and includes activities required to maintain situational understanding; monitor and evaluate running estimates, tasks, MOEs, and MOPs; and identify variances for decision support. These assessments generally provide commanders with a composite estimate of preoperational force readiness or status in time to make adjustments.

During preparation, operations to shape, and operations to prevent, the protection working group focuses on threats and hazards that can influence preparatory activities, including monitoring new Soldier integration programs and movement schedules and evaluating live-fire requirements for precombat checks and inspections. The protection working group may evaluate training and rehearsals or provide coordination and liaison to facilitate effectiveness in high-risk or complex preparatory activities, such as movement and sustainment preparation.

Assessment During Execution

The protection working group monitors and evaluates the progress of current operations to validate assumptions made in planning and to continually update changes to the situation. The protection working group continually meets to monitor threats to protection priorities, and they recommend changes to the protection plan, as required. They also monitor the conduct of operations, looking for variances from the operations order that affect their areas of expertise. When variances exceed a threshold value developed or directed in planning, the protection cell may recommend an adjustment to counter an unforecasted threat or hazard or to mitigate a developing vulnerability. It also tracks the status of protection assets and evaluates the effectiveness of the protection systems as they are employed. Additionally, the protection working group monitors the actions of other staff sections by periodically reviewing plans, orders, and risk assessments to determine if those areas require a change in protection priorities, posture, or resource allocation. The protection working group monitor and evaluate—

- Changes to threat and hazard assessments.
- Changes in force vulnerabilities.
- Changes to unit capabilities.
- The relevancy of facts.
- The validity of assumptions.
- Reasons that new conditions affect the operation.
- Running estimates.
- Protection tasks.
- System failures.
- Resource allocations.
- Increased risks.
- Supporting efforts.
- Force protection implementation measures, including site-specific AT measures.
- OPSEC measures and countermeasures.

8-26 (Protection) IV. Assessment

Index

A
Accepting Risk, 2-11
Achieving Convergence, 1-42
Actions by Friendly Forces, 4-14
Adversary, 1-24
Adversary Methods, 1-24, 1-64, 1-78
 - during armed conflict, 1-88
 - during competetion, 1-84
 - during crisis, 1-78
 - in maritime environments, 1-134
Agility, 1-38
Air and Missile Defense Planning/Integration, 6-24
Air Domain, 1-19
Air Force Assets, 6-8
Air Interdiction, 6-9
Air Operations, 1-34
Airspace Coordinating Measures (ACMs), 3-34
Airspace Planning and Integration, 6-17
Air-to-Surface Fires, 6-6
All-Source Intelligence, 5-15
Amphibious Operations, 1-136
Analyze, 5-14
Antiaccess and Area Denial (A2 & AD), 1-27
 - during armed conflict, 1-100
 - during competetion, 1-63
 - during crisis, 1-77
Antiterrorism Working Group, 8-22
Area Defense, 1-110
Area of Influence, 1-55
Area of Interest, 1-55
Area of Operations, 1-56
Area Reconnaissance, 4-19
Area Security, 4-24

Armed Conflict, 1-15
Army Airspace Command and Control (A2C2), 3-33
Army Command & Support Relationships, 3-25
Army Design Methodology (ADM), 3-15
Army Echelons, 1-32
Army Force Posture, 1-36
Army Operations, 1-1
Army Planning Methodologies, 3-15
Army Pre-Positioned Stocks (APS), 7-6
Army Strategic Contexts, 1-14
Army Support to the Joint Force during Crisis, 1-80
Army Surface-to-Surface Capabilities, 6-7
Army Targeting Process (D3A), 6-20
Art and Science of Tactics, 4-12
Art of Command, 2-7
Aspects of Mobility, 4-28
Assess, 3-11, 3-13, 5-14
Assigned Areas, 1-54
Assured Mobility, 4-26
Attack, 1-119

B
Basing, 7-7
Battle Rhythm, 3-22
Brigade Combat Teams (BCT), 1-33

C
Cannon Artillery, 6-7
Chain of Command, 3-23
Challenges for Army Forces, 1-4
Characteristics of Warfare, 1-9

Chemical, Biological, Radiological, and Nuclear Working Group, 8-22
China, 1-23, 1-90
Close Air Support (CAS), 6-9
Collect, 5-13
Combat Power, 2-1
Combined Arms, 1-11
Command and Control During Degraded or Denied Communications, 2-12
Command and Control System, 3-8
Command and Control Warfighting Function, 2-2, 3-1
Command and Support Relationships, 3-23
Command Posts, 3-29
Commander Presence on the Battlefield, 2-9
Commander's Guidance, 5-10
Commander's Intent, 2-9
Common Operational Picture (COP), 5-8
Competition (Below Armed Conflict), 1-15, 1-63
Conflict Type Determination, 1-75
Consolidating Gains, 1-16
 - during armed conflict, 1-104
 - during competetion, 1-73
 - during crisis, 1-86
Contested Deployment, 1-139
Conventional & Special Operations Forces Integration, 1-28
Conventional Warfare, 1-9
Convergence, 1-39
Corps, 1-32
Counter Improvised Explosive Device Working Group, 8-22

Index-1

Counterintelligence (CI), 5-16

Countermobility, 4-30

Country Team, 1--69

Cover, 4-24

Crisis, 1-15

Critical and Creative Thinking, 3-14

Critical Asset List (CAL), 8-16

Criticality Assessment, 8-10

Criticality, 8-16

Cyberspace Domain, 1-19

Cyberspace Electromagnetic Activities (CEMA), 6-23

Cyberspace Operations (CO), 2-16, 6-13

Cyberspace Operations, 1-35

D

Decide, Detect, Deliver, Assess (D3A), 6-20

Deep, Close, and Rear Operations, 1-58

Defeat Mechanisms, 1-47, 1-106

Defeating Enemy Forces in Detail, 1-46

Defended Asset List (DAL), 8-16

Defense Support Of Civil Authorities (DSCA), 1-144

Defense, 4-6

Defensive Operations, 1-108

Degraded or Denied Communications, 2-12

Deployment Operations, 4-10

Depth, 1-43

Describe, 3-11

Destroy, 1-48

Dimensions, 1-20

Direct, 3-11

Discipline, 2-10

Disintegrate, 1-48

Dislocate, 1-48

Disseminate, 5-13

Distribution, 7-14

Division, 1-33

Doctrinal Template of Depths & Frontages, 1-95

Domain Interdependence, 1-34

Domains, 1-18

Drive the Operations Process, 3-10

Dynamics of Combat Power, 2-4

E

Echelon Roles & Responsibilities, 1-97

Effects of Operations on Units and Soldiers, 2-9

Effects on Enemy Forces, 4-15

Electromagentic Warfare (EW), 2-16

Elements of Operational Art, 1-50

Enabling Operations, 1-107, 4-7

Endurance, 1-42, 7-12

Enemy Approaches to Armed Conflict, 1-88

Enemy Defense, 1-118

Enemy Offense, 1-112

Enemy, 1-24

Engineer Support, 4-34

Establishing Command and Control, 1-96

Execute, 3-12

Exploitation, 1-119

Extended Deep Area, 1-52

F

Field Army, 1-32

Fighting for Intelligence, 5-1

Financial Management, 7-4

Firepower, 2-5

Fires Across the Domains, 6-4

Fires in the Operations Process, 6-15

Fires Warfighting Function, 2-3, 6-1

Fixed-Wing Aircraft, 6-6

Force Posture, 1-36

Force Projection, 1-82, 4-10

Force Projection and Threat Capabilities, 1-139

Force Projection Planning, 1-84

Force Protection, 1-75

Fort to Port, 1-140

Foundations of Operations, 1-1

Freedom of Action, 7-12

Fundamentals of Operations, 1-37

G

Geospatial Intelligence, 5-16

Guard, 4-22

H

Health Service Support, 7-42

High-to-Medium Air Defense (HIMAD), 6-10

Homeland Defense, 1-144

Human Dimension, 1-21

Human Intelligence (HUMINT), 5-16

Humanitarian Assistance, 1-71

I

Imperatives, 1-44

Information, 2-5, 2-13, 2-14

Information Collection, 3-20, 5-7

Information Dimension, 1-21

Information Operations (IO), 2-13, 6-14

Information Use Across the Competition Continuum, 2-18

Information Warfare, 1-25

Informational Considerations, 1-22

Information-Related Capabilities (IRCs), 2-15

Initial Assessments, 8-9

Initiative, 2-9

Integrate Army, Multinational and Joint Fires, 6-15

Integrating Processes, 3-20

Integration, 1-40

Intelligence Capabilities, 5-15

Intelligence Disciplines, 5-16

Intelligence Estimate, 5-8

Intelligence Integrating Functions, 5-6

Intelligence Preparation of the Battlefield (IPB), 3-20, 5-6

Intelligence Process, 5-10

Index-2

Index

Intelligence Process, 5-9

Intelligence Products, 5-8

Intelligence Running Estimate, 5-8

Intelligence Summary, 5-8

Intelligence Support to Commanders and Decision-makers, 5-5

Intelligence Warfighting Function, 2-2, 5-1

Interagency Coordination, 1-28, 1-69

Interoperability, 1-67

Iran, 1-23

Irregular Warfare, 1-10

Isolate, 1-48

Isolation, 1-27

J

Joint Command Relationships, 3-24

Joint Intelligence Process, 5-9

Joint Interdependence, 1-36

Joint Operations and Activities, 1-28

Joint Security Area, 1-52

Joint Targeting, 6-22

K

Knowledge Management, 3-21

L

Land Domain, 1-18

Large-Scale Combat Operations, 1-11, 1-86, 7-15

Lead, 3-11

Leadership, 2-4, 2-7

Lethality: Overcoming Challenges, 1-6

Levels of Warfare, 1-12

Leveraging Information, 2-14

Logistics, 7-1

M

Main Effort, Supporting Effort, & Reserve, 1-62

Maritime Domain, 1-18

Maritime Environment, 1-127

Maritime Operations, 1-35

Measurement and Signature Intelligence (MASINT), 5-17

Methods of Warfare, 1-9

Military Decision-Making Process (MDMP), 3-15

Military Engagement, 1-70

Missiles, 6-7

Mission Command, 3-4

Mission Symbols, 4-16

Mission Variables - METT-TC (I), 1-22

Mobile Defense, 1-110

Mobility, 2-5, 4-25

Mortars, 6-7

Movement & Maneuver Warfighting Function, 2-2, 4-1

Movement Phase, 1-140

Movement to Contact, 1-119

Mulitdomain Operations, 1-2, 1-3, 1-29, 1-37

Multifunctional and Functional Brigades, 1-33

Mutual Support, 1-57

N

National Strategic Level of Warfare, 1-12

Nature of War, 1-7

Negotiations and Agreements, 7-13

Noncombatant Evacuation Operations (NEO), 1-75

Non-state Actors, 1-23

North Korea, 1-23

Notional Echelon Roles & Responsibilities, 1-97

Nuclear Deterrence and Countering Weapons of Mass Destruction, 1-71

O

Obstacle Control Measures, 4-33

Obstacle Planning, 4-32

Offense, 4-6

Offense, Defense, and Stability, 1-11

Offensive Operations, 1-118

Open-Source Intelligence (OSINT), 5-17

Operating as Part of the Joint Force, 1-94

Operational and Mission Variables, 1-22

Operational Approach, 1-46

Operational Art, 1-50

Operational Categories, 1-1

Operational Environment, 1-17

Operational Framework, 1-54

Operational Level of Warfare, 1-13

Operational Reach, 7-6

Operational Variables - PMESII-PT, 1-22

Operations, 1-1, 1-37
- *during armed conflict, 1-86*
- *during competetion, 1-63*
- *during crisis, 1-77*
- *in maritime environments, 1-127*

Operations Process, 3-9, 2-10, 6-21

Operations Security, 1-79

Other Relationships, 3-28

P

Peer threat, 1-24

Personnel Services, 7-4

Physical Dimension, 1-20

Plan, 3-13

Plan and Direct, 5-12

Port to Port, 1-141

Preclusion, 1-26

Prepare, 3-13

Principles of Joint Operations, 1-8

Principles of War, 1-8

Processing, Exploitation, and Dissemination (PED), 5-18

Produce, 5-13

Protection Cell and Working Group, 8-17

Protection Integration in the Operations Process, 8-4

Protection Planning, 8-9

Protection Priorities, 8-16

Protection Prioritization List, 8-16

Index-3

Index

Protection Warfighting Function, 2-3, 8-1
Pursuit, 1-119

R

Reception, Staging, Onward Movement, and Integration (RSOI), 1-85, 4-10
- *During Contested Deployments, 1-143*
Reconnaissance, 4-17
Reconnaissance in Force, 4-19
Reconnaissance Objective, 4-17
Reconstitution Operations, 7-24
Regeneration, 7-24
Relative Advantages, 1-44, 4-4
- *during armed conflict, 1-89*
- *during competetion, 1-72*
- *during crisis, 1-79*
Reorganization, 7-24
Requirements Management, 5-11
Retrograde, 1-110
Risk, 1-49
Risk Management, 3-21
Risks during Large-Scale Combat, 7-21
Rockets, 6-7
Rotary-Wing Aircraft, 6-6
Route Reconnaissance, 4-19
Russia, 1-23, 1-91

S

Sanctuary, 1-27
Scheme of Protection Development, 8-15
Screen, 4-21
Sector, 1-56
Security Operations, 4-21
Set the Theater, 1-66
Short-Range Air Defense (SHORAD), 6-11
Signals Intelligence (SIGINT), 5-17
Single-Source Intelligence, 5-18
Situational Understanding, 3-14

Space Domain, 1-19
Space Operations, 1-34, 6-12
Special Operations, 1-28, 6-12
Special Reconnaissance, 4-19
Stability Mechanisms, 1-49
Stability Operations, 4-8
Strategic Contexts, 1-14
Strategic Environment, 1-23
Strategic Framework, 1-52
Strategic Support Area, 1-52
Subordinate Forms of the Attack, 1-120
Support Area, 1-58, 7-22
Surface-to-Air Fires, 6-10
Surface-to-Surface Fires, 6-5
Survivability, 2-6, 8-6
Sustain Large-Scale Combat Operations, 7-15
Sustaining Offensive Operations, 7-18
Sustaining Offensive Operations, 7-20
Sustainment Execution, 7-12
Sustainment of Unified Land Operations, 7-5
Sustainment Preparation of the Operational Environment, 7-13
Sustainment Preparation, 7-12
Sustainment Rehearsals, 7-22
Sustainment Synchronization, 7-16
Sustainment Warfighting Function, 2-3
Sustainment Warfighting Function, 7-1
Synchronization, 1-41
Systems Warfare, 1-25

T

Tactical Doctrinal Taxonomy, 4-13
Tactical Level of War, 1-13, 4-6, 4-11
Tactical Mission Tasks, 4-12

Tactics, 4-11
Targeting, 3-21
Targeting Process (D3A), 6-20
Technical Intelligence (TECHINT), 5-17
Tenets, 1-38
Theater Army, 1-32
Theater Closing, 7-6
Theater Opening, 7-6
Theater Strategic Level of Warfare, 1-13
Threat, 1-24, 8-17
Threat and Hazard Assessment, 8-10
Threat Methods, 1-24
- *during armed conflict, 1-88*
Threat Probability, 8-17
Threat Vulnerability, 8-17
Threats and Hazards, 8-11
Threats to Sustainment Units, 7-17
Transitions, 3-18
- *during armed conflict, 1-104, 1-117, 1-126*
- *during competition, 1-74*
- *during crisis, 1-86*
Troop Leading Procedures (TLP), 3-15

U

UAS, 6-6
Understand, 3-11
Understanding an Operational Environment, 1-17
Unified Action, 1-28
United States Diplomatic Mission, 1-69

V

Visualize, 3-11
Vulnerability Assessment, 8-14

W

War and Warfare, 1-6
Warfighting Functions, 2-1

Z

Zone, 1-56
Zone Reconnaissance, 4-19

SMARTbooks
INTELLECTUAL FUEL FOR THE MILITARY

Recognized as a **"whole of government"** doctrinal reference standard by military, national security and government professionals around the world, SMARTbooks comprise a **comprehensive professional library** designed with all levels of Soldiers, Sailors, Airmen, Marines and Civilians in mind.

The SMARTbook reference series is used by **military, national security, and government professionals** around the world at the organizational/institutional level; operational units and agencies across the full range of operations and activities; military/government education and professional development courses; combatant command and joint force headquarters; and allied, coalition and multinational partner support and training.

Download FREE samples and SAVE 15% everyday at:
www.TheLightningPress.com

 The Lightning Press is a **service-disabled, veteran-owned small business,** DOD-approved vendor and federally registered — to include the SAM, WAWF, FBO, and FEDPAY.

SMARTbooks
INTELLECTUAL FUEL FOR THE MILITARY

MILITARY REFERENCE: SERVICE-SPECIFIC

Recognized as a "whole of government" doctrinal reference standard by military professionals around the world, SMARTbooks comprise a comprehensive professional library.

MILITARY REFERENCE: MULTI-SERVICE & SPECIALTY

SMARTbooks can be used as quick reference guides during operations, as study guides at professional development courses, and as checklists in support of training.

JOINT STRATEGIC, INTERAGENCY, & NATIONAL SECURITY

The 21st century presents a global environment characterized by regional instability, failed states, weapons proliferation, global terrorism and unconventional threats.

The Lightning Press is a **service-disabled, veteran-owned small business,** DOD-approved vendor and federally registered — to include the SAM, WAWF, FBO, and FEDPAY.

RECOGNIZED AS THE DOCTRINAL REFERENCE STANDARD BY MILITARY PROFESSIONALS AROUND THE WORLD.

THREAT, OPFOR, REGIONAL & CULTURAL

In today's complicated and uncertain world, the military must be ready to meet the challenges of any type of conflict, in all kinds of places, and against all kinds of threats.

 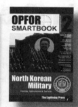   

HOMELAND DEFENSE, DSCA, & DISASTER RESPONSE

Disaster can strike anytime, anywhere. It takes many forms—a hurricane, an earthquake, a tornado, a flood, a fire, a hazardous spill, or an act of terrorism.

DIGITAL SMARTBOOKS (eBooks)

In addition to paperback, SMARTbooks are also available in digital (eBook) format. Our digital SMARTbooks are for use with Adobe Digital Editions and can be used on up to **six computers and six devices**, with free software available for **85+ devices and platforms**—including PC/MAC, iPad and iPhone, Android tablets and smartphones, Nook, and more! Digital SMARTbooks are also available for the **Kindle Fire** (using Bluefire Reader for Android).

Download FREE samples and SAVE 15% everyday at:
www.TheLightningPress.com

Purchase/Order

SMARTsavings on SMARTbooks! Save big when you order our titles together in a SMARTset bundle. It's the most popular & least expensive way to buy, and a great way to build your professional library. If you need a quote or have special requests, please contact us by one of the methods below!

View, download FREE samples and purchase online:
www.TheLightningPress.com

Order SECURE Online
Web: www.TheLightningPress.com
Email: SMARTbooks@TheLightningPress.com

24-hour Order & Customer Service Line
Place your order (or leave a voicemail)
at 1-800-997-8827

Phone Orders, Customer Service & Quotes
Live customer service and phone orders available
Mon - Fri 0900-1800 EST at (863) 409-8084

Mail, Check & Money Order
2227 Arrowhead Blvd., Lakeland, FL 33813

Government/Unit/Bulk Sales

The Lightning Press is a **service-disabled, veteran-owned small business**, DOD-approved vendor and federally registered—to include the SAM, WAWF, FBO, and FEDPAY.

We accept and process both **Government Purchase Cards** (GCPC/GPC) and **Purchase Orders** (PO/PR&Cs).

Keep your SMARTbook up-to-date with the latest doctrine! In addition to revisions, we publish incremental "**SMARTupdates**" when feasible to update changes in doctrine or new publications. These SMARTupdates are printed/produced in a format that allow the reader to insert the change pages into the original GBC-bound book by simply opening the comb-binding and replacing affected pages. Learn more and sign-up at: **www.thelightningpress.com/smartupdates/**